Prentice Hall Series in Geographic Information Science

KEITH C. CLARKE,
Series Editor

Arnold, *Interpretation of Airphotos and Remotely Sensed Imagery*

Berlin/Avery, *Fundamentals of Remote Sensing and Airphoto Interpretation, 6th edition*

Clarke, *Analytical and Computer Cartography, 2nd edition*

Clarke, *Getting Started with Geographic Information Systems, 4th edition*

Clarke, Parks and Crane (eds), *Geographic Information Systems and Environmental Modeling*

Foresman, *The History of Geographic Information Systems*

Jensen, *Introductory Digital Image Processing: A Remote Sensing Perspective, 2nd edition*

Jensen, *Remote Sensing of the Environment: An Earth Resource Perspective, 1st edition*

Lo/Yeung, *Concepts and Techniques in Geographic Information Systems*

Peterson, *Interactive and Animated Cartography*

Slocum, *Thematic Cartography and Visualization*

Vincent, *Fundamentals of Geological and Environmental Remote Sensing*

Prentice Hall Series in
Geographic Information Science

Getting Started with Geographic Information Systems

Fourth Edition

Keith C. Clarke
University of California, Santa Barbara

Pearson Education, Inc., *Upper Saddle River, New Jersey 07458*

Library of Congress Cataloging-in-Publication Data
Clarke, Keith, C.,
Getting started with geographic information systems / Keith C. Clarke–4th ed.
p. cm.- (Prentice Hall series in geographic information science)
Includes bibliographical references.
ISBN 0-13-046027-3
1. Geographic information systems. I. Title. II. Series.

G70.212.C57 2003
910′.285–dc21 2002032954
CIP

Executive Editor: *Dan Kaveney*
Associate Editor: *Amanda Griffith*
Editorial Assistant: *Margaret Ziegler*
Executive Managing Editor: *Kathleen Schiaparetti*
Production Editor: *Kim Dellas*
Senior Marketing Manager: *Christine Henry*
Managing Editor, AV Production & Management: *Patty Burns*
Production Manager, Artworks: *Ronda Whitson*
Manager, Production Technologies, Artworks: *Matt Haas*
A/V Art Editor: *Jessica Einsig*
Illustrator, Artworks: *Todd Frey, Jo Thompson*
Quality Assurance, Artworks: *Kenneth Mooney, Timothy Nguyen, Pamela Taylor*
Art Director: *Jayne Conte*
Cover Design: *Bruce Kenselaar*
Cover Art: *Dawn Wright OrSt/The Fagatele Bay National Marine Sanctuary,*
a division of the Office of National Marine Sanctuaries,
National Oceanic and Atmospheric Administration,
U.S. Department of Commerce
Manufacturing Manager: *Trudy Pisciotti*
Manufacturing Buyer: *Alan Fischer*
Assistant Managing Editor: *Nicole Bush*
Media Editor: *Chris Rapp*
Media Production Editor: *Elizabeth Klug*

Pearson Education, Inc.
Upper Saddle River, NJ 07458

Printed in the United States of America
10 9 8 7 6 5 4 3 2 1

ISBN 0-13-046027-3

Pearson Education LTD., *London*
Pearson Education Australia PTY, Limited, *Sydney*
Pearson Education Singapore, Pte. Ltd.
Pearson Education North Asia Ltd, *Hong Kong*
Pearson Education Canada, Ltd., *Toronto*
Pearson Educación de Mexico, S.A. de C.V.
Pearson Education—Japan, *Tokyo*
Pearson Education Malaysia, Pte. Ltd.

In memory

Raymond Harry Clarke
Helmut E. Ehrenspeck
John (Jack) E. Estes
Nicholas Bourdakis

The world is smaller without them.

Brief Contents

1 What Is a GIS?
2 GIS's Roots in Cartography
3 Maps as Numbers
4 Getting the Map into the Computer
5 What Is Where?
6 Why Is It There?
7 Making Maps with GIS
8 How to Pick a GIS
9 GIS in Action
10 The Future of GIS
Glossary
Index

Contents

Preface x

1 What Is a GIS? **1**
1.1 Getting Started 2
1.2 Some Definitions of GIS 2
1.3 A Brief History of GIS 8
1.4 Sources of Information on GIS 12
1.5 Study Guide 19
1.6 Exercises 22
1.7 References and Bibliography 22
1.8 Key Terms and Definitions 26

2 GIS's Roots in Cartography **34**
2.1 Map and Attribute Information 34
2.2 Map Scale and Projections 36
2.3 Coordinate Systems 45
2.4 Geographic Information 54
2.5 Study Guide 57
2.6 Exercises 60
2.7 Bibliography 60
2.8 Key Terms and Definitions 61

3 Maps as Numbers **66**
3.1 Representing Maps As Numbers 67
3.2 Structuring Attributes 71
3.3 Structuring Maps 72
3.4 Why Topology Matters 78
3.5 Formats for GIS Data 80
3.6 Exchanging Data 88
3.7 Study Guide 91
3.8 Exercises 95
3.9 References 95
3.10 Key Terms and Definitions 96

4 Getting the Map into the Computer **100**
4.1 Analog-to-Digital Maps 100
4.2 Finding Existing Map Data 101
4.3 Digitizing and Scanning 106
4.4 Field and Image Data 112
4.5 Data Entry 115
4.6 Editing and Validation 119

4.7 Study Guide 121
4.8 Exercises 122
4.9 References 123
4.10 Key Terms and Definitions 124

5 What Is Where? 128
5.1 Basic Database Management 128
5.2 Searches by Attribute 133
5.3 Searches by Geography 135
5.4 The Query Interface 140
5.5 Study Guide 141
5.6 Exercises 143
5.7 References 143
5.8 Key Terms and Definitions 144

6 Why Is It There? 146
6.1 Describing Attributes 146
6.2 Statistical Analysis 149
6.3 Spatial Description 157
6.4 Spatial Analysis 162
6.5 Study Guide 175
6.6 Exercises 177
6.7 References 177
6.8 Key Terms and Definitions 178

7 Making Maps with GIS 182
7.1 The Parts of a Map 182
7.2 Choosing a Map Type 185
7.3 Designing the Map 191
7.4 Study Guide 195
7.5 Exercises 196
7.6 Bibliography 197
7.7 Key Terms and Definitions 197

8 How to Pick a GIS 201
8.1 The Evolution of GIS Software 201
8.2 GIS and Operating Systems 204
8.3 GIS Functional Capabilities 204
8.4 GIS Software and Data Structures 214
8.5 Choosing the "Best" GIS 214
8.6 Study Guide 224
8.7 Exercises 226
8.8 References 226
8.9 Key Terms and Definitions 226

9 GIS in Action 232
9.1 Introducing GIS in Action 233
9.2 Case Study 1: GIS Fights the Gypsy Moth 233

9.3 Case Study 2: GIS and Road Accidents in Connecticut 237
9.4 Case Study 3: GIS at the World Trade Center after September 11, 2001 241
9.5 Case Study 4: Resource Management for California's Coastal Islands: The Channel Islands GIS 252
9.6 Case Study 5: Using GIS and GPS to Map the Sliding Rocks of Racetrack Playa 256
9.7 Study Guide 261
9.8 Exercises 264
9.9 References 265
9.10 Key Terms and Definitions 265

10 The Future of GIS **267**
10.1 Why Speculate? 267
10.2 Future Data 268
10.3 Future Hardware 276
10.4 Future Software 281
10.5 Some Future Issues and Problems 288
10.6 Conclusion 291
10.7 Study Guide 294
10.8 Exercises 296
10.9 References 296
10.10 Key Terms and Definitions 297

Glossary **304**

Index **329**

Preface

After six years and four editions, *Getting Started with Geographic Information Systems* remains a basic-level textbook for the beginning student in the expanding field of geographic information science. Books in GIS have tended to be rather advanced, for the specialist rather than the beginner. GIS is not just for the specialist, but for everyone. **Geographic information science** is the discipline that uses geographic information systems as tools to understand the world, by describing and explaining humankind's relationship to that world. The usual order of intellectual discovery has been reversed by GIS. In the past, geography students in their advanced studies met the tools of spatial description and analysis for the first time. Today, students from many disciplines and professionals find their way into the newly evolving academic discipline of geographic information science *through* their hands-on use of geographic information systems and through the medium of real-world problems. Geographic information systems are an important new entry point into fields where location in geographic space makes a difference, what might be called the *mapping sciences*.

Nevertheless, it is reassuring to find that as geographic information technologies have evolved, necessitating revision after revision of this book, the same old principles have reemerged to assert their significance as the roots of the new discipline. Much of this book is simply an old story retold, one that most geographers will find very familiar. Bernhard Varenius's 1650 *Geographica Generalis*, for example, contained much of the basic cartography in this book. Yet technology has brought change, and the evolution of the GIS field has now reached maturity, and the benefits to all are self-evident.

This book evolved from a tried and trusted approach to basic education. This approach is to first revisit the **basics**, such that all students will have the same foundation in underlying principles—both students who have covered them *and* those who skipped them during their grade school education. Next, the **scope** of the field is covered, and the critical underlying issues are highlighted in the context of the learned principles. Finally, the approach works toward the development of **critical thinking**, using the knowledge base and the basic concepts to develop educated thinking in context.

Getting Started with Geographic Information Systems uses these three stages of learning. In the early chapters, the basics of cartography, geodesy, and geography are covered. The following chapters cover the breadth and a little of the depth of GIS. In the course of this coverage, critical thinking is developed by visiting themes and challenges around issues and applications. Accuracy, data models, how data structure dictates capability, the demands of analysis—all are considered in context.

Chapter 1, *What Is a GIS?*, is an introduction to the concepts of GIS by the examination of alternative definitions, a glimpse at the historical context and heritage of the field, and a guide to the many information sources available, including those on the Internet and the World Wide Web. Chapter 2, *GIS's Roots in Cartography*, is a basic concepts chapter, introducing the cartographic necessities of map projections, coordinate systems, and geodesy. Chapter 3, *Maps as Numbers*, begins a consideration of map data representation, necessary for storage of the data within a GIS. The survey approach to data structures and formats is supplemented by consideration of how data structures both facilitate and limit GIS data use.

Chapter 4, *Getting the Map into the Computer*, also covers the basics of computer cartography and database systems, and getting maps into the computer in digital form, the process of geocoding. The broader issue introduced is the relationship between map accuracy and resolution, and the cartographic process of generalization. When an existing map is the source of data for a GIS, we often make faithful reproduction in the digital world of cartographic errors in the real world. Chapter 5, *What Is Where?*, is information rarely covered in GIS books: the database management capabilities of GIS and how they have evolved. Most database systems have a great deal in common. The attribute database is used as a vehicle to understand the concept of retrieval from a spatial database, a very different process indeed, and one right at the core of GIS power. Chapter 6, *Why Is It There?*, looks at data analysis, building from attribute data description and analysis toward spatial analysis. Here less depth is used because many curricula in geography already teach advanced methods of spatial analysis. A strength is the use of a real-world GPS data set. As my grandfather often said, it makes no sense to teach carpentry with blunt tools—you are just as likely to get cut. The same goes for GIS.

Chapter 7, *Making Maps with GIS*, is the last of the basic review chapters, in this case covering the map design component of cartography. So often, GIS is taught without a link to the substantial literature and body of experience on map design. With just a few of the basics, a GIS novice will be better able to understand when mistakes are being made. Chapter 8, *How to Pick a GIS*, is intended to allow the GIS novice to make an informed choice of systems, all too often the first step required in a GIS user's education. Although no coverage can be complete, the view here is that the educated shopper is the best consumer. Far more information is available to the sophisticated user; here the GIS novice gets the essentials. Chapter 9, *GIS in Action*, explores five original contributed case studies in GIS, one of them new to the fourth edition. The chapter highlights the full scope of GIS work, the inter- and multidisciplinary nature of the research in the field, and the many problems that are addressable (and created) by GIS. Chapter 10, *The Future of GIS*, speculates on future hardware, software, and issues, providing the GIS novice with a road map to the intellectual issues driving GIS research.

Each chapter includes four essential learning aids. First, each chapter's content is summarized in "bullet" form for easy use as classroom summaries and study guides. Next, each technical term is treated as a keyword, and definitions are given chapter by chapter. This assists in review, in learning concepts along with lectures, and in learning along with reading. A summary dictionary-style glossary is also provided at the end of the book. Finally, two sets of questions are included. The study questions are specific to the chapter and can serve as useful enrichment. Each chapter ends with exercises that can be completed using software on a PC. These exercises are generic enough that almost any GIS can be used to complete them. New to this edition, I include the labs developed for my class at UC Santa Barbara and the ArcView GIS software to complete them with.

The final feature of the book is a collection of interviews, included as sections called *People in GIS*. These sections are included because, first, it is hard to relate to a subject that does not have a human face, and, second, because the snippets of information each of these GIS users relate reinforce concepts highlighted elsewhere in the book. We are all, just because we are reading this book, people in GIS!

As with any book, there are many people to thank. Ray Henderson, my first editor at Prentice Hall and creator of the *Prentice Hall Series in Geographic Information Science*, originally talked me into writing this book. Dan Kaveney saw the project to completion, and has now stuck with me through four editions and the whole GIS series, as has his assistant Margaret Ziegler. Applications were provided by Ellen Cromley, Sean Ahearn, Leal Mertes, Bryan Pijanowski, and Paula Messina. The interviewees, Nils Larsen, Mark Bosworth, Susan Benjamin, Assaf Anyamba, Michael Goodchild and Brenda Faber, gave generously of their time and energy. Assistance with some of the graphics came from David Lawson, Barbara Tempalski, Brett Gilman, Westerly Miller, Jeff Hemphill and many others. Susan Baumgart did a highly professional job of redrafting many of the original figures for the third edition.

The following people read and reviewed all or parts of the manuscript: Len Gaydos, Joshua Lerner, E. Lynn Usery, Robert Sechrist, William Lawrence, Thomas Hodler, Benjamin Richason III, Leland Dexter, Mark Jakubauskas, Steven Walsh, Robert Churchill, Michael Peterson, and others. Len Gaydos was kind enough to torture-test the book in the classroom at San Jose State University in the spring of 1996. Sarah Battersby worked on the laboratory exercises on the web site, and Jordan Hastings did the actual site creation. The cartoons at the beginnings of each chapter are the inspired work of Englishman Jon Paul Fahy. Thanks JP!

I received a great deal of feedback on previous editions, from the reviews in the GIS literature, to E-mail from instructors around the nation and the world, to student evaluations from my own class, Geography 176A, at UCSB. Despite the fact that I have acted upon almost all of the suggested improvements, I humbly ask again that you keep them coming! I am committed to ongoing update, correction and improvement of this book—something that is especially important in a field as volatile and fast-paced as GIS. I have also received positive feedback on the teaching materials that accompany this book on the World Wide Web at `http://www.prenhall.com/clarke`. Encouraged by this, I have refined and expanded these materials to include lecture notes, exams and quizzes; a series of HTML-based labs using the ArcView software can be found on the CD in the back of this book. The outstanding efforts to create the CD and the Web site were by Ann Ricchiazzi and Jordan Hastings, with content provided by Violet Gray for the third edition and Sarah Battersby for the fourth.

There is much new material in the fourth edition. The graphics are much improved by color, and there is a new case study plus the laboratory and Web material. I hope this makes the basic GIS class, especially in geography programs around the country, of increased value to students. Geographers now find their discipline in demand intellectually, and they have the opportunity to conduct work of increased relevance. This is possible because of the power that GIS has placed into the hands of the spatial analyst. The challenge to use the power well is as vivid as ever.

I dedicate this fourth edition to my father, Raymond Harry Clarke, who died in England in September of 2000, shortly after the third edition appeared. I will always remember our last trip together, to the Greenwich Observatory. He took the photograph there that I have used in Chapter 2. And last, of course, and yet again, I thank Margot, Chantal, Elizabeth, Anne, and Caroline. Not only did they put up with a husband and father who travels way too much, they also modeled for some of the photographs in the book.

Keith C. Clarke
Santa Barbara

CHAPTER 1

What Is a GIS?

1.1 GETTING STARTED
1.2 SOME DEFINITIONS OF GIS
1.3 A BRIEF HISTORY OF GIS
1.4 SOURCES OF INFORMATION ON GIS
1.5 STUDY GUIDE
1.6 EXERCISES
1.7 REFERENCES AND BIBLIOGRAPHY
1.8 KEY TERMS AND DEFINITIONS

What in the world is a "GIS"?

—Item on the Internet's `comp.infosystems.gis` *FAQ list (FAQ = frequently asked question)*

GISs are simultaneously the telescope, the microscope, the computer, and the Xerox machine of regional analysis and synthesis of spatial data.

—(Ron Abler, 1988)

SQUAWK! "Who needs a GIS?"
SQUAWK! "Who needs a GIS?"

1.1 GETTING STARTED

If you are getting started with geographic information systems (GISs), or perhaps are curious to know what a GIS is and what it can do for you—then this book is for you. Whatever your field of interest, the chances are strong that you will at least come across and probably use a GIS in some way in the years ahead. So getting started now is a good idea! Perhaps you have already checked into some of the sources of information about GIS covered in this chapter. If you have, you may have noticed that there are already a great many GIS textbooks. Why, you may ask, do I need another? What is different about *this* GIS book?

Getting Started with Geographic Information Systems is intended to supplement the many new GIS books, not by updating or adding to them, but by *gently easing new GIS users into their community of understanding* without that long, slow, expensive, and sometimes painful climb up the GIS learning curve. As your first book in GIS, this one will set the foundation for a more breadth-first tour through the discipline than what the more advanced books can offer. By keeping the text up to date, the author and editors are working hard to ensure that your first GIS experience is timely, pleasant, and constructive.

First, we get started with GIS definitions, outline the development of the field, and map out some of the sources of information that can teach you more about GIS. It should be clear at the outset that GIS is not a new "killer app," namely a "must-have" innovative and essential computer application like a spreadsheet, a word processor, or a database manager. GIS is partly a killer app, but the upward shift in capability that its users receive is not due to computer software alone. Instead, GIS has built on the collective knowledge of the academic fields of geography and cartography, with some geodesy, database theory, and mathematics thrown in for good measure. As Ron Abler's definition shows, GIS is not just one but *many* simultaneous technological revolutions. *Getting Started with Geographic Information Systems* introduces a distilled version of the theory and content from the fields of these technologies—the minimum necessary to get you started—and then offers some signposts pointing toward where the revolutions will lead next. If you choose to go further, there are plenty of paths forward.

Using GIS requires you to think like a geographic information scientist. *Geographic information science* is a new field, born by merging skills and theory across many different disciplines, and it has now reached maturity after years of development. Like all new fields, geographic information science requires some mental readjustment. The purpose of this book is to gently guide you, the reader, through this readjustment. Fortunately, because you are reading this book, the chances are that you are already used to thinking graphically, mapping out information, and building analytical solutions around these maps and graphics. If not, I hope that this text will serve both to get you started and to unleash a part of the brain you may never have used before—the spatial part—that holds real power as a new way of problem solving.

1.2 SOME DEFINITIONS OF GIS

Good science starts with clear definitions. In the case of geographic information systems, however, definitions have sometimes been as clear as mud. As a result, different definitions have evolved over the years as they were needed. It is no surprise, then, that "geographic information system" can be defined in many different ways. Which definition you choose depends on what you seek. Common to all the definitions is that one type of data, *spatial data*, is unique because it can be linked to a geographic map (Figure 1.1).

Spatial means "related to the space around us, in which we live and function." Our own definition of GIS can start with a simple description of the three parts of a GIS, which are (1) the database, (2) the spatial or map information, and (3) some way to link the two. Necessary parts are a computer, some software, and people to use the system. We also need an underlying problem or task for which the GIS will be used, such as choosing a site for a nature preserve, routing an ambulance to a house, or maintaining a set of data that citizens of a town can use to become informed. Then, of course, we need both understanding and experience, both of the system and of the problem. As you will quickly learn, the last two items are the hardest to come by.

1.2.1 A GIS Is a Toolbox

A GIS can be seen as a set of tools for analyzing spatial data. These are, of course, computer tools, and a GIS can then be thought of as a software package containing the elements necessary for working with spatial data. If we want to write a book, we might visit a computer store and buy a word processing package in a box to install on our computer. Similarly, if we seek to work with spatial data, one definition of a GIS is the software in the box that gives us the geographic capabilities we need.

Peter Burrough, in his pioneering textbook, defined GIS as "**a powerful set of tools for storing and retrieving at will, transforming and displaying spatial data from the real world for a particular set of purposes**" (Burrough, 1986, p. 6). The key word in this definition is "powerful." Burrough's definition implies that GIS is a tool for geographic analysis. This is often called the "toolbox definition" of GIS because it stresses a set of tools each designed to solve specific problems.

Part Number	Quantity	Description
1034161	5	Wheel spoke
1051671	1	Ball bearing
1047623	6	Wheel rim
1021413	2	Tire
1011210		

Crimes during 2003		
Date	Location	Type
22-Jan	123 James St.	Robbery
24-Jan	22 Smith St.	Burglary
10-Feb	9 Elm St. #4A	Assault
13-Feb	12 Fifth Avenue	Breaking and Entering
14-Feb	17 Del Playa	Drunk and Disorderly

FIGURE 1.1: Two databases. A database contains columns (attributes) and rows (records). The bicycle parts list on the left is not spatial. The parts could be located anywhere. The list of crimes on the right is spatial because one of the attributes, the street address, locates the crimes on a map. This list could be used in a GIS.

If a GIS is a toolbox, a logical question is, What types of tools does the box contain? Several authors have tried to define a GIS in terms of what it does, offering a *functional definition* of GIS. Most agree that the functions fall into categories and that the categories are subtasks that are arranged sequentially as data move from the information source to a map and then to the GIS user and decision maker. Another GIS definition, for example, states that GISs are "**automated systems for the capture, storage, retrieval, analysis, and display of spatial data**" (Clarke, 1995, p. 13). This has been called a "process definition" because we start with the tasks closest to the collection of data and end with tasks that analyze and interpret the information. This book's chapters are structured around this sequence of functions, and each will be discussed in detail as the book progresses.

1.2.2 A GIS Is an Information System

Jack Estes and Jeffrey Star defined a GIS as "**an information system that is designed to work with data referenced by spatial or geographic coordinates. In other words, a GIS is both a database system with specific capabilities for spatially-referenced data, as well as a set of operations for working with the data**" (Star and Estes, 1990, p. 2).

This definition stresses that a GIS is a system for delivering answers to questions or queries, what might be called an *information system* sort of definition. This means that a GIS collects data, sifts and sorts them, and selects and rebuilds them to find precisely the right piece of information to answer a specific question. The reference to geographic coordinates is an important one, because the coordinates are literally how we are able to link data with the map. This theme is examined in detail in Chapter 2.

Another information system definition of a GIS is one that has stood the test of time remarkably well. As such, this definition is worth considerable thought. In 1979, during the infancy of the technology, Ken Dueker defined a GIS as "**a special case of information systems where the database consists of observations on spatially distributed features, activities or events, which are definable in space as points, lines, or areas. A geographic information system manipulates data about these points, lines, and areas to retrieve data for ad hoc queries and analyses**" (Dueker, 1979, p. 106).

The phrase "special case of information systems" implies that GIS has a heritage in information systems technology, which it indeed does. GIS did not invent database management, and there exists in computer science a 40-year tradition in this field all the way from the earliest spreadsheet programs, through relational database management, to the object-oriented database management of today. Information systems are used extensively in library science, in business, and around the Internet.

In Dueker's definition of GIS, the database itself consists of a set of observations, which implies a scientific approach to measurement. Scientists take measurements and record those measurements in some kind of system to help them analyze the data. The observations are *spatially distributed*; that is, they occur over space at different times and at different locations at the same time.

The observations are those of *features*, *activities*, and *events*. A *feature* is a term from cartography meaning an item to be placed on a map. *Point features*, such as an elevation bench mark (Figure 1.2), have only a location. *Line features* have several locations strung out along the line in sequence like a bead necklace, an example being

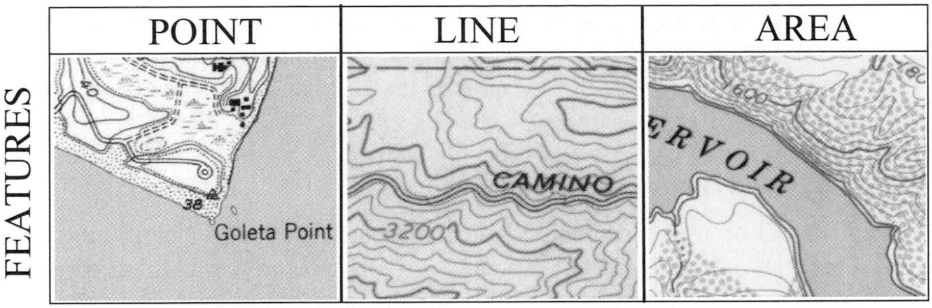

FIGURE 1.2: The Feature Model: Examples of a point feature (38 foot elevation bench mark), a line feature (road, contours), and area features (reservoir, vegetation).

a road or a stream. *Area features* consist of one or more lines that form a loop, such as the shoreline of a reservoir or lake, or the edge of a patch of vegetation. Traditionally, the source of geographic information is the map, and the information on a map consists of a set of graphic symbols, such as colors, lines, patterns, and shades.

"Activities" implies a link to the social sciences. Human activities create geographic patterns and distributions. They lead to the population map, census map, distribution of disease incidences, location of infrastructure, and so on—all related to how people live their daily lives. The "event" part of GIS implies that geographic data fall not only into space but also into time. Time gives us a fourth dimension and becomes a part of the data because events happen in time and features exist over a duration. The reservoir in Figure 1.2, for example, was created by damming a stream and so was not shown on a map made 100 years ago.

Dueker's GIS definition assumes that events also have expressions as points, lines, or areas in space and on the map. An example of a point event is the location of a traffic accident. A line activity could be the flow of electricity along a segment of a power cable. An area event could be the freezing of a body of water, such as the Central Park reservoir in New York City. The information element becomes useful to the GIS user because it exists, it has data associated with it, and it has cartographic reality as a feature on a map.

We use the information mapped in the GIS for doing exactly what an information system should do: solve problems, do queries, come up with the answer, or try out a possible solution. So we manipulate the data, not by hand, but digitally. We manipulate data about events or activities by using the digital map features that represent them as "handles." In other words, *the points, lines, and areas in this map database are used to manage the data.* Another key part of Dueker's GIS definition is that the queries must be ad hoc or context-specific queries. We don't have to know in advance when building a GIS exactly what we want to use it for. This means that GIS is a generic problem-solving tool; it is not something built just for that project or to get this week's assignment done. The value of GIS comes from its ability to apply general geographic methods to specific geographic regions.

Finally, in Dueker's definition a GIS can also do analysis. Usually, the purpose of having data in GIS form is so that an analyst can extract what is necessary to make predictions and explanations about geographic phenomena. A focus on GIS technology ignores the fact that the ultimate purpose of the system is to solve problems. Geographic information science goes beyond description, to include analysis, modeling, and prediction. The information systems definition, then, leads back to the role of a GIS as a problem solver. It begs the question: Is this just one more scientific method, or is this a new scientific approach?

1.2.3 GIS Is an Approach to Science

As a tool or as an information system, GIS technology has changed the entire approach to spatial data analysis. GIS has already been compared to not one but several simultaneous revolutionary changes in the way that data can be managed. The convergence of GIS with allied technologies, those of surveying, remote sensing, air photography, the global positioning system (GPS), and mobile computing and communications has fed a spectacular growth of these technologies.

As a result, the way of doing business—the standard operating procedure of geographic and spatial information handling—has rapidly restructured itself. First, the technology of GIS has become much simpler, more distributed, cheaper, and has crossed the boundary into disciplines such as anthropology, epidemiology, facilities management, forestry, geology, and business. Second, this mutation has led to a culling of the body of knowledge that constitutes geography so that it is suitable for use in these parallel fields as a new approach to science. Goodchild called this "geographical information science" (Goodchild, 1992). In the United States the preferred term is *geographic information science*.

Goodchild defined geographic information science as "**the generic issues that surround the use of GIS technology, impede its successful implementation, or emerge from an understanding of its potential capabilities**" (Goodchild, 1992). He also noted that this involved both research *on* GIS and research *with* GIS. Supporting the science are the uniqueness of geographic data, a distinct set of pertinent research questions that can only be asked geographically, the commonality of interest of GIS meetings, and a supply of books and journals. On the other hand, Goodchild noted that the level of interest depends on innovation, that it is hard to sustain a multidisciplinary (rather than interdisciplinary) science, and that at the core of the science, in geography, a social science tradition has to some extent an antipathy toward technological approaches.

This book is an effort to distill from the discipline of geography exactly those components that are derived from the areas of research outlined by Goodchild. As such, this book adopts Goodchild's approach. The chapters that compress the principles of cartography are Chapters 2 and 7; analytical cartography's contributions fall into Chapters 3–5; and spatial analysis is discussed in Chapter 6. Added to these are doses of general geography, database management, and applied GIS. This knowledge base constitutes the new and strengthening field of geographic information science.

1.2.4 GIS Is a Multibillion-Dollar Business

Groups monitoring the GIS industry estimate the total value of the hardware, software, and services conducted by the private, governmental, educational, and other sectors that handle spatial data to be billions of dollars a year. Furthermore, for the last half decade of the 1990s, and into the current decade, the industry has seen double-digit annual growth. Anyone who attends a national or international conference in the field can feel an overwhelming sense of rapid growth, sophistication, and the sheer magnitude of the transformation that GIS has led.

Largely responsible for this situation were the massive cost reductions in technology dating from about 1982, when computers moved out from behind glass windows as tended by people in white coats and onto the desktop. This decline in cost, aided by the success of the workstation as a tool in engineering settings, has led to a rapid increase in what is usually called the "installed base" of GIS. Just about every major academic institution in the United States and in many other countries now teaches at least one class in GIS. Most local, state, and federal government agencies use GIS, as do businesses, planners, architects, foresters, geologists, archeologists, and so on. This growth in pure numbers, added to the increase in sophistication of the systems, is what has led to the big business aspect of GIS.

However, other steps have been critical to the booming (and blooming) of GIS. First, the industry was founded on vast amounts of inexpensive federal government data, mostly data of the U.S. Census Bureau and the U.S. Geological Survey. Second, the community has been a successful advocate of the field and has rapidly developed an infrastructure for self-support, user groups, network conference groups, and so on. Third, the addition of graphical user interfaces and the addition of extremely useful features such as help screens and automatic installation routines have played an essential role. Fourth, GIS has merged successfully with parallel technologies and has benefited from the resultant multiplier effect.

The growth of GIS has been a marketing phenomenon of amazing breadth and depth and will remain so for many years to come. Clearly, GIS will continue to integrate its way into our everyday life to such an extent that it will soon be impossible to imagine how we functioned before. The GIS operations could become so transparent to the public that we would not even realize that GIS was there, just as we give no thought to the microprocessor that calculates our change at the cash register. Then again, many would argue that it already has done just this.

1.2.5 A GIS Plays a Role in Society

Many people doing research on GIS have argued that defining GIS narrowly, as a technology, as software, or as a science, ignores the role that GIS plays in changing the way people live and work. Not only has GIS radically changed how we do day-to-day business, but also how we operate within human organizations. Nick Chrisman (1999) has defined GIS as "**organized activity by which people measure and represent geographic phenomena then transform these representations into other forms while interacting with social structures**."

This definition has emerged from an area of GIS research that has examined how GIS fits into society as a whole, including its institutions and organizations, and how GIS can be used in decision making, especially in a public setting such as a town meeting, or on a community group Web site. This latter field is termed PPGIS, for Public Participation GIS.

Few people would doubt that as GIS has become part of the way of doing business within many organizations, such as planning offices or state planning agencies, the result has been a shifting in the work assignments, job descriptions, responsibilities, and even the power relationships of the organization. For example, when GIS is first introduced into a work environment, it is very important to have a "champion," someone who is a GIS advocate from within the group. Many in the study of GIS have focused on describing and analyzing these impacts rather than looking technically at GIS, or at GIS in its application. So far, the field has generated a history of the discipline (Foresman, 1997) and an increasingly popular set of meetings and conferences. Several books have stimulated interest in this approach, including *Ground Truth* (Pickles, 1995), which introduced the somewhat more humanistic and social science dimension to GIS research work.

Nick Chrisman's definition of GIS includes all of the social process of GIS functions. For example, a GIS may be used to capture data about land holdings as ownership parcels. However, the use and purpose of the data and their dissemination will vary according to the philosophy and traditions of the community in which the data are being used. In a growth-oriented community, for example, a GIS might be seen as a mechanism for expediting building permits and increasing land sales. In a more conservation-oriented community, a GIS might be seen and used as a vehicle for raising public awareness about environmental issues, supporting community planning, or enforcing pollution controls. Although essentially the same GIS software, hardware, and data may be in place in the two settings, the staff, their work assignments, and the degree of administrative control might be very different. It is the human factors involved that determine much about the GIS, rather than the technical capabilities.

Another component that Chrisman's definition recognizes is the importance of a basis in measurement. In the abstract sense, a GIS supports measurements about the land with many different levels of accuracy and reliability. In most cases, the GIS is based on the "best available data," but virtually always some of the data are incomplete, outdated, or missing. How GIS users come to terms with this problem is often as large a factor in the GIS's capabilities and effective use as are the software, hardware, and processes involved. As we state later, a GIS, like a map, is often a set of errors that have been agreed on. This definition notes that not just the errors and the system supporting them define GIS, but also those critical agreements about the data that result among the people involved.

1.3 A BRIEF HISTORY OF GIS

Many of the principles of the new geographic information science have been around for quite some time. General-purpose maps date back centuries and usually focused on topography, the lay of the land, and transportation features such as roads and rivers. More recently, in the last century, thematic maps came into use. Thematic maps contain information about a specific subject or a theme, such as surface geology, land use, soils, political units, and data collection areas. Although both types of maps are used in GIS, it is the thematic map that led cartography toward GIS. Some themes on maps are clearly linked. For example, a map of vegetation is closely tied to a map of soils.

It was the field of planning that first began to exploit thematic maps by extracting data from one map to place them on another. As an early example, the geographic extent of the German city of Dusseldorf was mapped at different time periods in this way in 1912, and a set of four maps of Billerica, Massachusetts, were prepared as part of a traffic circulation and land-use plan in the same year (Steinitz et al., 1976). By 1922, these concepts had been refined to the extent that a series of regional maps were prepared for Doncaster, England, which showed general land use and included contours or isolines of traffic accessibility. Similarly, the 1929 "Survey of New York and Its Environs" clearly shows that overlaying maps on top of each other was an integral part of the analysis, in this case of population and land value.

In 1950, the publication of the *Town and Country Planning Textbook* in Britain included a landmark chapter, "Surveys for Planning," by Jacqueline Tyrwhitt (Steinitz et al., 1976). Various data themes, including land elevation, surface geology, hydrology/soil drainage, and farmland, were brought together and combined into a single map of "land characteristics" (Figure 1.3). The author described how the maps were drafted at the same scale, and how map features were duplicated so that the maps could be superimposed precisely, using these features as a guide. Just as many others had "discovered" America, it was Columbus who is remembered because he was the first to write about it (and, incidentally, to draw a map!).

In 1950, the technique of map overlay, now so common in GIS packages, was "invented" by Tyrwhitt, although it is likely that there were earlier precedents. Nevertheless, it is clear that by 1950, maps were regularly being traced onto transparent overlays for use in land analysis and presentation. Twenty years later, Ian McHarg, in his 1969 book *Design with Nature*, described using blacked-out transparent overlays to assist in finding locations in New York's Staten Island that were solutions to multiple siting control factors (Figure 1.4).

As early as 1962, two planners at the Massachusetts Institute of Technology had evolved the map overlay idea to include weighting, by making the overlays different in their importance with respect to each other. The plan involved 26 maps showing the desirability of highways. Maps were ordered in a "procedural tree," and different combinations were made by reordering the map layers photographically.

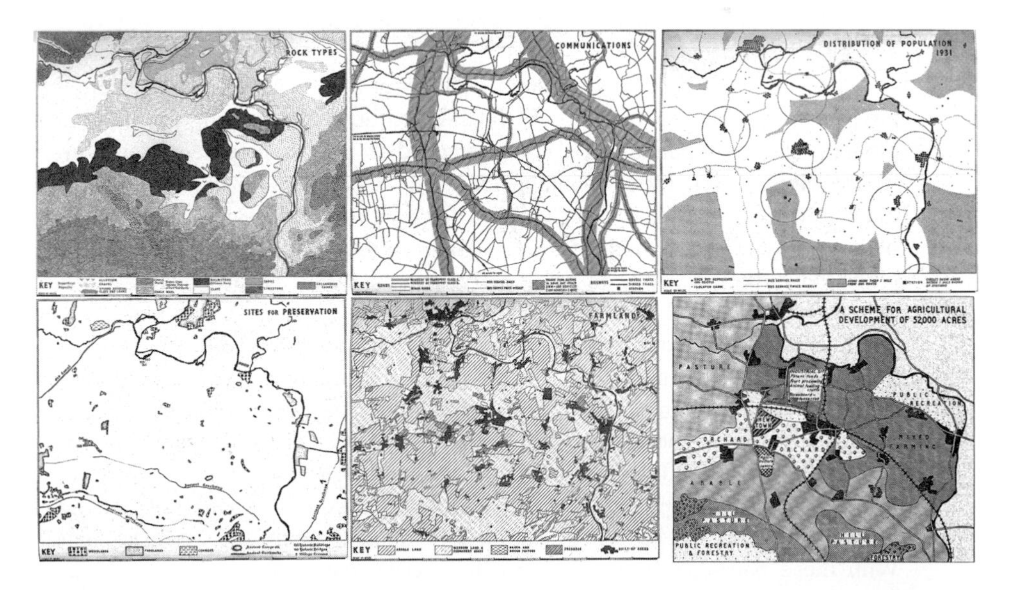

FIGURE 1.3: Map overlay as presented in Town and Country Planning Textbook by Jacqueline Tyrwhitt. Map at lower right is a composite overlay of several others, and is used to plan a community by applying basic map layer overlay principles.

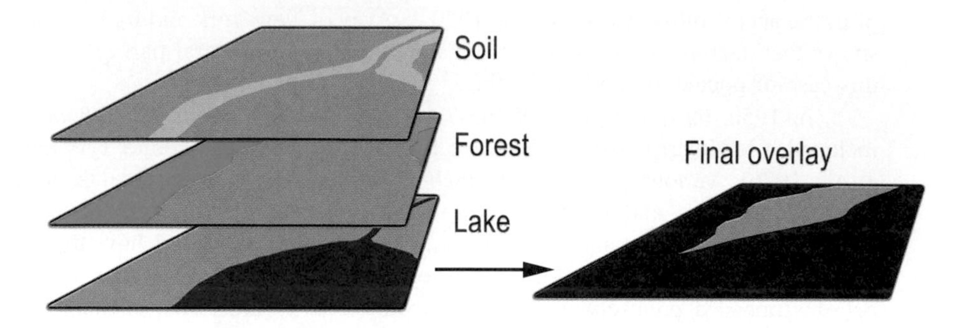

FIGURE 1.4: Map overlay as presented in *Design with Nature* by Ian McHarg. Each transparent layer map "blacked out" areas excluded as unsuitable locations.

During the 1960s, many new types of thematic maps were becoming available in standardized scales, such as topographic and land cover maps from the U.S. Geological Survey and soil maps from the U.S. Department of Agriculture's Soil Conservation Service (now the Natural Resource Conservation Service). It became fairly straightforward to select the right maps, trace off a layer, or photographically build a "separation" for one type of feature on the map, and then to combine the layers mechanically.

The scene was set for the arrival of the computer. In 1959, Waldo Tobler, then a graduate student, published a paper in *Geographical Review* outlining a simple model for applying the computer to cartography (Tobler, 1959). His model, often referred to as a MIMO (map in–map out) system, had three elements: a map input, map "manipulation," and a map output stage. These three simple steps were the distant origins of the geocoding and data capture, data management and analysis, and data display modules now part of every GIS package.

Within just a few years, many people were busy writing computer programs using programming languages such as FORTRAN to draw maps using primitive printers and plotters. The new demands on computing led to the development of the first digitizer by the New Haven group planning the 1960 census and to the development of many other new devices. As new capabilities for mapping came along, the first experiments with entirely new mapping methods, such as animation and automated hill shading, took place. Nevertheless, none of these early systems could be described as a GIS. During the early years, development of computer mapping resulted in less and less dependence on individual computer programs and more and more on *software packages*, sets of linked computer programs that had common formats, structures, and files. When *modular computer programming languages* came along during the 1960s, the process of writing integrated software became easier. Among the early computer mapping packages were SURFACE II, IMGRID, CALFORM, CAM, and SYMAP.

Most of these programs were sets of modules for the analysis and manipulation of data and the production of choropleth (shaded area) and isoline (contour) maps. With these packages it was possible to overlay data sets, reducing the hard work of doing this only with transparencies. Closely related to the mapping software was the development of the first systematic map databases. First came the Central Intelligence Agency's (CIA) World Data Bank, a global map of coastlines, rivers, and national boundaries still in use today, along with the CAM software that projected it onto maps at different scales.

After many prototype systems, the DIME (dual independent map encoding) coding system was devised by the U.S. Census Bureau as an experiment in digital mapping and data handling. The DIME and the resultant files, called geographic base files (GBFs), were a major breakthrough in the history of geographic information representation. The GBF/DIME recognized that attribute information, in this case all the data collected by the census, and the computer maps used in planning the census could be integrated not just for mapping but also to search for geographic patterns and distributions. Some landmark early systems were the Canada Geographic Information System (CGIS) in 1964, the Minnesota Land Management System (MLMIS) in 1969, and the Land Use and Natural Resources Inventory System in New York (LUNR) in 1967. Both MLMIS and LUNR were derivatives of the GRID system that replaced SYMAP at Harvard University.

During the mid and late 1960s a cluster of faculty and students at Harvard University's Laboratory for Computer Graphics and Spatial Analysis made some major theoretical contributions and developed and implemented several new systems. Most influential among these was the GIS program Odyssey. With program modules named for sections of Homer's *The Odyssey*, the team pioneered a set of data structures that came into common use after their publication in 1975 (Peucker and Chrisman, 1975) called the *arc/node* or *vector* data structure. The computer routines that sorted digitized chains and lines and assembled topologically connected polygons, for example, was called the *Whirlpool*. Odyssey was a highly influential arc/node-based GIS and influenced much of the software that followed.

In Chapter 4 we will examine this structure in some detail, but what was different then was that the data structure captured polygon information using a series of nodes; there was a beginning node and an ending node with an arc between them. The arcs could be assembled to construct a polygon because the structure contained information about adjacency and connectivity between features. Many GIS packages, including Arc/Info, have been based on this simple model of geographic features.

In 1974 the International Geographical Union surveyed software in the mapping sciences and found enough GIS software to publish an entire inventory volume entitled *Complete Geographical Information Systems*. While in the early days many different terms were used to describe a GIS, this report began the convergence on the term GIS as a generic name for this new applications and research field. Reporting on the results of the survey, Kurt Brassel noted that "we understand that a mapping system is mainly designed for display purposes, even though it may fulfill some secondary functions that are not graphical. A geographical information system is designed for a broader range of applications, even though mapping functions may represent an important subset of its activities" (Brassel, 1977, p. 71). Both GIS and computer mapping continue to have this significant and constructive overlap in their content.

Development of GIS persisted into the 1980s, with large computers and FORTRAN continuing to dominate. In 1982, IBM introduced its PC, or personal computer, following from the Apple II microcomputer of a few years earlier. The impact of this single advance cannot be understated. Within just a few years, some of the large GIS packages, such as Arc/Info, had made the difficult transition to the microcomputer. Others, such as IDRISI, owe their origin to the low cost and high degree of efficiency that characterized the first generation of PCs. Other packages migrated instead to the new workstation platform that had developed from the minicomputer and networking trend. Again, other packages, such as GRASS, owe their origins to this transition.

The 1980s and early 1990s saw GIS mature as a technology. Many older packages that failed to move to the new languages and platforms died out, to be replaced by newer systems that could exploit the capabilities of the more powerful equipment. Costs of storage fell remarkably, computer power increased many-fold, and the first generation of GUIs or graphical user interfaces, among them X-Windows, Microsoft Windows, and Apple's Macintosh, made the software considerably easier to use, adding features such as menus, online manuals, and context-sensitive help. During the 1980s, the Internet arose out of the collection of early networks, such as Arpanet and NSFNet, that were beginning to link scientists and became a significant new component of computing.

The 1980s also saw the origins of the infrastructure for GIS: the books, journals, conferences, and other resources that are so critical to finding out about GIS. During this era, the National Science Foundation created the National Center for Geographic Information and Analysis (NCGIA), which devised a national college curriculum and developed broad research agendas for academic research on GIS.

The 1990s saw remarkable growth in the GIS world. Several new factors emerged. First, GIS spread far beyond its origins in the mapping science to encompass developments in new fields such as geology, archeology, epidemiology, and criminal justice. Also, the cost of GIS fell markedly after a series of desktop GIS products emerged. The increasing market penetration of personal computers, and the more mobile laptop and portable digital assistants, took GIS into many new work environments. Object-oriented programming approaches made radical improvements in the software engineering that could be applied to GIS software and allowed the portability of programs across many computer platforms.

In addition, GISs became fully integrated with the global positioning system, greatly enhancing the system's data capture capability. High-resolution imagery became common as a reference base for GIS data. Finally, the emergence of the Internet and e-commerce has placed GIS onto the World Wide Web as Web-GIS. Many now talk of a new era of g-commerce or g-trade, based on geographically enabled Web search capability rather than simply map display.

And so we arrive at the present. Although GIS's lineage dates back to the roots of cartography and although thematic cartography and map overlay date from the nineteenth century, what is today known as GIS owes its birth to a cluster of interrelated events and human interactions in the 1960s, and its spectacular growth to the microcomputer, the workstation, and the Internet. It is, indeed, a rather short history and one that is still being written.

1.4 SOURCES OF INFORMATION ON GIS

This section is designed to help you find more information about GIS topics not covered or covered in insufficient depth in this basic book. Historically, GIS has been a somewhat disjoint field from a reader's standpoint, and most of the major books, journals, and online resources date from only the last few years. This is far less an issue today, however, and there are now some excellent sources of GIS information. These fall into groups and are covered here under journals and magazines, books, professional societies, the Internet and the World Wide Web, GIS conferences, and educational organizations and universities.

The amount of information available about GIS is somewhat overwhelming. An excellent place to begin one's search is at a library, or perhaps by connecting to the Internet and using one of the World Wide Web search tools. This is possible even at one's home computer, but slow enough that a visit to the library may be more productive. Some libraries have facilities to connect to network search systems and even specialized staff with training in geographic information.

As in our definition of geographic information science, the information sources on GIS fall into the broad categories of research *with* GIS and research *on* GIS. As a beginner, try restricting your search to basic material rather than going straight to the research frontier. This can come later. A good way to research a topic is to find publications that came out at about the time a new idea was being introduced. In the older papers, articles, or book chapters, the authors had to write for an audience that would be unfamiliar with the language and concepts under discussion. This is the case in several classic papers in the GIS arena. The writing remains today as a good first step toward understanding and an excellent place to get started with GIS.

1.4.1 Journals and Magazines

Today many journals and magazines publish articles and papers on GIS, and a large number occasionally publish a few papers or a special issue. Journals that publish exclusively on GIS include the academic research journals *International Journal of Geographical Information Systems*, *Geographical Systems*, and *Transactions in GIS*, and the news and applications periodicals *Geospatial Solutions, Geoinformatics,* and *GeoWorld.*

Some journals are specialized in their audience. For example, *Business Geographics* catered to the business world until it stopped publishing in June 2001, *GIS Law* to the legal profession, and *GrassClippings* to users of the GRASS GIS package. There are also regional journals, such as *GIS Asia/Pacific* and *GeoEurope*. Many foreign-language journals exist.

Among the scholarly journals that publish academic work about GIS and its uses are the *Annals of the Association of American Geographers*; *Cartographica*; *Cartography and GIS*; *Computer*; *Computers, Environment, and Urban Systems*; *Computers and Geosciences*; *IEEE Transactions on Computer Graphics and Applications*; the *URISA Journal* (Urban and Regional Information Systems Association); and *Photogrammetric Engineering and Remote Sensing.* The latter devotes one issue a year entirely to GIS.

Some journals carry occasional articles, including *Cartographic Perspectives, Cartographica, Journal of Cartography*, *Geocarto International*, *IEEE Geosciences*, the *International Journal of Remote Sensing*, *Landscape Ecology*, *Remote Sensing Review* and *Infoworld.* GIS is also occasionally national news, with articles in leading newspapers and weekly magazines.

1.4.2 Books

Over the last few years, many books have been written that relate to GIS, although not all have been aimed at the new GIS user. The first generation of GIS books was targeted more toward the advanced user or expert in GIS who wished to see where research in a specialty was going. Some early sources of information for those in GIS were the set of collected readings by Marble et al. (1984), Ripple (1987; 1989), and Worrall (1991a).

At one time in most college classrooms, faculty taught the courses in GIS using reproductions of classic journal papers, so books containing collections of readings served an important role. The first comprehensive textbook in GIS appeared in 1986: Peter Burrough's book *Principles of Geographical Information Systems for Land Resources Assessment.* Burrough's text was used almost exclusively until the appearance of several other works in the late 1980s and 1990s, including those by Star and Estes (1990), Tomlin (1990), Laurini and Thompson (1992), Aronoff (1989), Huxhold (1991), Chrisman (1997), DeMers (1997), and the second edition of Burrough's *Principles* (Burrough and McDonnell, 1998). Recent additions include Longley et al. (2001) and Lo (2002).

The textbooks share the fact that they are written primarily for the advanced student in a college classroom. The professional market includes a large number of people in small offices, planning divisions, and so on, who also need information and training. These individuals have been adequately served by the professional book market, from both a tutorial and a reference perspective.

Aimed at particular segments of users, GIS books have covered such disciplines as health (DeLepper et al., 1995), business (Grimshaw, 1994), surveying (Onsrud and Cook, 1990), management (Aronoff, 1989; Obermeyer and Pinto, 1994), defense (Ball and Babbage, 1989), geology (Bonham-Carter, 1994), social theory (Pickles, 1995), archeology (Allen et al., 1990), urban planning (Huxhold, 1991), public health (Cromley and McLafferty, 2002), history (Knowles, 2002), environmental modeling (Clarke et al., 2002), and landscape ecology (Haines-Young et al., 1993).

Strictly from the professional's perspective, several books serve as comprehensive guides to the industry and technology of GIS as a whole, in some cases listing and reviewing the sources covered in this section. Among these are the AGI Source Book (AGI, 1995), Berry (1993), Antenucci (1991), Korte (1994), and Montgomery and Schuch (1993). Others are aimed at training in particular and are specific to one or another of the various GIS software packages on the market, such as ESRI (1995).

Some books look at a particular component of the more advanced end of GIS research, among them the inclusion of time as a data element (Langran, 1992), three-dimensional visualization (Raper, 1989), data accuracy (Goodchild, 1989), data structures (Samet, 1990), and analytical cartography (Clarke, 1995). Finally, one source has tried to conduct a comprehensive overview of the entire field of GIS, including surveys of applied, theoretical, and research frontier contributions. This rather lengthy and expensive set of volumes (Longley et al., 1999), now in its second edition, is recommended as a reader and source of more detailed information for all GIS scholars. Another comprehensive reference source is Bossler's manual (Bossler, 2002).

It is almost impossible to keep up with the GIS literature from a static source such as this book. Searches under "Geographic Information System" using Google (`www.google.com`) yielded 265,000 hits in February of 2000 and 218,000 in June of 2002 (assuming more stringent search rules). Even Amazon.com (`www.amazon.com`) showed 29 items. Two recent projects have attempted separate online bibliographies on GIS, covering books, journals, and proceedings volumes from meetings. These two Web-based projects are the GIS Master Bibliography (Web location `http://liinwww.ira.uka.de/bibliography/Database/GIS/index.html`) and the Spatial Odyssey (`http://wwwsgi.ursus.maine.edu/biblio/`). Clearly, GIS as a "literature" is a moving target.

1.4.3 Professional Societies

The major GIS journals follow the professional societies closely, and many have book distribution lists with member discounts. The professional societies associated with the technology are the *American Congress of Surveying and Mapping*, the *American Society for Photogrammetry and Remote Sensing* (ASPRS), the *Association of American Geographers*, the *Geospatial Information and Technology Association,* and the *Urban and Regional Information Systems Association* (URISA).

The ACSM (American Congress of Surveying and Mapping) has member organizations, each of which has an interest in GIS, including the Cartographic and Geographic Information Society. Journals produced are *Cartography and Geographic Information Systems*, formerly the *American Cartographer*, and *Surveying and Land Information Systems*. The American Society for Photogrammetry and Remote Sensing covers the mapping science fields broadly. Its journal, *Photogrammetric Engineering and Remote Sensing*, is monthly and has taken on a very strong GIS theme over the last few years. Once a year the journal publishes a special issue on GIS. The journal itself publishes GIS articles in an even balance with traditional mapping and remote sensing. The Association of American Geographers (AAG) has a GIS specialty group, which constitutes the largest specialty group within the organization. The association has regular regional and annual national meetings, and the organization supports a newsletter with job listings.

URISA is a large organization aimed primarily at professionals in planning, government, infrastructure, and utilities. The organization holds an annual national conference and hosts many activities, including job listings, it publishes a journal, and it distributes newsletters. Another professional organization is the Geospatial Information and Technology Association. This group hosts an annual national conference, publishes conference proceedings and other publications, issues a newsletter, and provides scholarships and internships for college students to work in GIS firms.

1.4.4 The Internet and the World Wide Web

An extraordinary, indeed overwhelming, amount of information about GIS can be found on the Internet using the World Wide Web (WWW). Everything from newsgroup frequently asked questions (FAQs) to commercial GIS software vendors' Web sites, to entire online and downloadable GIS packages, such as GRASS, is available.

The best way to search is to load a suitable Web browser such as Internet Explorer or Netscape, and then follow your own interest. Although slow, this is possible from a home computer using a modem and an online service such as America Online. The bibliography at the end of this chapter lists some of the places where GIS information resides, but there are many, many more, and even more are added every day. The network news group GIS-L (comp.infosystems.gis) is a long-standing source of technical information on GIS (Figure 1.5). Users post questions to the list, and people answer back. Replies are archived, and when common threads emerge, they are compiled into a FAQ list, sometimes echoed and hosted on sites across the World Wide Web. GIS-L is currently hosted by the URISA professional organization at `http://www.hdm.com/urisa3.htm`. Following discussions on GIS-L is an excellent way to get an introduction to the software and environment of GIS applications.

One very useful GIS online resource is a network-accessible copy of the U.S. Geological Survey's brochure *Geographic Information Systems* (Figure 1.6) and accessible at

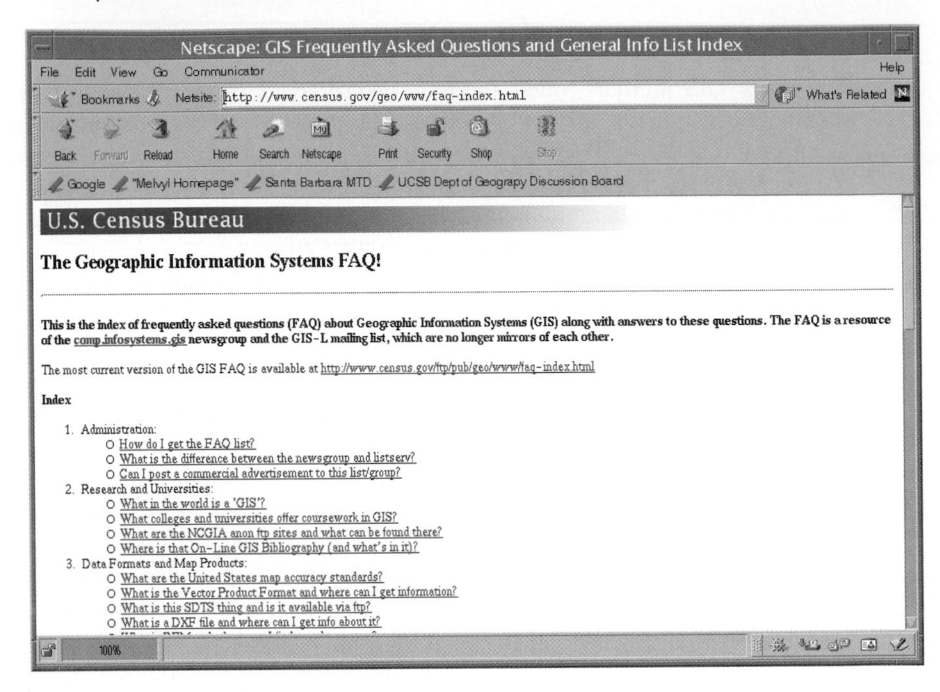

FIGURE 1.5: Web page for GIS-L FAQ index maintained by the U.S. Census Bureau. (See: `http://www.census.gov/geo/www/faq-index.html`.

`http://www.usgs.gov/research/gis/title.html`. This Web document was originally a wall-size poster, also available free, containing all sorts of GIS samples, examples, and definitions.

A recent addition to the Internet GIS information sources are news services that update frequently, some daily, information about GIS. Among these are GIS Monitor `www.gismonitor.com`, Spatial News `www.spatialnews.com`, the GIS Café `www.giscafe.com`, and Geoplace, home of several of the news and information journals (`www.geoplace.com`). Some of these will send you daily updates on the Web site's contents by e-mail.

More information about the GIS data available on the WWW is included in Chapter 3. Newly available are the various data clearinghouses now forming part of the Spatial Data Clearinghouse, an online "library catalog" of available GIS format data available free or at cost. In addition to acting as a library, the WWW also serves as an information source, a software source, a data source, and even as a place to publish results. In Chapter 10, the future role of the WWW and the Internet is discussed from a GIS viewpoint.

1.4.5 Conferences

As a growing and new industry, especially in the early days when there was as yet no major journal where research and applications were published, the various professional conferences for GIS served as "literature." As a result, some of the key papers in GIS technology and theory appear, at least in their early and most readable form, as papers

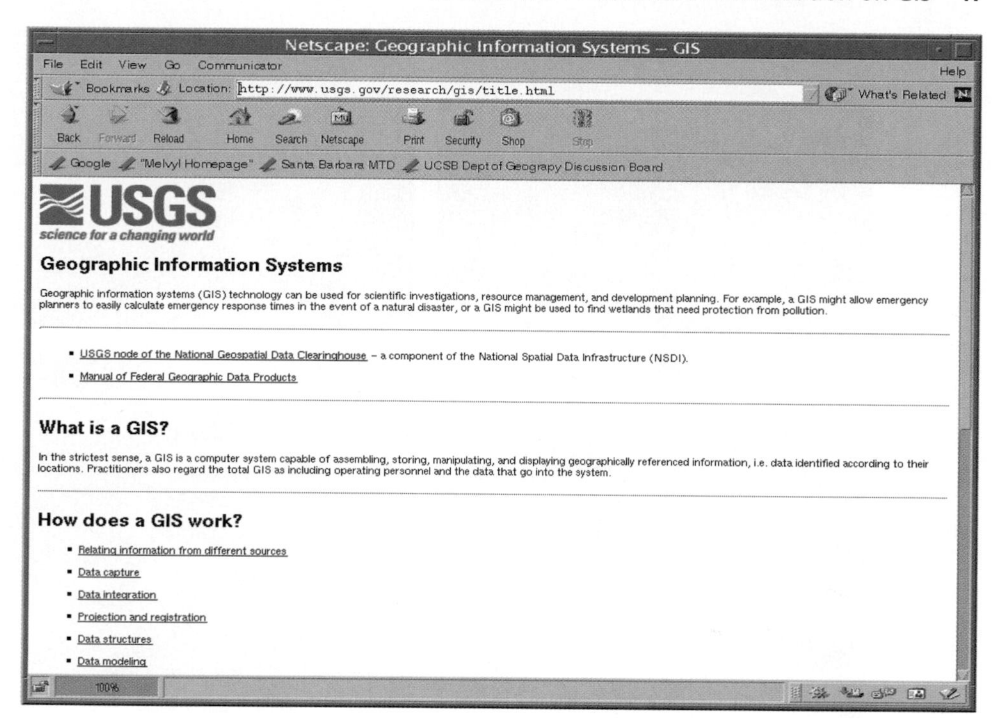

FIGURE 1.6: Entry point on the World Wide Web for the USGS brochure *Geographic Information Systems*, an excellent first glimpse of GIS capabilities. `http://www.usgs.gov/research/gis/title.html`. (Used with permission.)

in conference proceedings. Unfortunately, these papers are often hard to find. In many cases, the professional societies sell back copies of proceedings at a discount.

The earliest conference in GIS was probably the original Harvard Conference on Topological Data Structures. Very soon, the AutoCarto (International Symposium on Automated Cartography) took over as a key place for the publishing of papers. This series now has 12 volumes, with the most recent conference in 1995. During the 1980s, the GIS/LIS conference became a leading focus of GIS activity but was terminated in 1998, having completed its major task. The proceedings remain as a valuable GIS resource.

Other major conferences have been the URISA annual conference, which has more of a GIS application focus; the ACSM/ASPRS technical meetings, with both a research and applications orientation; and the GITA conference, which is the one many municipalities and industrial GIS users attend (Figure 1.7). The biannual spatial data handling conference, held alternately in the United States and internationally, has become a major concentration of people working on GIS research and development. Several states, including New York, Texas, California, and North Carolina, also hold annual meetings. In addition, the various GIS packages or local areas hold their own user group meetings, some of which even approach the professional conferences in size. Largest of them all is the ESRI User Conference, held annually in San Diego, California, with over 10,000 attendees.

FIGURE 1.7: Selected photographs taken in the exhibits area at the ACSM/ASPRS/FIG conference in Washington, D.C. in April, 2002. (Photographs by the author.)

1.4.6 Educational Organizations and Universities

Many colleges and universities teach classes in GIS, and some offer complete programs with course sequences and certificates. No national body as yet certifies people in GIS, but some vendors offer certification as instructors. Some universities and extension services offer short courses, and most of the major GIS vendors offer short training programs lasting anywhere from a few hours at a national or regional conference to several days or weeks.

Within universities and colleges, GIS classes are taught in many departments. Most are in geography, but many are also in departments and programs in geology, environmental science, forestry, civil engineering, computer and information science, and many others. There is little consensus among those teaching GIS as to what the content for a course in GIS should be, although standardization efforts are under way. Many programs around the country offer just a single class, structured in much the same way as this book. Others use the national GIS curriculum of the National Center for Geographic Information and Analysis (NCGIA) (Figure 1.8). This center is a National Science Foundation–funded program designed to channel GIS research and learning toward an improvement for the discipline of geographic information science. The center, a consortium of three universities, maintains a Web site at `http://www.ncgia.ucsb.edu`. The group has conducted a comprehensive set of research initiatives in GIS covering many different areas. Publications, research reports, outreach activities, and sponsorship of conferences and visitors to the center have been the main activities of the NCGIA.

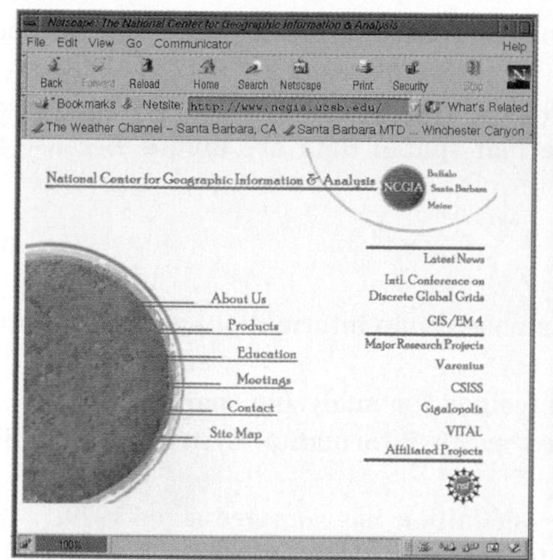

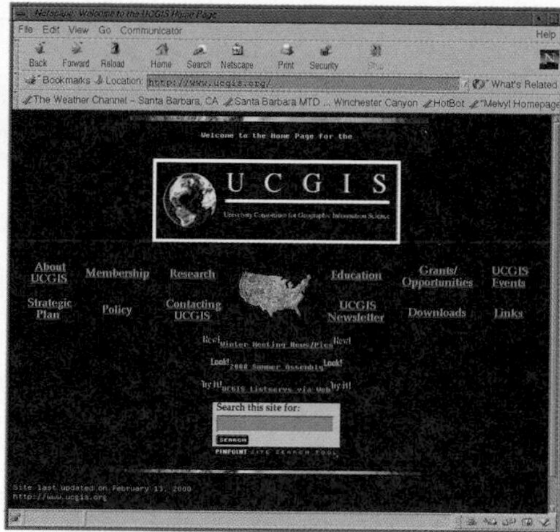

FIGURE 1.8: GIS University Consortia Websites. Left: NCGIA (`www.ncgia.uscb.edu`). Right: UCGIS (`www.ucgis.org`).

The primary mission of the NCGIA is to conduct basic research, but the organization also coordinated the formation of a far broader geographic information science community in the United States. In 1994, a total of thirty-three universities, research institutions, and the Association of American Geographers met to establish the University Consortium for Geographic Information Science (UCGIS) (Figure 1.8). UCGIS is a nonprofit organization of universities and other research institutions dedicated to advancing the understanding of geographic processes and spatial relationships through improved theory, methods, technology, and data. As of spring 2000, there were 64 members of the consortium. Several meetings and resource collections have proven very useful as GIS information sources, and the updated list of UCGIS initiatives gives an indication of research directions in GIS.

A college or university near you may be able to provide information about GIS courses or help you to find out more. University libraries hold many GIS publications and conference proceedings, and these are also a good starting point. Perhaps, after reading this book, you will be tempted to take a college course, or maybe you are using this book as part of one. If so, don't forget that learning never ends and that increasing your GIS education also increases your effectiveness as a GIS user, your ability as a geographic information scientist, and your employability as a GIS specialist.

1.5 STUDY GUIDE

1.5.1 Summary

CHAPTER 1: What Is a GIS?

Getting Started (1.1)

- **GIS is built on knowledge from geography, cartography, computer science, and mathematics.**

- **Geographic information science is a new interdisciplinary field built out of the use and theory of GIS.**
- **Different definitions of a GIS have evolved in different areas and disciplines.**
- **All GIS definitions recognize that spatial data are unique because they are linked to maps.**

Some Definitions of GIS (1.2)

- **A GIS at least consists of a database, map information, and a computer-based link between them.**
- **A GIS has been defined as a toolbox for analyzing spatial data.**
- **A GIS has also been defined as an information system for handling spatial data.**
- **Dueker's information systems definition has survived since 1979.**
- **Dueker's definition uses the feature model of geographic space.**
- **The standard feature model divides a mapped landscape into features, which can be points, lines, or areas.**
- **Using a GIS involves capturing the spatial distribution of features by measurement of the world or of maps.**
- **Almost all human activity and natural phenomena are spatially distributed, so they can be studied using GIS.**
- **A GIS uses map features to manage data.**
- **A GIS is flexible enough to be used for ad hoc query and analysis.**
- **A GIS can do analysis, modeling, and prediction.**
- **GIS is an approach to science that crosses several technologies.**
- **Geographic information science is research both on and with GIS.**
- **GIS is a multimillion-dollar business.**
- **Chrisman defines GIS as including the people and institutions that make decisions using geographic measurements and data transformations.**
- **GIS is integrating its way into many aspects of contemporary life.**

A Brief History of GIS (1.3)

- **The origins of GIS lie in thematic cartography.**
- **Many planners employed the method of map overlay using manual techniques.**
- **Manual map overlay as a method was first described comprehensively by Jacqueline Tyrwhitt in a 1950 planning textbook.**
- **McHarg used blacked out transparent overlays for site selection in *Design with Nature*.**
- **The 1960s saw many new forms of geographic data and mapping software.**
- **Within computer cartography the first basic GIS concepts were developed during the late 1950s and 1960s.**
- **Linked software modules, rather than stand-alone programs, preceded GISs.**
- **Early influential data sets were the World Data Bank and the GBF/DIME files.**
- **Early systems were CGIS, MLMIS, GRID, and LUNR.**
- **The Harvard University Odyssey system was influential because of its topological arc/node (vector) data structure.**

- **GIS was significantly altered by both the personal computer and the workstation.**
- **During the 1980s, new GIS software could better exploit more advanced hardware.**
- **User interface developments led to GIS's vastly improved ease of use during the 1990s.**

Sources of Information on GIS (1.4)

- **The large amount of information available about GIS can be overwhelming.**
- **Sources of GIS information include journals and magazines, books, professional societies, the World Wide Web, and conferences.**
- **GIS has Web home pages, network conference groups, professional organizations, and user groups.**
- **Most colleges and universities now offer GIS classes in geography departments.**

1.5.2 Study Questions

GIS Definitions

Summarize the various definitions of a GIS given in this chapter. What are the sources of the definitions? Underline in the definitions terms that are common across the definitions. Why would the definitions reflect different user needs and expectations of what a GIS is and what it can accomplish? How would you expand this set of definitions today?

The History of GIS

Take a look at the online GIS-timeline project, and then draw your own diagram showing a timeline of GIS development. What are the developments that point to a heritage for GIS in various academic disciplines, such as geography, cartography, computer science, environmental science, and planning? Give two examples of ways in which GIS has integrated knowledge and problem solving across these disciplinary boundaries. Search the World Wide Web to find any other GIS timelines. How well does yours agree with any others you may have found?

Sources of Information on GIS

Make a list of the key GIS information sources and find as many as you can in your local public library. If your library offers data services or the Internet, use these to find out as much information about GIS information sources (GIS metadata, or "data about data") as possible. What local, regional, or national GIS meetings or conferences are taking place in your area in the near future? Can you get onto pertinent mailing lists? If possible, arrange a field trip to one of these meetings to see any exhibits and to collect vendor information.

Which colleges, universities, or other educational establishments in your area offer GIS classes? What does the educational establishment teach in the GIS curriculum?

Prepare a two-page "GIS Guide" on finding out about GIS for a complete GIS novice.

1.6 EXERCISES

1. *Use the Internet to search for information about GIS usage and for digital map data online covering your town or city. Are there any attribute data for the maps, perhaps in gazetteers, almanacs, or data books? If you are not connected to the Internet, visit a library and use its facilities, or look in the reference section for information. If you live near a map library, perhaps at a university, see if you can use this facility in your search.*
2. *After a few searches, make an inventory of data you were successful in locating. Put the inventory in the form of a list. Add a column to show which agencies supplied the data you found and how recent the data were. How are the data made available to the public?*
3. *Using the information in this chapter, obtain as much information as possible about the GIS you are or will be using. When was the package created? What is the package's history? Who wrote the software? Are there manuals, research papers, or other sources of information, such as newsletters, user groups, or other manuals, relating to the software of which you are unaware?*
4. *Retrieve the GIS-L FAQ (frequently asked questions) list. How would you get answers to any questions you have at this stage that are not on the list?*
5. *Review the operating system of the computer on which you will work with GIS. Whether it be Unix, Windows, Linux, DOS, or any other, what operations are necessary to create and delete files, create and delete directories, copy and move files, edit the contents of an ASCII file, and use peripheral devices such as CD-ROM drives and plotters or digitizers? Become familiar with the system's manuals and/or online help facility.*
6. *Take a look at the interactive GIS Timeline project on the World Wide Web at* `http://www.casa.ucl.ac.uk/gistimeline`. *What would you consider to be the six most important events in the history of GIS?*

1.7 REFERENCES AND BIBLIOGRAPHY

1.7.1 Books

AGI (1995) *The AGI Source Book for Geographic Information Systems*. London: Association for Geographic Information.

Allen, K. M. S., Green, S. W., and Zubrow, E. B. W. (eds.) (1990) *Interpreting Space: GIS and Archaeology*. London: Taylor & Francis.

Antenucci, J. C. (1991) *Geographic Information Systems: A Guide to the Technology*. New York: Van Nostrand Reinhold.

Aronoff, S. (1989) *Geographic Information Systems: A Management Perspective*. Ottawa: WDL Publications.

Ball, D. and Babbage, R. (eds.) (1989) *Geographic Information Systems: Defence Applications*. Rushchutters Bay, Australia: Brasses Australia.

Berry, J. K. (1993) *Beyond Mapping: Concepts, Algorithms and Issues in GIS*. Fort Collins, CO: GIS World.

Bonham-Carter, G. (1994) *Geographic Information Systems for Geoscientists: Modelling with GIS*. Tarrytown, NY: Pergamon Press.

Bossler, J. D. (ed.) (2002) *Manual of Geospatial Science and Technology*. New York: Taylor and Francis.

Burrough, P. A. (1986) *Principles of Geographical Information Systems for Land Resources Assessment*. Oxford: Clarendon Press.

Burrough, P. A. and McDonnell, R. A. (1998) *Principles of Geographical Information Systems*. Oxford: Oxford University Press.

Castle, G. H. (1993) *Profiting from a Geographic Information System.* Fort Collins, CO: GIS World.

Chrisman, N. (1997) *Exploring Geographic Information Systems.* New York: Wiley.

Clarke, K. C. (1995) *Analytical and Computer Cartography.* 2nd ed. Upper Saddle River, NJ: Prentice Hall.

Clarke, K. C., Parks, B. O., and Crane, M. P. (eds.) (2002) *Geographic Information Systems and Environmental Modeling*, Upper Saddle River, NJ: Prentice Hall.

Cromley, E. K. and McLafferty, S. L. (2002) *GIS and Public Health.* New York: Guilford.

De Lepper, M. J. C., Scholten H. J., and Stern R. M. (eds.) (1995) *The Added Value of Geographical Information Systems in Public and Environmental Health.* Boston: Kluwer Academic Publishers.

DeMers, M. N. (1997) *Fundamentals of Geographic Information Systems.* New York: Wiley.

ESRI (1995) *Understanding GIS: The Arc/Info Method.* New York: Wiley.

ESRI (1997) *Getting to Know ArcView GIS.* Cambridge, UK: Geoinformation International/Prentice Hall.

Fischer, M. F. and Nijkamp, P. (eds.) (1993) *Geographic Information Systems, Spatial Modelling and Policy Evaluation.* New York: Springer-Verlag.

Fotheringham, S. and Rogerson, P. (1994) *Spatial Analysis and GIS.* Bristol, PA: Taylor & Francis.

Garson, G. D. and Biggs, R. S. (1992) *Analytic Mapping and Geographic Databases.* Newbury Park, CA: Sage Publications.

GIS World. (1989, 1990) *The GIS Sourcebook.* Fort Collins, CO: GIS World.

GIS World. *International GIS Sourcebook.* (1991/92, 1994). Fort Collins, CO: GIS World.

GIS World. (1995) *GIS World Sourcebook.* Fort Collins, CO: GIS World.

Goodchild, M. F. (ed.) (1989) *Accuracy of Spatial Databases.* London: Taylor & Francis.

Goodchild, M. F., Parks, B. O., and Steyaert, L. T. (eds.) (1993) *Environmental Modeling with GIS.* New York: Oxford University Press.

Grimshaw, D. J. (1994) *Bringing Geographical Information Systems into Business.* Harlow, Essex, UK: Longman Scientific & Technical; New York: Wiley.

Haines-Young, R., Green, D. R., and Cousins, S. (1993) *Landscape Ecology and Geographic Information Systems.* Bristol, PA: Taylor & Francis.

Hearnshaw, H. W. and Unwin, D. J. (1994) *Visualization in Geographical Information Systems.* New York: Wiley.

Heit, M. and Shortreid, A. (eds.) (1991) *GIS Applications in Natural Resources.* Fort Collins, CO: GIS World.

Huxhold, W. E. (1991) *An Introduction to Urban Geographic Information Systems.* New York: Oxford University Press.

Johnson, A. I., Pettersson, C. B., and Fulton, J. L. (eds.) (1992) *Geographic Information Systems (GIS) and Mapping: Practices and Standards.* Philadelphia: ASTM.

Kennedy, M. (1996) *The Global Positioning System and GIS: An Introduction.* Ann Arbor, MI: Ann Arbor Press.

Knowles, A. K. (ed.) (2002) *Past Time, Past Place: GIS for History.* Redlands, CA: ESRI Press.

Korte, G. (1994) *The GIS Book: A Practitioner's Handbook.* 3rd ed. Santa Fe, NM: OnWord Press.

Langran, G. (1992) *Time in Geographic Information Systems.* Bristol, PA: Taylor & Francis.

Laurini, R. and Thompson, D. (1992) *Fundamentals of Spatial Information Systems.* London: Academic Press.

Lo, C. P. and Yeung, A. K. W. (2002) *Concepts and Techniques in Geographic Information Systems.* Upper Saddle River, NJ: Prentice Hall.

Longley, P. A., Goodchild, M. F., and Maguire, D. J. (1999) *Geographical Information Systems: Principles, Techniques, Applications and Management.* 2nd ed. New York: Wiley.

Longley, P. A., Goodchild, M. F., Maguire, D. J., Rhind, D. W., and Lobley, J. (2001) *Geographic Information Systems and Science*. New York: Wiley.

Maguire, D. J., Goodchild, M. F., and Rhind, D. W. (1991) *Geographical Information Systems: Principles and Applications*. Harlow, Essex, UK: Longman Scientific & Technical; New York: Wiley.

Marble, D. F., Calkins, H. W., and Peuquet, D. J. (1984) *Basic Readings in Geographic Information Systems*. Williamsville, NY: SPAD Systems.

Masser, I. and Blakemore, M. (eds.) (1991) *Handling Geographical Information: Methodology and Potential Applications*. Harlow, Essex, UK: Longman Scientific & Technical: New York: Wiley.

Montgomery, G. E. and Schuch, H. C. (1993) *GIS Data Conversion Handbook*. Fort Collins, CO: GIS World.

Morgan, J. M., III (1990) *Directory of Colleges and Universities Offering Geographic Information Systems Courses*, prepared by J. M. Morgan III and G. R. Bennett; edited by J. H. Treadwell. Bethesda, MD: ASPRS: ACSM; Washington, DC: AAG.

Obermeyer, N. J. and Pinto, J. K. (1994) *Managing Geographic Information Systems*. New York: Guilford Press.

Onsrud, H. J. and Cook, D. W. (eds.) (1990) *Geographic and Land Information Systems for Practicing Surveyors: A Compendium*. Bethesda, MD: American Congress on Surveying and Mapping.

Peuquet, D. J. and Marble, D. F. (1990) *Introductory Readings in Geographic Information Systems*. Bristol, PA: Taylor & Francis.

Pickles, J. (1995) *Ground Truth: The Social Implications of Geographic Information Systems*. New York: Guilford Press.

Raper, J. (ed.) (1989) *Three-Dimensional Applications in Geographic Information Systems*, London: Taylor & Francis.

Ripple, W. J. (ed.) (1987) *Geographic Information Systems for Resource Management: A Compendium*. Falls Church, VA: American Society for Photogrammetry and Remote Sensing.

Ripple, W. J. (ed.) (1989) *Fundamentals of Geographic Information Systems: A Compendium*. Bethesda, MD: American Congress on Surveying and Mapping; Falls Church, VA: American Society for Photogrammetry and Remote Sensing.

Samet, H. (1990) *Design and Analysis of Spatial Data Structures*. Reading, MA: Addison-Wesley.

Star, J. and Estes, J. E. (1990) *Geographic Information Systems: An Introduction*. Upper Saddle River, NJ: Prentice Hall.

Tomlin, D. (1990) *Geographic Information Systems and Cartographic Modelling*. Upper Saddle River, NJ: Prentice Hall.

Worrall, L. (ed.) (1991a) *Geographic Information Systems: Developments and Applications*. New York: Routledge.

Worrall, L. (ed.) (1991b) *Spatial Analysis and Spatial Policy Using Geographic Information Systems*. New York: Belhaven Press.

Zweig White & Associates (1999) *Geographic Information System (GIS) Trends for A/E/P & Environmental Consulting Firms*. Zweig White & Associates, Inc.; ISBN: 1885002890.

1.7.2 Magazines and Journals

Popular Distribution Magazines

Geospatial Solutions. (formerly Geo Info Systems) vol. 1, no. 1 (Nov./Dec. 1990). Avanstar Communications, 859 Willamette Street, Eugene, Oregon 97401-6806 USA. E-mail: sbarnes@advanstar.com Web: `http://www.geospatial-online.com`.

GIS Law. GIS Law and Policy Institute, Suite 501, NationsBank Building, Harrisonburg, VA 22801. Quarterly since 1992.

Geoworld (formerly GIS World), vol. 1, no. 1 (July 1988). Adams Business Media, 250 S. Wacker Drive, Suite 1150, Chicago, IL 60606. E-mail: mball@mail.aip.com Web: `http://www.geoplace.com`.

GPS World Magazine. Avanstar Communications, 859 Willamette Street, Eugene, Oregon 97401-6806 USA. E-mail: editorial-gps@gpsworld.com Web: `http://www.gpsworld.com`.

Mapping Awareness. vol. 7, no. 7 (Sept. 1993). Oxfordshire, UK: Continued as *Mapping Awareness and GIS in Europe*.

Scientific and Academic Journals

Cartographica

Cartographic Perspectives

Cartography and GIS

Computer (algorithms and visualization)

Computers, Environment, and Urban Systems

Computers and Geoscience

Geocarto International

Geographical Systems

GrassClippings

IEEE Transactions on Computer Graphics and Applications (visualization)

IEEE Geosciences

International Journal of GIS

International Journal of Remote Sensing

Landscape Ecology

Photogrammetric Engineering and Remote Sensing

Remote Sensing Review

Transactions in GIS

URISA Journal

1.7.3 Proceedings of Conferences

AUTOCARTO 12, ACSM/ASPRS Annual Convention and Exposition Technical Papers, Charlotte, NC. Feb. 27–Mar. 2, 1995, vol. 4. Continues International Symposium on Automated Cartography.

First International Advanced Study Symposium on Topological Data Structures for Geographic Information Systems. Dedham, MA, 1977. Harvard Papers on Geographic Information Systems: G. Dutton, ed. Cambridge, MA: Laboratory for Computer Graphics and Spatial Analysis, Graduate School of Design, Harvard University, 1978.

GIS/LIS Annual Conference and Exposition. Sponsored by AAG, ACSM, AM/FM, ASPRS, and URISA. Held annually until 1999.

Proceedings, 5th International Symposium on Spatial Data Handling, August 1992, Charleston, SC: IGU Commission on GIS. Held annually.

SSD '95 Advances in spatial databases: 4th International Symposium, Portland, ME, Aug. 6–9, 1995. *Proceedings*, M. J. Egenhofer, and J. R. Herring (eds.). New York: Springer-Verlag. Held biannually.

1.7.4 Professional Organizations

URISA: Urban and Regional Information Systems Association, 900 Second Street NE, Suite 304, Washington, DC 20002. (202) 289-1685. E-mail: urisa@macc.wisc.edu.

AM/FM International: Automated Mapping and Facilities Management, 14456 East Evans Avenue, Aurora, CO 80014. (303) 337-0513.

AAG: The Association of American Geographers, 1710 Sixteenth St. NW, Washington, DC 20009-3198. Also publishes *AAG Newsletter*. (202) 234-1450. E-mail: gaia@aag.org.

ACSM: American Congress on Surveying and Mapping, 5410 Grosvenor Lane, Suite 100, Bethesda, MD. 20814-2122. (301) 493-0200. Web: `http://www.acsm.net`.

ASPRS: American Society for Photogrammetry and Remote Sensing. 5410 Grosvenor Lane, Suite 210, Bethesda, MD 20814-2162. (301) 493-0290.

NACIS: North American Cartographic Information Society, AGS Collection, P.O. Box 399, Milwaukee, WI 53201, (414) 229-6282, fax: (414) 229-3624, E-mail: nacis@nacis.org. See Web Page: `http://www.nacis.org`.

1.7.5 World Wide Web Sites

There are many WWW sites providing information about GIS. These tend to change quite frequently. Links to some of the most useful sites are maintained on the WWW Home Page for Getting Started with GIS at `http://www.prenhall.com/clarke`.

1.7.6 Chapter References

Abler, R. F. (1988) "Awards, rewards and excellence: keeping geography alive and well," *Professional Geographer*, vol. 40, pp. 135–40.

Brassel, K. E. (1977) "A survey of cartographic display software," *International Yearbook of Cartography*, vol. 17, pp. 60–76.

Chrisman, N. R. (1999) "What does 'GIS' mean?" *Transactions in GIS*, vol. 3, no. 2, pp. 175–186.

Dueker, K. J. (1979) "Land resource information systems: a review of fifteen years' experience," *Geo-Processing*, vol. 1, no. 2, pp. 105–128.

Dutton, G. (ed.) (1979) *Harvard Papers on Geographic Information Systems.* First International Advanced Study Symposium on Topological Data Structures for Geographic Information Systems. Reading, MA: Addison-Wesley.

Foresman, T. W. (ed.) (1997) *The History of Geographic Information Systems: Perspectives from the Pioneers*. Upper Saddle River, NJ: Prentice Hall.

Goodchild, M. F. (1992) "Geographical information science," *International Journal of Geographical Information Systems*, vol. 6, no. 1, Jan.–Feb. 1992.

McHarg, I. L. (1969) *Design with Nature*. New York: Wiley.

Peucker, T. K. and Chrisman, N. (1975) "Cartographic data structures," *American Cartographer*, vol. 2, no. 1, pp. 55–69.

Pickles, J. (ed.) (1995) *Ground Truth: The Social Implications of Geographic Information Systems*. New York: Guilford Press.

Steinitz, C., Parker, P., and Jordan, L. (1976) "Hand-drawn overlays: their history and prospective uses," *Landscape Architecture*, vol. 66, no. 5, pp. 444–455.

Tobler, W. R. (1959) "Automation and cartography," *Geographical Review*, vol. 49, pp. 526–534.

1.8 KEY TERMS AND DEFINITIONS

academic research: New learning created by the activity of university and other scholars.

ad hoc: For the particular case at hand.

adjacency: The topological property of sharing a common boundary or being in immediate proximity.

analysis: The stage in science when measurements are sorted, tested, and examined visually for patterns and predictability.

arc/node: Early name for the vector GIS data structure.

arc: A line represented as a set of sequential points.

area feature: A geographic feature recorded on a map as a sequence of locations or lines that, taken together, trace out an enclosed area or ring that represents the feature. Example: a lake shoreline.

attribute: A characteristic of a feature that contains a measurement or value for the feature. Attributes can be labels, categories, or numbers; they can be dates, standardized values, field measurements or other data. An item for which data are collected and organized. A column in a table or data file.

AUTOCARTO (International Symposium on Automated Cartography): A sequence of computer cartography and GIS conferences.

cartography: The science, art, and technology of making, using, and studying maps.

CGIS (Canadian Geographic Information System): An early national land inventory system in Canada that evolved into a full GIS.

choropleth map: A map showing numerical data (but not simply "counts") for a group of regions by (1) grouping the data into classes and (2) shading each class on the map.

computer mapping: Producing maps using the computer as the primary or only tool.

connectivity: The topological property of sharing a common link, such as a line connecting two points in a network.

context-sensitive help: A component of a user interface that can reveal to the user information that assists with the current status of other elements of the user interface.

data structure: The logical and physical means by which a map feature or an attribute is digitally encoded.

database: The body of data that can be used in a database management system. A GIS has both a map and an attribute database.

database manager: A computer program or set of programs allowing a user to define the structure and organization of a database, to enter and maintain records in the database, to perform sorting, data reorganization, and searching, and to generate useful products such as reports and graphs.

digitizing tablet: A device for geocoding by semi-automated digitizing. A digitizing tablet looks like a drafting table but is sensitized so that as a map is traced with a cursor on the tablet, the locations are picked up, converted to numbers, and sent to the computer.

Dueker's definition (of GIS): "A special case of information systems where the database consists of observations on spatially distributed features, activities or events, which are definable in space as points, lines, or areas. A geographic information system manipulates data about these points, lines, and areas to retrieve data for ad hoc queries and analyses."

FAQ: A list of frequently asked questions, usually posted on a network newsgroup or conference group to save new users the trouble of asking old questions over again.

feature: A single entity that makes up part of a landscape.

file: Data logically stored together at one location on the storage mechanism of a computer.

format: The specific organization of a digital record.

FORTRAN: An early computer programming language, initially for converting mathematical formulas into computer instructions.

fourth dimension: A common way of referring to time; the first three dimensions determine location in space, the fourth dimension determines creation, duration, and destruction in time.

functional definition: Definition of a system by what it does rather than what it is.

g-trade: (also g-commerce): Web-oriented use of GIS capability to spatially enable the search and browse processes during online business activity or e-trade.

GBF (Geographic Base File): A database of DIME records.

general-purpose map: A map designed primarily for reference and navigation use.

geocoding: The conversion of analog maps into computer-readable form. The two usual methods of geocoding are scanning and digitizing.

geographic information science: Research on the generic issues that surround the use of GIS technology, impede its implementation, or emerge from an understanding of its capabilities.

geographic(al) information system: (1) A set of computer tools for analyzing spatial data; (2) a special case of an information system designed for spatial data; (3) an approach to the scientific analysis and use of spatial data; (4) a multibillion-dollar industry and business; (5) a technology that plays a role in society.

geographic pattern: A spatial distribution explainable as a repetitive distribution.

geography: The science concerned with all aspects of the earth's surface, including natural and human divisions, the distribution and differentiation of regions, and the role of humankind in changing the face of the earth.

GIS/LIS: A U.S. national conference on geographic information and land information systems, sponsored by most GIS professional organizations and held annually.

GUI (graphical user interface): The set of visual and mechanical tools (such as window, icons, menus, and toolbars, plus a pointing device such as a mouse) through which a user interacts with a computer.

information: The part of a message placed there by a sender and not known by the receiver.

information system: A system designed to allow the user to be delivered the answer to a query from a database.

installed base: The number of existing implemented systems.

Internet: A network of many computer networks. Any computer connected to the Internet can access any of the computers accessible through the network.

isoline map: A map containing continuous lines joining all points of identical value.

killer app: A computer program or "application" that by providing a superior method for accomplishing a task in a new way becomes indispensable to computer users. Examples are word processors and spreadsheets.

land-cover map: A map showing the type of actual surface covering at a given time. Categories could be grassland, forest land, cropland, bare rock, and so on.

land-use map: A map showing the human use to which land is put at a given time. Categories could be pasture, national forestland, agricultural land, wasteland, and so on.

landscape: The part of geographic space shown on a map, including all its features.

learning curve: The relationship between learning and time. A steep learning curve means that much is learned quickly (usually thought to be the opposite). A difficult learning curve is one where learning takes place slowly, over a long time period.

line feature: A geographic feature recorded on a map as a sequence of locations tracing out a line. An example is a stream.

LIS (land information system): Surveying profession's term for GIS where that data are for land ownership.

location: A position on the earth's surface or in geographic space definable by coordinates or some other referencing system, such as a street address or space indexing system.

LUNR (Land Use and Natural Resources Inventory System): An early GIS in New York.

map: A depiction of all or part of the earth or other geographic phenomenon as a set of symbols and at a scale whose representative fraction is less than 1 : 1. A digital map has had the symbols geocoded and stored as a data structure within the map database.

map overlay: Placing multiple thematic maps in precise registration, with the same scale, projections, and extent, so that a compound view is possible.

measurement: A quantitative assessment of a phenomenon.

menu: A component of a user interface that allows the user to make selections and choices from a preset list.

MIMO system: A term used to describe a first-generation computer mapping system designed to capture the map by computer and reproduce it (map in–map out).

MLMIS (Minnesota Land Management System): An early statewide GIS for the state of Minnesota.

modeling: The stage in science when a phenomenon under test is sufficiently understood that an abstract system can be built to simulate the real system.

modular computer program: Computer programs composed of integrated sections of reusable functions rather than a single program.

National GIS Curriculum: An NCGIA-sponsored national college curriculum for GIS, used in many colleges and universities worldwide and with available teaching materials.

National Spatial Data Clearinghouse: A World Wide Web resource that serves as a cross-reference point for the distributed database of all U.S. government public-domain and other geographic information.

NCGIA (National Science Foundation's National Center for Geographic Information and Analysis): A three-university consortium funded to assist in GIS education, research, outreach, and information generation.

newsgroup: An area on the Internet for asynchronous many-to-many discussions.

node: At first, any significant point in a map data structure. Later, only those points with topological significance, such as the ends of lines.

observation: The process of recording an objective measurement.

Odyssey: A first-generation GIS developed at Harvard to implement the original arc/node vector data structure.

online manual: A digital version of a computer application manual available for searching and examination as required.

overlay weighting: Any system for map overlay in which the separate thematic map layers are assigned unequal importance.

PC (Personal Computer): A self-contained microcomputer, providing the necessary components for computing, including hardware, software, and a user interface.

point feature: A geographic feature recorded on a map as a location. Example: a single house.

prediction: The scientific ability to forecast the outcome of a process in advance.

proceedings: The formal record of the papers and other prepared presentations at a conference. Usually available to conference attendees and later distributed as a softcover book.

professional publication: Books, journals, or other information designed primarily for those using GIS technology as part of their job.

query: A question, especially if asked of a database by the user via a database management system or GIS.

record: A set of values for all attributes in a database. Equivalent to a row in a data table.

scientific approach: A method for rationally explaining observations about the natural and human world.

search engine: A software tool designed to search the Internet and the WWW for documents meeting the user's query. Examples: Yahoo and Alta Vista.

software package: A computer program application.

spatial data: Data that can be linked to locations in geographic space, usually via features on a map.

spatial distribution: The locations of features or measurements observed in geographic space.

spreadsheet: A computer program that allows the user to enter numbers and text into a table with rows and columns and then maintain and manipulate those numbers using the table structure.

thematic map: A map designed primarily to show a "theme," a single spatial distribution or pattern, using a specific map type.

topographic map: A map type showing a limited set of features but including, at the minimum, information about elevations or landforms. Example: contour maps. Topographic maps are common for navigation and for use as reference maps.

topology: The numerical description of the relationships between geographic features, as encoded by adjacency, linkage, inclusion, or proximity. Thus a point can be inside a region, a line can connect to others, and a region can have neighbors.

transparent overlay: An analog method for map overlay, where maps are traced or photographed onto transparent paper or film and then overlain mechanically.

U.S. Census Bureau: An agency of the Department of Commerce that provides maps in support of the decennial (every 10 years) census of the United States, especially the census of population.

user group: Any formal or informal organization of users of a system who share experiences, information, news, or help among themselves.

USGS (U.S. Geological Survey): A part of the Department of the Interior and a major provider of digital map data for the United States.

vector: A map data structure using the point or node and the connecting segment as the basic building block for representing geographic features.

workstation: A computing device that includes, as a minimum, a microprocessor, input and output devices, a display, and hardware and software for connecting to a network. Workstations are designed to be used together on local area networks, and to share data, software, and so on.

World Data Bank: One of the first digital maps of the world, published in two versions by the Central Intelligence Agency in the 1960s.

World Wide Web (WWW or W3): A distributed database of information stored on servers connected by the Internet.

PEOPLE IN GIS

Nils Larsen GIS Coordinator and Staff Geologist

Nils Larsen has been GIS Coordinator and Staff Geologist at IWT, Inc. in Santa Barbara for the last seven years. A native of the Seattle area, his background is in geology, with a Bachelor of Science degree at Western Washington University. Nils moved to California and started working as a soils technician in a soils lab before taking a position at a hydrologic consulting firm, IWT. As the firm became more and more involved in GIS, Nils became responsible for making GIS work in the consulting activity of the company, with a continuing focus on water and the use of digital information in resource exploration.

KC: Nils, did you use computers in college?

NL: There was one programming class in Pascal. I took geography courses in map reading and analysis, where I spent time with paper maps, learning what's on them and how they're put together, what the information means. I've always been interested in maps.

KC: What sort of clients does your company have?

NL: Two main types of clients: private companies that want to develop or to invest in water resource assets or water rights. We do the water resource investigations, finding out how much there is associated with a particular piece of property, then they can decide if they want to purchase that land. Our public clients are typically water districts, water service agencies, and so forth. We help them manage the water resources that they have,

find out where the water's coming from, how much is going into the system, where it's getting taken out, where there are contamination problems or overdraft problems.
KC: Those problems are related to the subsurface geology and flow within that complicated structure?
NL: Definitely. The GIS is very helpful for visualizing those problems relative to the geologic structure. We worked for a Water Conservation District in the local area who operate some artificial recharge basins, and also have sea water intrusion and water quality problems. The GIS was instrumental in showing the relationships between them.
KC: What brought you into GIS in your field?
NL: Well, at IWT, we wanted to try and get a jump on some of the competition so we decided to invest in GIS. We had what we thought was somewhat of a specialized field—water resource investigations in fractured bedrock, beyond drilling wells into alluvial basins. The option was whether to go PC-based or workstations. We use NT workstation PCs and some Windows 95 PCs; lots of memory, lots of storage space. We use a large format plotter, a large digitizing tablet and a flatbed scanner.
KC: And what GIS software did you use?
NL: The original suite of software was ArcCad and AutoCad 12, all in DOS then ArcView 1 and an AutoCad third party package called Quicksurf.
KC: With a scanner and a digitizing tablet you must digitize maps. What kind?
NL: Just about anything that's not in digital form. We've captured data from geologic maps and consulting firms' reports, and from blue-line maps typical of engineering firms.
KC: Is it hard to georegister so many different maps?
NL: The difficulty is when there's no projection or coordinate system on the map. Those are typically blue-line, engineering type maps.
KC: What sort of attribute data do you bring into GIS that you might use in combination with the maps?
NL: If I'm working with well data, then depth, well diameter, water levels, water quality, well perforation interval, aquifer, well ID, owner. These are all typical attributes. For geologic data we use things like the formation name, rock type, age, and different levels of classification, such as formation subunits.
KC: If you were going to give advice to an intern or an incoming freshman, what would it be?
NL: Get to know the paper maps, get to know projection systems, datums, the different kinds and why certain ones are used for certain reasons. Area versus shape versus direction issues regarding different projection systems. As well, learn how to read manuals and spend time reading manuals.
KC: Thanks very much Nils.

(Interview and photograph used with permission of Nils Larsen)

CHAPTER 2

GIS's Roots in Cartography

2.1 MAP AND ATTRIBUTE INFORMATION
2.2 MAP SCALE AND PROJECTIONS
2.3 COORDINATE SYSTEMS
2.4 GEOGRAPHIC INFORMATION
2.5 STUDY GUIDE
2.6 EXERCISES
2.7 BIBLIOGRAPHY
2.8 KEY TERMS AND DEFINITIONS

Blow winds, and crack your cheeks! rage! blow!
You cataracts and hurricanoes, spout
Till you have drenched our steeples, drowned the cocks!
Your sulphurous and thought-executing fires,
Vault couriers to oak-cleaving thunderbolts,
Singe my head white! And thou, all-shaking thunder,
Strike flat the thick rotundity o' the world!
—Shakespeare, King Lear, *act 3, sc. 2.*

It was difficult to understand exactly what was happening, but the general impression was favourable, and I remember being struck by a message from a young officer in a tank of the 7th Armoured Division, "Have arrived at the second B in Buq Buq."
—Sir Winston Churchill, The Second World War, Vol. 2, Their Finest Hour, *1949.*

Completely due to a lack of attendance in "Introduction to GIS" lectures, Bradley's science fair project enters the school "Cartography Hall of Shame."

2.1 MAP AND ATTRIBUTE INFORMATION

Information permeates our society, but fortunately, it takes on only a few tangible forms. Without the preordering of information, much of it would not be usable by humans in their everyday lives. Among these are the everyday methods for organizing information, visible by everyday examples such as the Yellow Pages, baseball box scores, magazines, or the television listings. Most information is usually preordered into lists, numbers, tables, text, pictures, maps, or indexes. Clusters or chunks of similar information, usually numbers and text, are called *data*. When data are entered into the computer, we store them as *files* and refer to them collectively as a *database*. In database language, the items that we gather information about are referred to as *attributes*, and individual data items as *records*. For example, we may have a shoebox filled with baseball cards. On each card (record) in the *database* (shoebox) we have a picture and some *values* for the attributes. The attributes may be the player's name, team, batting average, or the years playing on which teams in the major leagues. Values of the attributes could be "*Jason Giambi*," "*The New York Yankees*," *0.3084*, and *2001–02* (Figure 2.1).

A basic difference between these types of information and the information that is collected into geographic information systems is that GIS information has associated with it an underlying *geography*, or descriptions of *locations* on the face of the earth. This means that pictures and especially maps can be a database, too. A link to the earth must somehow be placed into the GIS database, so that we can refer to the data by the location—and the location by the data. With this feature comes the fact that we can now manage the data using the underlying geography, the attributes, or both.

This is possible for our baseball card example. On each card we have the name of the team and the city where the team plays. If we went to an atlas and looked up the *latitude* and *longitude* of the city, we would be *geocoding* the baseball cards. If we then entered the latitudes and longitudes in pencil on each card, we would have a geographic information system of sorts, although we would have to enter the cards into a computer to really have a GIS. The data are now more usable, because if we have the capability of mapping in our GIS, we can place any baseball card information on a map. Later, we will see that this is only the first of many new abilities that georeferencing the data brings. For now, however, location is everything!

The power of the GIS is in allowing the attribute and the geographic or map information to be linked together in a useful way. For example, we can search the data both by the attributes (find all players with a batting average over 0.300) and by using the map (find all the players who play within 200 miles of Yankee Stadium). Obviously, if the two sorts of information are linked, we can use either one to search the other, or we can use them together. So we could ask the GIS, for example, to select from the database all players who bat left-handed, have better than a 0.300 average, and play no more than 200 miles from New York. Furthermore, as we have locations, we can answer the query about the data with a map rather than a list.

Central to this map and attribute data use is finding a way to *link* the map with the attributes. As we are using a computer, obviously the link should be in the form of numbers. When we locate people and houses, we usually use street addresses rather than numbers. Later in the book, we will see that a GIS gives us the power to move from one to the other of these descriptions of location with numbers. For now, however, we need a simple number description for a location. In the example here, we used

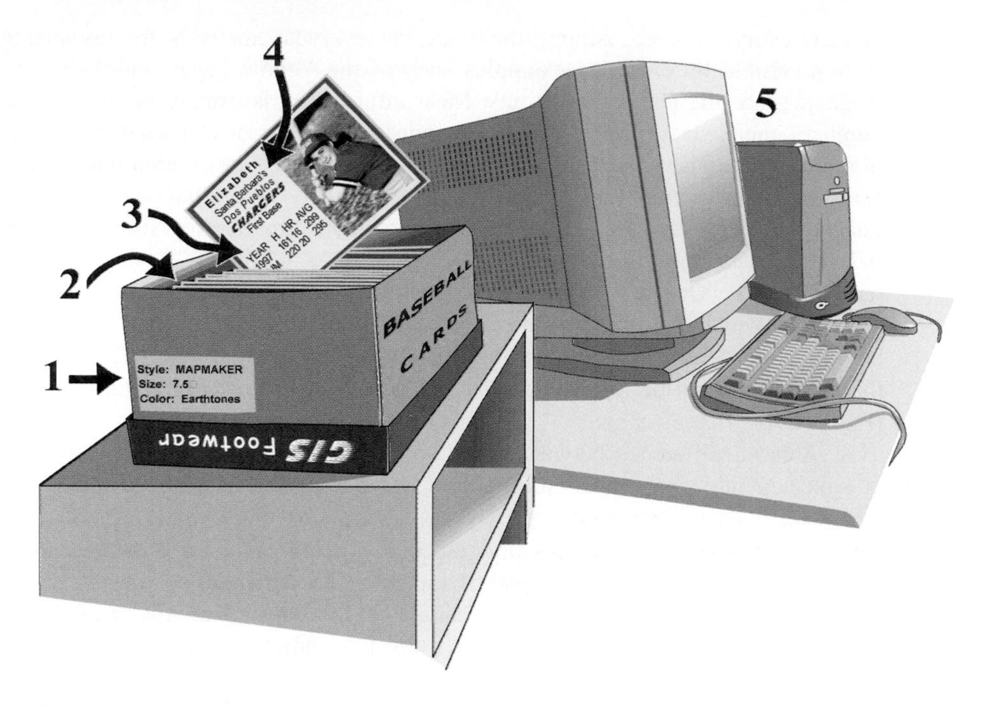

FIGURE 2.1: The elements of a GIS. (1) the database (shoebox); (2) the records (baseball cards); (3) the attributes (the categories on the cards, such as a batting average, (4) the geographic information (locations of the team's stadium in latitude and longitude); (5) a means to use the information (the computer).

latitude and longitude. Many GIS packages use latitude and longitude, so this is quite appropriate.

Before we move on, however, it is important to get a feel for what these geographic numbers mean and how they correspond to places on both the earth and the map. It is a little more complex than it first seems, but with a little digression, we can quickly come up to speed, and even be experts. This means that to understand GISs, we need to know a little *cartography*, which is the science that deals with the construction, use, and principles behind maps and map use. The basics here go back a long way, to the work of the ancient Greek Ptolemy, the father of latitude and longitude and of map projections. A little digression, therefore, is called for on the way that we "strike flat the thick rotundity o' the world."

2.2 MAP SCALE AND PROJECTIONS

2.2.1 The Shape of the Earth

A rather astounding fact, and a true one, so I am told, is that there are many more members of the Flat Earth Society than of the American Cartographic Association (now the Cartography and Geographic Information Society). Nevertheless, it would be hard to maintain that the earth is flat, appearances aside. A trip to the beach to observe a ship sailing into the distance reveals not a visible dot that gets smaller and smaller, but a dot that eventually just disappears over the curve of the earth. When high in a plane at cruising altitude, hold a ruler at arm's length up to the horizon and judge for yourself whether the horizon line is straight. Two issues suggest themselves if the idea is to "contain" the earth, or parts of its surface, onto a map inside a GIS. First, how big is the earth? Second, how can a flat map (and simple numbers) be used to describe locations on the earth's surface?

First of all, how big? This question becomes one of what shape we use as a description of the earth. Although for many mapping applications the earth can be assumed to be a *perfect sphere*, there is a small but significant difference between the distance around the earth pole to pole (39,939,593.9 meters) versus the distance around the equator (40,075,452.7 meters). This is because the earth resembles more closely than the sphere a figure called an *oblate ellipsoid* or *spheroid*, the three-dimensional shape you get by rotating an ellipse about its shorter axis (Figure 2.2).

There have been many attempts to measure the size and shape of the earth's ellipsoid. In 1866 the mapping of the United States was based on the ellipsoid measured by Sir Alexander Ross Clarke, which had a basis in measurements taken in Europe, Russia, India, South Africa, and Peru. The Clarke 1866 ellipsoid had an equatorial radius of 6,378,206.4 meters and a polar radius of 6,356,538.8 meters. This gave a "flattening" of the ellipsoid of 1/294.9787 (Figure 2.3). In 1924, a simpler measure of 1/297 with a longer radius of 6,378,388 was adopted as an international standard. As mapping had already begun in the United States, the older values were used, adopted as the North American Datum of 1927 (NAD27).

The satellite era has brought with it more accurate means of measurement, including the global positioning system (GPS). An estimate of the ellipsoid allows calculation of the elevation of every point on earth, including sea level, and is often called a *datum*. Recent datums have been calculated using the center of the earth as a reference point instead of a point on the ground as was the case before. In 1983 a new datum was adopted for the United States, called the North American Datum of 1983 (NAD83), based on measurements taken in 1980 and accepted internationally as the geodetic

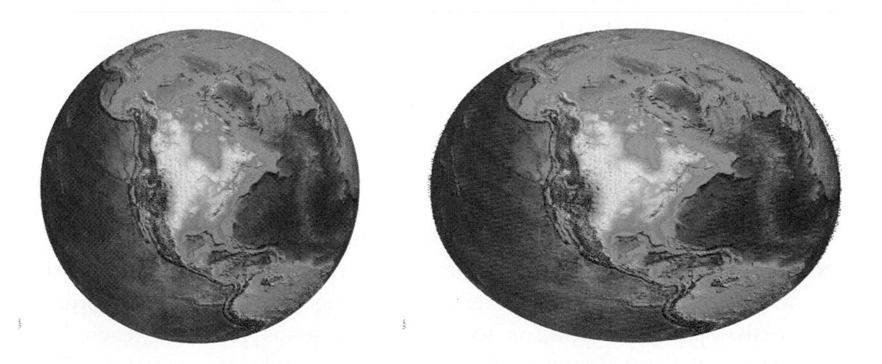

FIGURE 2.2: Sphere and ellipsoid (or spheroid). Earth's ellipsoid is actually only about 1/300 off from the sphere.

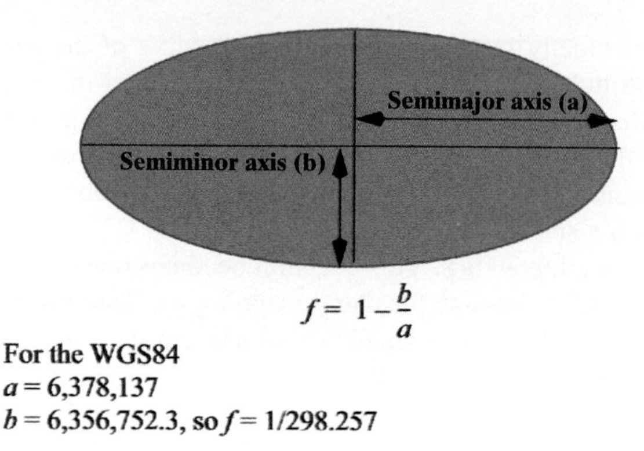

FIGURE 2.3: The ellipsoid. The long axis is the major axis, the short the minor axis. Half of each of these lengths is used to calculate the flattening of the ellipsoid.

reference system (GRS80). Efforts have been under way since then to make the slight necessary corrections to maps of the United States, which amount to about 300 meters in places.

The U.S. military has also adopted the GRS80 ellipsoid but refined the values slightly in 1984 to make the world geodetic system (WGS84). It is important that when maps are to be used that the datum and ellipsoid reference information be known, as at large scales there can be major differences, especially in elevations (Figure 2.4). The datum and ellipsoid are also essential to know when using a GPS receiver, as coordinates will be different in each, sometimes by a significant distance.

As a final complication, the science of geodesy, which measures the earth's size, shape, and gravitational fields exactly, has mapped out all of the local variations from the ellipsoid and calls the resultant surface a *geoid*. Only under highly demanding circumstances would a geoid be used in a GIS. In fact, in cartography, a common reference base is the sphere. The ellipsoid becomes necessary when we deal with finer, more detailed, or "large" map scales, and differences caused by not using it can become significant at scales coarser (smaller) than about 1 : 100,000 but are noticeable even at scales less detailed than 1 : 50,000.

2.2.2 Map Scale

All maps, whether on a sheet of paper or inside a computer, are reductions in size of the earth. A map at one-to-one scale (1 : 1) would be virtually useless; you would barely be able to unfold it. In cartography, the term *representative fraction* is used for the amount of scaling. A representative fraction is the ratio of distances on the map to the same distances on the ground. A model airplane or train is usually at about a 1 : 40 scale. This means that every distance on the model is one-fortieth of its size on the real object. The world is so big that for maps we often reach some pretty small values for the representative fraction. Just a couple of examples will show why this is necessary.

First, let's use the WGS84 numbers for earth's size. The two ellipsoid distances average to 6,367,444.65 meters. To calculate the circumference of a circle of this size, we multiply by two times Π.

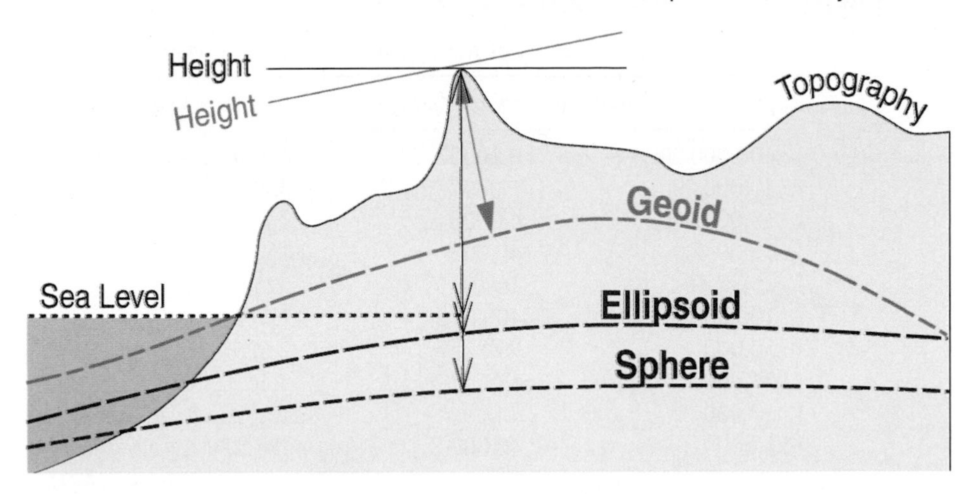

FIGURE 2.4: Elevations defined with reference to a sphere, ellipsoid, geoid, or local sea level will all be different. Even locations as latitude and longitude will vary somewhat. When linking field data such as GPS with a GIS, the user must know what base to use.

This gives a distance "around the average world" of 40,078,346.23 meters. Table 2.1 shows what this number becomes when multiplied by various representative fractions to give map distances associated with the earth's circumference. A quick look at Table 2.1 reveals a suspicious number. At 1 : 40,000,000 the earth's circumference maps onto a meter almost exactly. This is because the original definition of the meter was one ten-millionth of the distance from the equator to the north pole measured along the meridian passing through Paris, France. It is fairly obvious that the metric system makes these computations far easier, as we don't have to convert feet to inches and miles.

Over the range of scales shown in Table 2.1, the earth's equator would map onto about the circumference of a gumball at 1 : 400,000,000, (Figure 2.5) a baseball at 1 : 177,000,000, a basketball at 1 : 40,000,000, but at 1 : 50,000 would map onto about 10 Manhattan city blocks! At 1 : 1,000, a very detailed scale used in engineering and construction maps, the earth's equator would map onto twice the length of Manhattan Island. A convenient scale to hold in mind is 1 : 40,000,000, with the equator mapped across a 1-meter poster-size world map. Obviously, we don't use all scales in cartography. Most national mapping for GIS use is between 1 : 1,000,000 and 1 : 10,000. In the United States, key scales are 1 : 100,000 and 1 : 24,000, at which national coverages are available.

Another important factor to keep in mind is that a GIS is largely *scaleless*. The data can be multiplied up or reduced to any size that is appropriate. However, as we get farther and farther from the scale at which a map was made before it was captured into the GIS, problems of scale appear. As we enlarge maps, detail does not appear as if by magic. A smooth coastline, for example, remains smooth and imprecise as we enlarge it. On the other hand, if we keep reducing the scale of a map without eliminating detail, the map becomes so "dense" with data that we cannot see the forest for the trees. The proper presentation of information at a particular scale is one of the most important goals of cartographic design.

A last point to keep in mind as we finish this short discussion of map scale is that only on a globe is a scale constant. As we move the map from the curved surface of the

TABLE 2.1: Lengths of the Equator at Different Map Scales

Representative Fraction	Map Distance (m)	Distance in Feet (approx.)
1 : 400,000,000	0.10002	0.328 (3.9 inches)
1 : 40,000,000	1.0002	3.28
1 : 10,000,000	4.0008	13.1
1 : 1,000,000	40.008	131
1 : 250,000	160.03	525
1 : 100,000	400.078	1,312
1 : 50,000	800.157	2,625
1 : 24,000	1,666.99	5,469 (1.036 miles)
1 : 10,000	4,000.78	13,126 (2.486 miles)
1 : 1,000	40,007.8	131,259 (24.86 miles)

FIGURE 2.5: Assuming a sphere, the earth maps onto a gumball at 1 : 470,000,000; a baseball at 1 : 177,000,000 and a basketball at 1 : 40,000,000. At 1 : 50,000 the earth's circumference maps on 10 city blocks.

sphere or ellipsoid to the flat surface of paper or the computer screen, we necessarily have to distort the map in some way. The part of cartography that deals with this problem of putting a round earth onto flat paper is called *map projections*.

2.2.3 Map Projections

Given that the earth can be approximated by a shape like the sphere or the ellipsoid, how can we go about converting data in latitude and longitude into a flat map, with x and y axes? The simplest way is to ignore the fact that latitude and longitude are angles at the center of the earth, and just pretend that they are x and y values. Figure 2.6 shows this arrangement. Obviously, the map will range from 90 degrees north to 90 degrees south, and from 180 degrees east to 180 degrees west.

The corresponding (x, y) values are from $(-180, -90)$ to $(+180, +90)$. This map is now a *map projection*, because the earth's geographical (latitude and longitude) coordinates have been "mapped" or projected onto a flat surface. Obviously this can be done in many ways. We can "project" the sphere (or the ellipsoid) onto any of three flat surfaces and then unfold them to make the map. These can be the plane (as previously), the cylinder, or the cone. Projections onto these three surfaces are

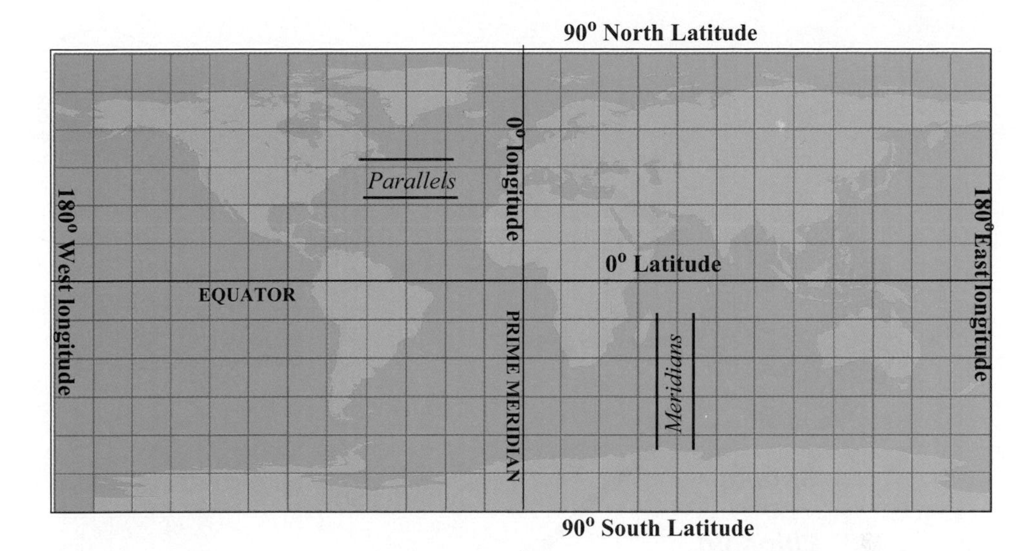

FIGURE 2.6: Geographic coordinates. The familiar latitude and longitude system, simply converting the angles at the earth's center to coordinates, gives the basic equirectangular projection. The map is twice as wide as high (360° east-west, 180° north-south).

called *azimuthal*, *cylindrical*, and *conic*, respectively. Examples of each are shown in Figure 2.7.

We can also choose how the mapping takes place with relationship to earth's surface. We can have the figure, such as the cone or the cylinder, "cut" through the earth. The resulting projection, shown in Figure 2.8, is called *secant*. So, for example, if a cone cuts the earth we would have a secant conic projection. The line on the map where the "cut" falls on the projection is important because it is a line along which the earth and the map match exactly, just as on a globe at the same scale, with no distortion. If this line coincides with a parallel of latitude, it is often called a *standard parallel*. Figure 2.8 shows a secant conic projection with two standard parallels. On or near these lines the map is most accurate. Similarly, there is no hard-and-fast rule that we have to orient the figure we are using in the projection with the earth's polar or rotation axis. If we align instead to a line at 90 degrees to this axis, we call the projection *transverse*. If we orient the projection axis at another angle, we call the projection *oblique* (Figure 2.9).

Cartographers have devised thousands of different map projections. Fortunately, they all fall into a set of "types" that are quite easily understood. The simplest way to evaluate a projection is by how it distorts the earth's surface during the transformation from a sphere or ellipsoid to a flat map. Some projections preserve the property of local shape, so that the outline of a small area like a state or a part of a coastline is correct. These are called *conformal* projections. They are easily identified, because on a conformal projection the lines of the latitude and longitude grid (called the *graticule*) meet at right angles, the way they do on a globe, although not all right-angle graticules mean conformal projections. Conformal projections are employed mostly

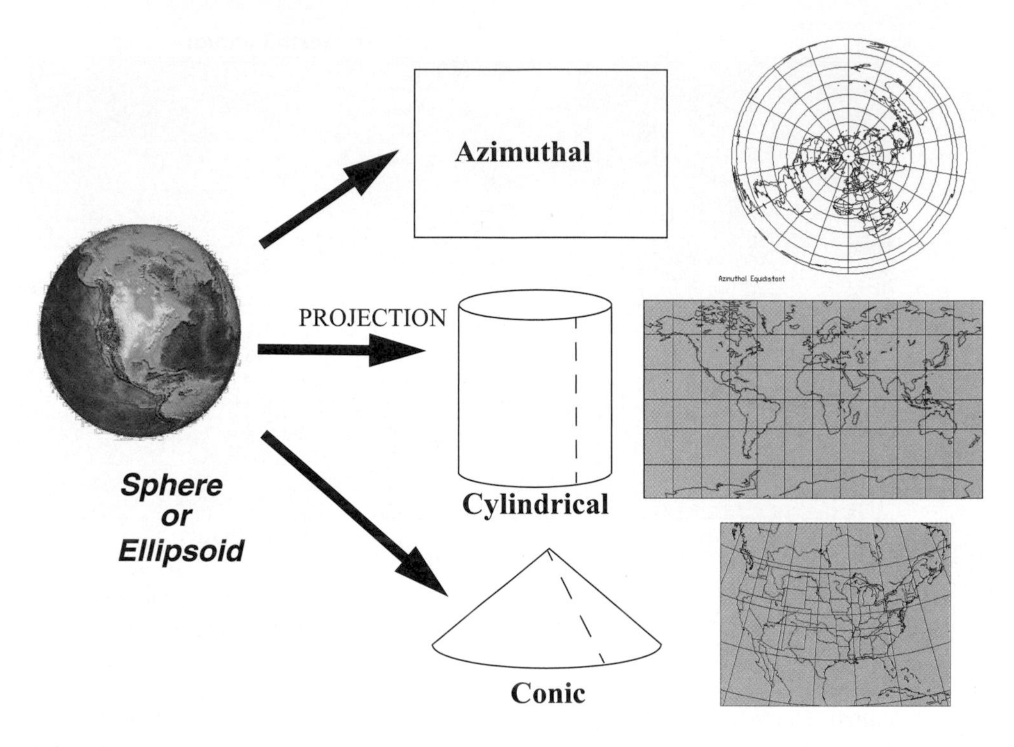

FIGURE 2.7: The earth can be projected in many ways, but basically onto three shapes that can be unrolled into a flat map: a flat plane, a cylinder, and a cone.

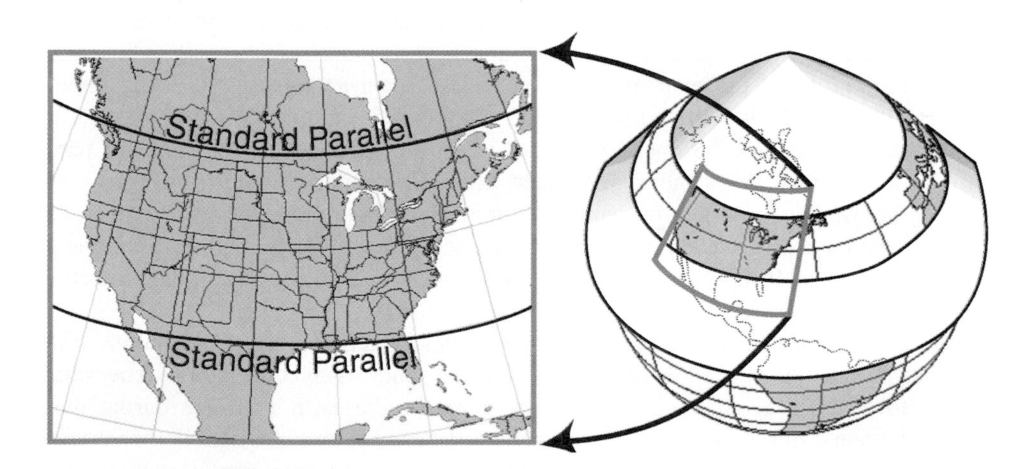

FIGURE 2.8: The projection in this figure is a secant conic projection. The figure also illustrates standard parellels. The conic projection cuts through the globe, and the earth is projected both in and out onto it. Lines of true scale, where the cylinder and sphere touch, become standard parallels. If the touching is along one line, the projection is tangent and has one standard parallel.

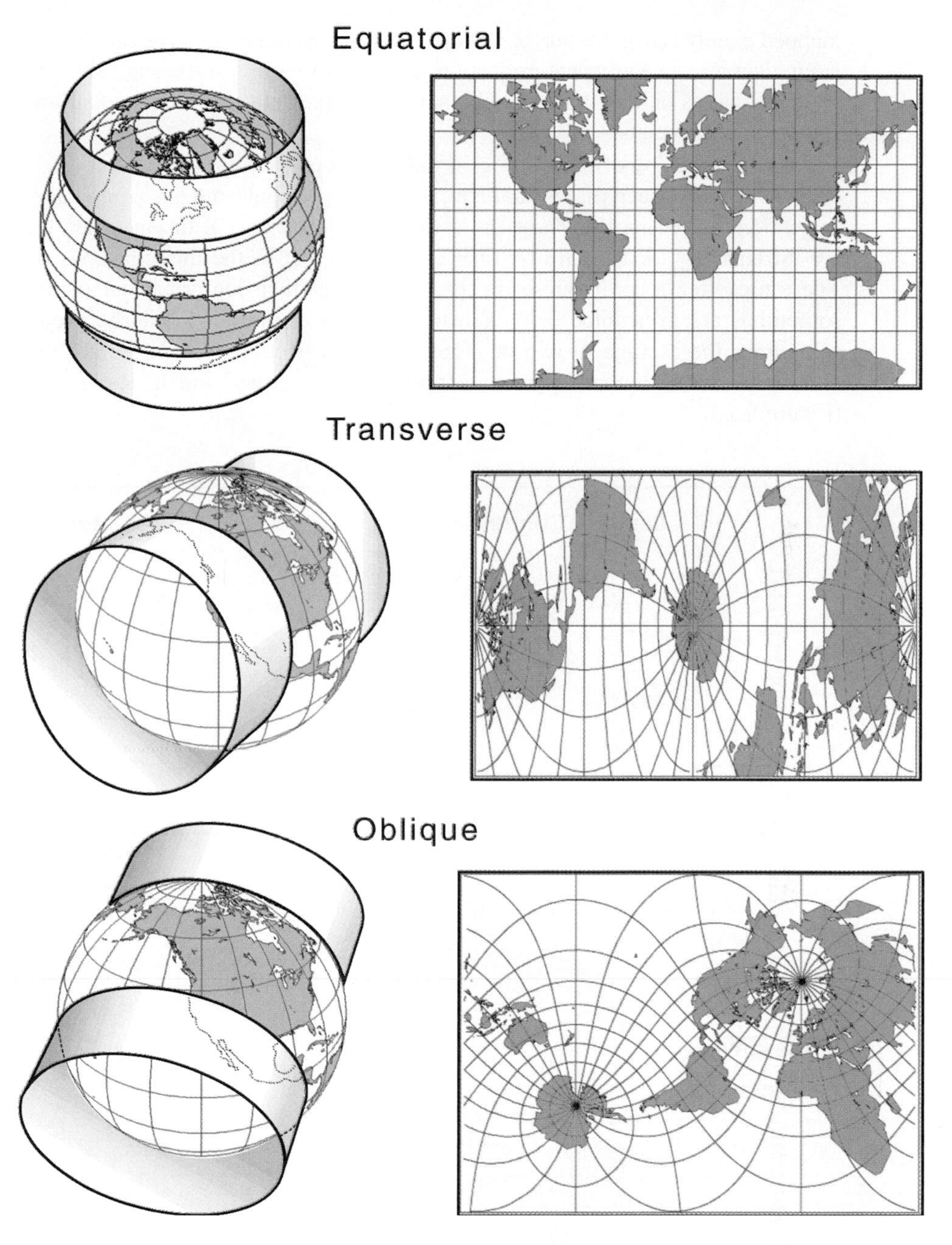

FIGURE 2.9: Variations on the Mercator (pseudocylindrical) projection shown as secant.

for maps that must be used for measuring directions, because they preserve directions around any given point. Examples are the Lambert conformal conic and the Mercator projections.

At the other extreme are projections that preserve the property of area. Many GIS packages compute and use area in all sorts of analyses, and as such must have area mapped evenly across the surface. Projections that preserve area are called *equal area* or *equivalent*. On an equivalent projection, all parts of earth's surface are shown with their correct area, as on the sphere or ellipsoid. Examples are the Albers equal area and the sinusoidal projections.

A third category of projections is the set that preserves distances but only along one or a few lines between places on the map. The simple conic and the azimuthal equidistant projections are examples. These projections are useful only if distances are critical, and are infrequently used in GIS. A final category is that of the miscellaneous projections. These are often a compromise, in that they are neither conformal nor equivalent, and sometimes are interrupted or broken to minimize distortion. Similarly, projections are sometimes the average of two or more similar projections. Examples are the Goode's homolosine (made by patching different projections together) and the Robinson projection (Figure 2.10).

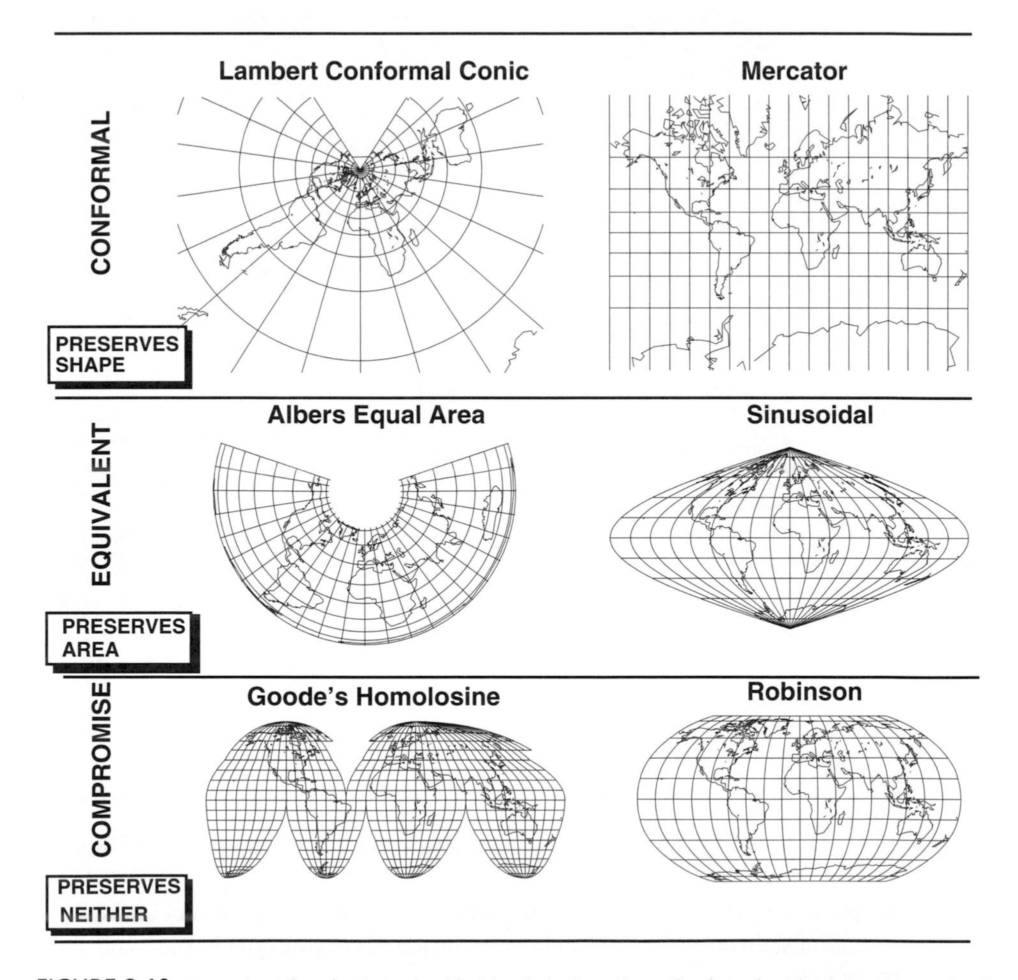

FIGURE 2.10: Examples of projections classified by their distortions. Conformal projections preserve local shape, equivalent projections preserve area, while compromise projections lie between the two. No projection can be both equivalent and conformal.

The most important implications of map projections for GIS are the following. First, the larger the area involved, the more important the mapping errors due to the projection become. At 1 : 24,000 scale, the errors are already significant, and at smaller scales like 1 : 1 million they are major. Second, the projection used should suit the GIS application. If directions or bearings from point to point are important, obviously a conformal projection is called for. If the analysis within the GIS consists of comparing or calculating areas or values based on areas, such as densities, then an equivalent projection is essential. Finally, to overlay or edge-match any two maps, they must be on the same map projection.

Many GIS packages have the ability to convert geographic coordinates to several different map projections. Some allow conversion backwards, from map coordinates in a projection to latitude and longitude. Obviously, this ability is rather important to the power and capability of a GIS, because usually GIS maps come from many different sources.

Finally, certain countries, and especially certain coordinate systems, rely entirely on the ability to work in a particular map projection, with a particular ellipsoid or a specific datum. In the United States, for example, the bulk of the 1 : 24,000 topographic map series of the US Geological Survey uses a polyconic projection, the Clarke 1866 ellipsoid, and the NAD27 datum. The recent movement to the NAD83 datum and its corresponding ellipsoid the GRS80, have "moved" features by as much as 300 meters on the ground or 12.5 millimeters (0.49 inch) on a 1 : 24,000 map. If the GIS user makes a basic mistake in comparing or assembling maps on different projections, based on different ellipsoids, and with different datums, many complex errors can result. This is especially important when data are to be captured from a map into the computer, as we will see in Chapter 3.

2.3 COORDINATE SYSTEMS

When we describe where we are, we usually give the place with reference to somewhere else. Giving directions, for example, we would say, "Go down to the second traffic light, turn right, then continue until you see the diner on the left, then take the second right." When we describe the location of a house or business, we might give the street address, "695 Park Avenue," for example. A *street address* is also a reference to another place, simply saying "Go to the street named Park Avenue and find the building labeled 695." Geography calls such references to locations *relative location*, because they give locations with respect to some other place. Later we will see that a GIS *can* handle some relative locations, such as street addresses. It can only do it, however, by fixing locations with respect to the earth as a whole. This is called an *absolute location*, because it is fixed with respect to an origin, a "zero point." For latitude and longitude, we use the earth's *equator* and the *prime meridian* as the system's origin. The location of the point, actually in the ocean off West Africa, is not critical, but locations fixed using the origin are indeed important.

Converting maps into numbers requires that we choose a standard way to encode locations on the earth. Maps are drawn (whether by computer or not) on a flat surface such as paper. Locations on the paper can be given in *map millimeters* or inches starting at the lower left-hand corner. A computer plotter or a printer can understand these dimensions also, and usually requires that the locations be given in (x, y) format; that is, an east-west distance or *easting*, followed by a north-south distance or *northing*. This pair of numbers is called a *coordinate pair* or, more usually, a *coordinate*. Standard ways of listing coordinates are then called *coordinate systems*. Maps on common coordinate systems are automatically aligned with each other.

A significant problem with coordinates is that while the map dimensions are simple and the (x, y) axes are at right angles to each other, locations on earth's surface are not so simply derived. The first and foremost problem is that a flat map of all or part of earth's surface is necessarily on a map projection. Something has been distorted to make the surface flat, usually scale, shape, area, or direction. On our flat map, we would like all of the earth's curvature removed. Just how this is done depends on which of the various coordinate systems we use, how big an area we seek to map, and what projection the system uses.

We consider four of the systems in common use in the United States in more detail in this section. As we cover each of the systems, take note of what projections are used and relate them back to the categories of projections introduced in Section 2.2. As you will quickly see, none of the coordinate systems in regular use is really ideal for computer mapping. Considering how complex a shape the earth is, however, many of the systems are perfectly adequate—indeed—extremely well suited for work with GIS.

The four systems we cover are the *geographic* coordinates themselves; then the worldwide *universal transverse Mercator* (UTM) coordinate system favored in many mapping efforts; the *military grid* system, an alternative form of the UTM that has been adopted in many countries outside the United States and for world mapping; and the *state plane* system, the basis of most surveying practice in the United States. Finally, we consider what other systems might be encountered in the GIS world and the implications of using these systems.

2.3.1 Geographic Coordinates

Many GIS systems store locations as numbers using latitude and longitude or geographic coordinates. This system was standardized by the International Meridian conference, held in Washington, D.C. in 1884. At this conference, it was decided to establish the origin for longitude for the earth at the Greenwich Observatory in England (Figure 2.11). In a GIS, latitude and longitude are almost always geocoded, or captured from the map into the computer, in one of two ways. These are degrees, minutes and seconds (DMS) and decimal degrees (DD). In both cases, latitudes go from 90 degrees south (−90) to 90 degrees north (+90) (Figure 2.6). Precision below a degree is geocoded as minutes and seconds, and decimals of seconds, in one of two formats: either in DMS as plus or minus DD.MMSS.XX, where DD are degrees, MM are minutes, and SS.XX are decimal seconds; or alternatively, in DD as DD.XXXX, or decimal degrees. Just as with time, a degree consists of 60 minutes, each with 60 seconds. Longitudes are the same, with the exception that the range is −180 to +180 degrees. In the second format, degrees are converted to radians and stored as decimal numbers with the appropriate number of significant digits.

For example, the file listed in Figure 2.12 is part of the World Data Bank, a listing of coordinates of the world's coastlines, rivers, islands, and political boundaries. The coordinates are decimal degrees, rounded to the nearest 0.001 degree. At the equator, one degree is about 40,000 km per 360 degrees = 111.11 km; 0.001 degree is then 111 meters. So, these data have a resolution of 111 meters on the ground. Their accuracy is determined by how well the line actually represents the real coastline of Africa (Figure 2.12).

FIGURE 2.11: The author standing on the prime meridian in Greenwich, England at the Royal observatory, Longitude 0°0′0″, Latitude 51°28′38″. Photo by Raymond H. Clarke.

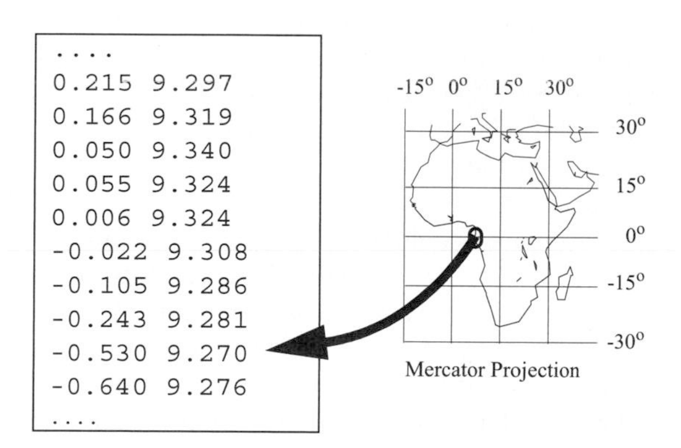

FIGURE 2.12: Part of the World Data Bank I listing of the coordinates of the coastline of Africa. Format is geographic coordinates in decimal degrees.

The advantage of using geographic coordinates in a GIS is that all maps can be transformed into a projection in the same way. If maps captured on a variety of projections are reprojected into geographic coordinates, there is some room for error. For example, the points in Figure 2.12 can never achieve a resolution better than 111 meters, regardless of the projection. If, however, the GIS does not support transformations between projections, then working in a common coordinate system such as the UTM or state plane system is very important if the maps are expected to overlay each other.

2.3.2 The Universal Transverse Mercator Coordinate System

The UTM coordinate system is commonly used in GIS because it has been included since the late 1950s on most USGS topographic maps. The choice of the transverse Mercator, probably now used more than any other projection for accurate mapping, has an interesting history. The story begins with the observation that the equatorial Mercator projection, which distorts areas so much at the poles, nevertheless produces minimal distortion laterally along the equator.

Johann Heinrich Lambert modified the Mercator projection into its transverse form in 1772, in which the "equator" instead runs north-south. The effect is to minimize distortion in a narrow strip running from pole to pole. Johann Carl Friedrich Gauss further analyzed the projection in 1822, and Louis Kruger worked out the ellipsoid formulas in 1912 and 1919 adjusting for "polar flattening." As a result, the projection is often called the Gauss conformal or the Gauss–Kruger, although the name *transverse Mercator* is used in the United States. Rarely, however, was the projection used at all until the major national mapping efforts of the post–World War II era.

The transverse Mercator projection, in various forms, is part of the civilian UTM system described here, the state plane system, and the military grid. It has been used for mapping most of the United States, many other countries, and even the planet Mars. The first version is the civilian UTM grid, used by the U.S. Geological Survey on its maps since 1977, and marked on many maps since the 1940s as blue tic marks along the edges of the quadrangle maps or grids over the surface. In 1977 the transverse Mercator projection replaced the polyconic for large-scale U.S. mapping.

The UTM capitalizes on the fact that the transverse Mercator is accurate in north-south strips by dividing the earth up into 60 pole-to-pole zones, each 6 degrees of longitude wide, running from pole to pole. The first zone starts at 180 degrees west (or east), at the international date line, and runs east, that is, from 180 degrees west to 174 degrees west. The final zone, zone 60, starts at 174 degrees east and extends east to the date line. The zones therefore increase in number from west to east. For the coterminous United States, California falls into zones 10 and 11, while Maine falls into zone 19 (Figure 2.13).

Within each zone we draw a transverse Mercator projection centered on the middle of the zone oriented north-south. Thus for zone 1, with longitudes ranging from 180 degrees west to 174 degrees west, the central meridian for the transverse Mercator projection is 177 degrees west. Because the equator meets the central meridian of the system at right angles, we use this point to orient the grid system (Figure 2.14). In reality, the central meridian is set to a map scale of slightly less than 1, making the projection for each zone secant along two lines at true scale parallel to the central meridian.

To establish a coordinate system origin for the zone, we work separately for the two hemispheres. For the southern hemisphere, the zero northing is the South Pole, and we give northings in meters north of this reference point. As the earth is about 40 million meters around, this means that northings in a zone go from zero to 10 million meters.

The numbering of northings starts again at the equator, which is either 10 million meters north in southern hemisphere coordinates or 0 meters north in northern hemisphere coordinates. Northings then increase to 10 million at the north pole. As we approach the poles, the distortions of the latitude-longitude grid drift farther and farther from the

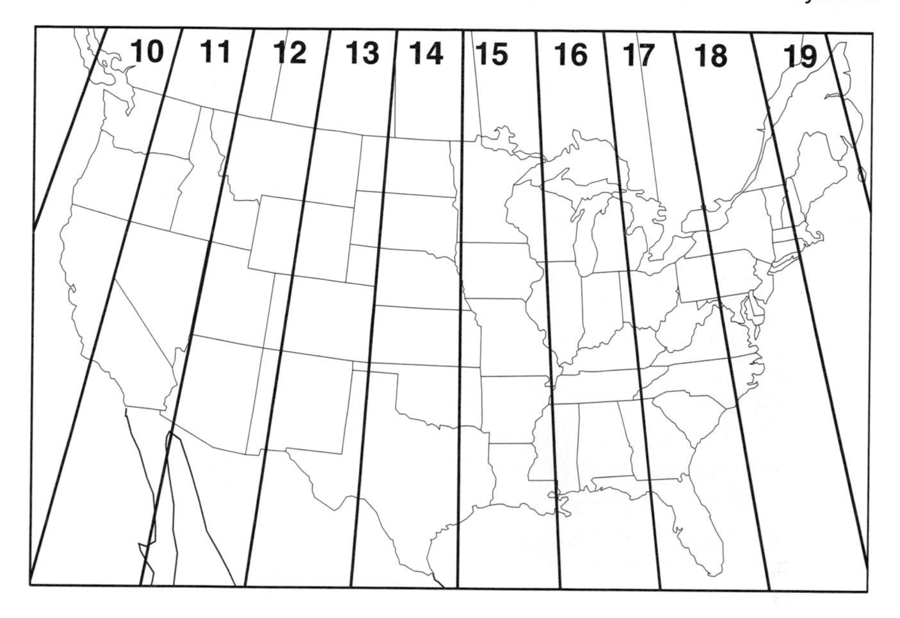

FIGURE 2.13: Universal transverse Mercator zones in the 48 contiguous states.

UTM grid. It is customary, therefore, not to use UTM beyond 84 degrees north and 80 degrees south. For the polar regions, the universal polar stereographic coordinate system is used.

For eastings, a false origin is established beyond the westerly limit of each zone. The actual distance is about half a degree, but the numbering is chosen so that the central meridian has an easting of 500,000 meters. This has the dual advantage of allowing overlap between zones for mapping purposes and of giving all eastings positive numbers. We can tell from our easting if we are east or west of the central meridian, and so the relationship between true north and grid north at any point is known. To give a specific example, Hunter College in New York City is located at UTM coordinate 4,513,410 meters north; 587,310 meters east; zone 18, northern hemisphere. This tells us that we are about four-tenths of the way up from the equator to the north pole, and are east of the central meridian for our zone, which is centered on 75 degrees west of Greenwich. On a map showing Hunter College, UTM grid north would therefore appear to be east of true north.

The variation from true scale is 1 part in 1000 at the equator. As a Mercator projection, of course, the system is conformal and preserves the shape of features such as coastlines and rivers. Another advantage is that the level of precision can be adapted to the application. For many purposes, especially at small scales, the last UTM digit can be dropped, decreasing the resolution to 10 meters. This strategy is often used at scales of 1 : 250,000 and smaller. Similarly, submeter resolution can be added simply by using decimals in the eastings and northings. In practice, few applications except for precision surveying and geodesy need precision of less than 1 meter, although it is often used to prevent computer rounding error and is stored in the GIS nevertheless.

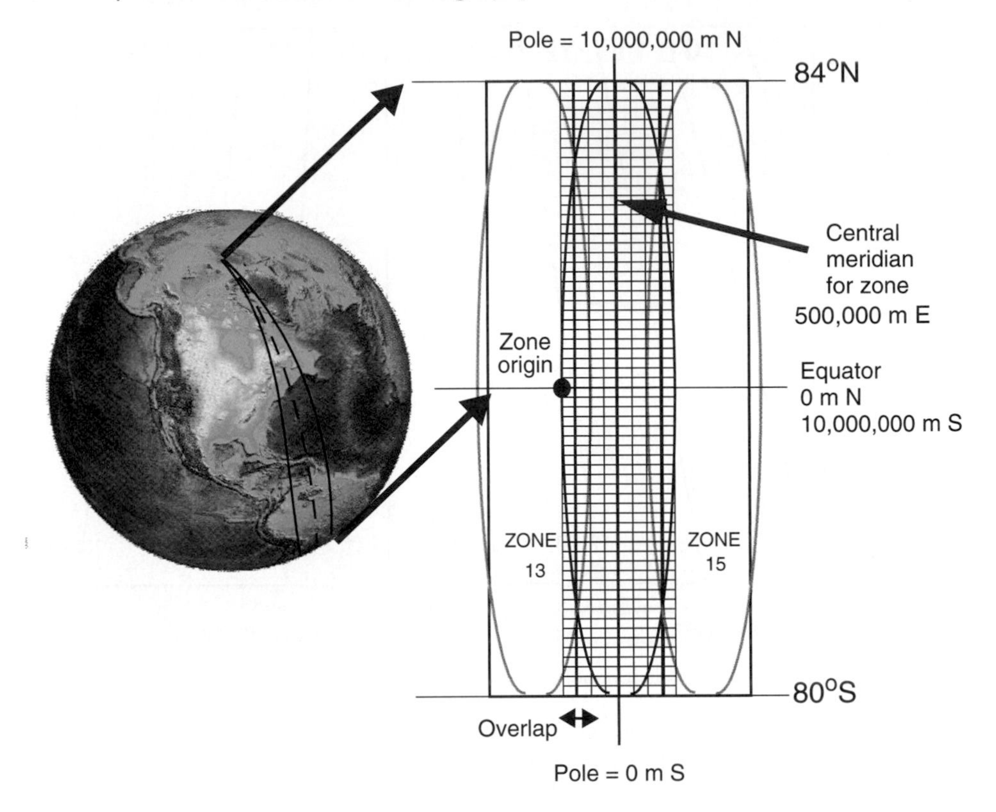

FIGURE 2.14: The Universal transverse Mercator coordinate system.

2.3.3 The Military Grid Coordinate System

The second form of the UTM coordinate system is the *military grid*, adopted for use by the U.S. Army in 1947 and used by many other countries and organizations. The military grid uses a lettering system to reduce the number of digits needed to isolate a location. Zones are numbered as before, from 1 to 60 west to east. Within zones, however, 8-degree strips of latitude are lettered from C (80 to 72 degrees south) to X (72 to 84 degrees north: an extended-width strip). The letter designations A, B, Y, and Z are reserved for Universal Polar Stereographic designations on the poles. A single rectangle, 6-by-8 degrees, generally falls within about a 1000-kilometer square on the ground. These squares are referenced by numbers and letters; for example, New York City falls into grid cell 18T (Figure 2.15).

Each grid cell is then further subdivided into squares 100,000 meters on a side. Each cell is assigned two additional letter identifiers (Figure 2.16). In the east-west (x) direction, the 100,000-meter squares are lettered starting with A, up to Z, and then repeating around the world, with the exception that the letters I and O are excluded, because they could be confused with numbers. The first column, A, is 100,000 meters wide and starts at 180 degrees west. The alphabet recycles about every 18 degrees and includes about six full-width columns per UTM zone. Several partial columns are given designations nevertheless, so that overlap is possible, and some disappear as the poles are approached.

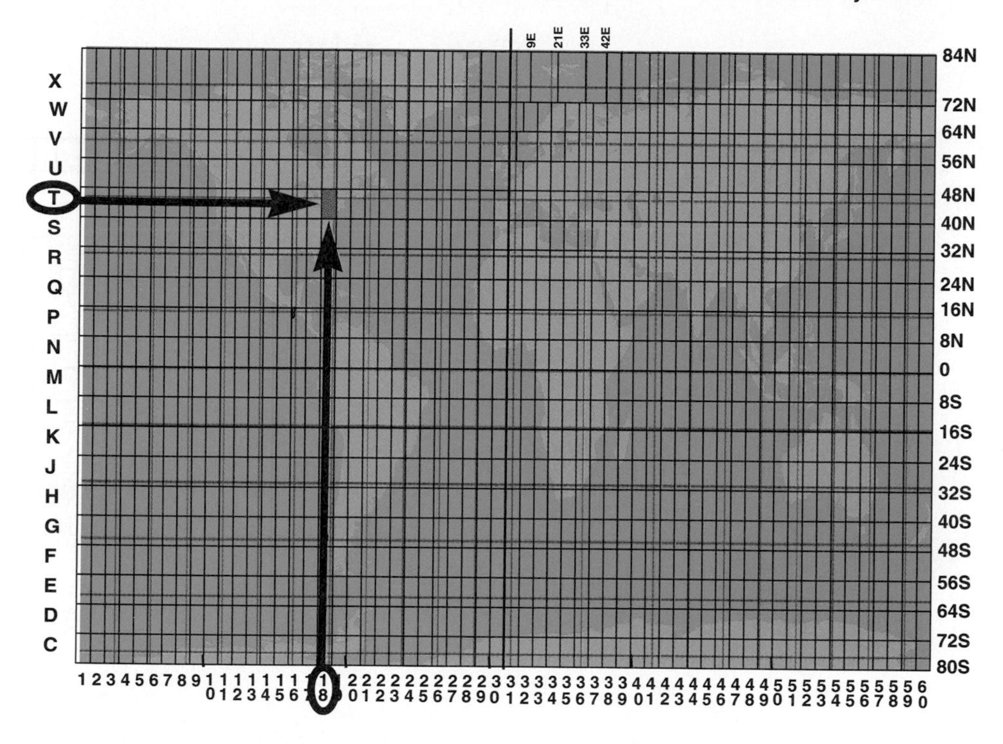

FIGURE 2.15: Six-by-eight-degree cells on the UTM military grid.

In the north-south (y) direction, the letters A through V are used (again omitting I and O), starting at the equator and increasing north, and again cycling through the letters as needed. The reverse sequence, starting at V and cycling backward to A, then back to V, and so on, is used for the southern hemisphere. Thus a single 100,000-meter grid square can be isolated using a sequence such as 18TWC. Within this area, successively accurate locations can be given by more and more pairs of x and y digits. For example, 18TWC 81 isolates a 10,000-meter square, 18TWC 8713 a 1000 meter square, and 18TWC 873134 a 100-meter square. These numbers are frequently stored without the global cell designation, especially for small countries or limited areas of interest. Thus WC873134, two letters and six numbers, would give a location to within 100-meter ground accuracy. Finally, the polar areas are handled completely separately on a different (UPS) projection.

2.3.4 The State Plane Coordinate System

Much geographic information in the United States uses a system called the *state plane coordinate system* (SPCS). The system is used primarily for engineering applications, especially by utility companies and local governments that need to do accurate surveying of facilities networks such as power lines and sewers, or for property designation. The SPCS is based on both the transverse Mercator and the Lambert conformal conic projections with units in meters (previously feet). The SPCS, which has been used for decades to write legal descriptions of properties and in engineering projects in many states, is based upon a different map of each state, except Alaska. States that are elongated north

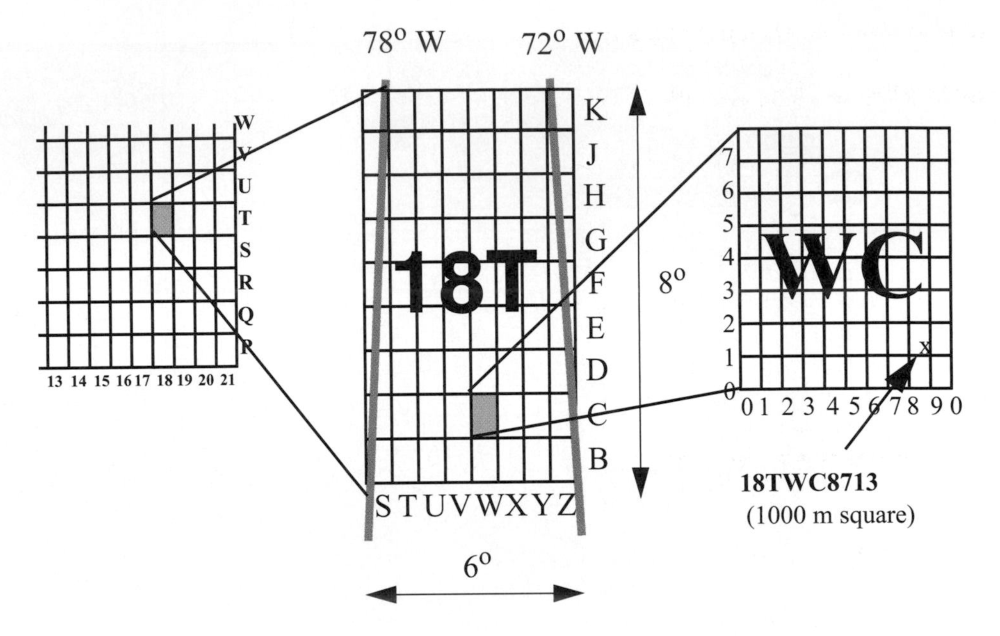

FIGURE 2.16: Military grid cell letters.

to south, such as California, are drawn on a Lambert conformal conic projection. States that are elongated east to west, such as New York, are drawn on a transverse Mercator projection, because the zones are divided into north-south strips.

The state is then divided up into zones, the number of which varies from small states, such as Rhode Island with one to as many as five zones. Some states have special case zones; for example, the state of California has one zone that consists of Los Angeles County alone. Some have more logic; for example, Long Island has its own zone for the state of New York. Because there are so many projections to cover the land area, generally the distortion attributable to the map projection is very small, much less than in UTM, where it can approach 1 part in 2000.

Each zone then has an arbitrarily determined origin that is usually some given number of feet west and south of the southwesternmost point on the map. This again means that the eastings and northings all come out as positive numbers. The system then simply gives eastings and northings in feet, often ending up with millions of feet, with no rounding up to miles. The system is slightly more precise than UTM because coordinates are to within a foot rather than a meter, and it can be more accurate over small areas. A disadvantage is the lack of universality. Imagine mapping an area covering the boundary between not only two zones, but two states. This means that you could be working with data that fall into four coordinate systems on two projections. Any calculation, such as computing an area, on that basis becomes a set of special-case solutions. On the other hand, SPCS is used universally by surveyors all over the United States.

A sample set of zone information is shown in Figure 2.17. New York State is somewhat of an exception. The bulk of the state, being east-west in extent, is divided into three north-south zones called "east," "central," and "west." Each of these three

New York State (3 zones on Transverse Mercator, 1 on Lambert Conic)
Based on NAD83

		Central Meridian	Origin	Scale reduction at Central Meridian	Easting at Origin
East	TM	74 30	40 00	1 in 10,000	150,000 m
Central	TM	74 30	40 00	1 in 10,000	250,000 m
West	TM	74 30	40 00	1 in 10,000	350,000 m
Long Island	LC	40 40 N 41 02 N	74 00W 40 10N		300,000 m

FIGURE 2.17: Statistics for New York State's state plane zones.

zones is drawn on a single transverse Mercator projection with a central meridian at 74°30′ West, with the scale factor at the center set at 0.9999, making the "cylinder" secant to the projection. Each zone then has an origin of zero meters (or feet) set at the northing of the origin (40° N) and at the easting given in Figure 2.17.

Coordinates in each zone are then numbered off in meters (or feet, without aggregation beyond feet). New York's Long Island zone uses a Lambert conic projection. For this one zone, the origin is a point and the standard parallels are given. For example, a position could have the following state plane coordinates:

870, 432 feet North; 730, 012 feet East, New York, Central Zone

Maps often show one or more of these coordinate systems, often by overprinted grids, and showing tic marks along the edges to show the coordinates. These are essential if the map data are to find their way into the GIS. Complicating the process are zone boundaries, and the fact that abbreviations of coordinates often look similar across coordinate systems (Figure 2.18).

2.3.5 Other Systems

There are, in addition, many other coordinate systems. Some are standardized, but many are not. Most countries have their own, although many use UTM or the military grid. The national grid of the United Kingdom uses the lettering system of the military grid but different-sized zones. In a few cases—Sweden, for example—the national census and other data are tied directly into the coordinates. Within the United States, many private companies and public services use unique systems, usually tied to specific functions such as power lines, or a specific region such as a municipality, or even a single construction project. There is also a tendency, especially when a base map is of unknown origin or when the map must be captured quickly, to throw away coordinates and just use "map millimeters" or "tablet coordinates." In this instance, unless we have the critical spatial fact of knowing at least two and preferably more points in both this and an accepted coordinate system, the map will be useless for matching with others or for overlay and analysis.

When using a coordinate system for geocoding in a GIS, we should be sure to remain consistent within that system and to record the relationship between the system and latitude and longitude or some other recognized system. Two points will suffice if

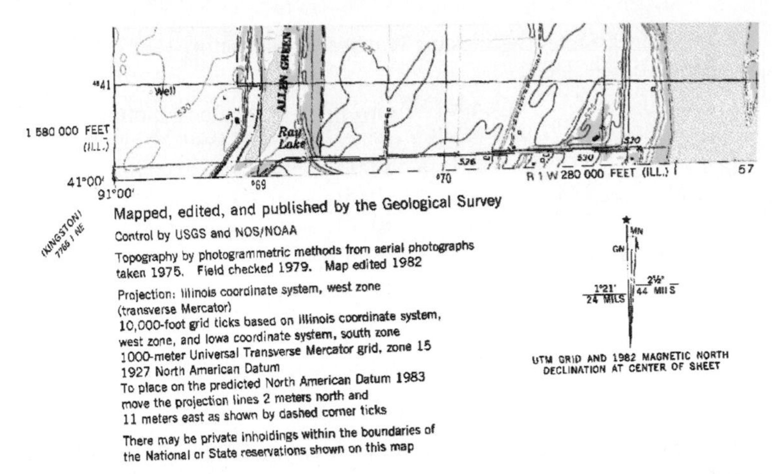

FIGURE 2.18: Map collar annotation from the Keithsburg USGS 7.5 minute Quadrangle (Illinois-Iowa). Note the grid systems in use are Geographic, UTM, and the intersection of two zones of the State Plane system. Note also the NAD27 to NAD83 correction. Note that UTM grid north is 1 degree 21 minutes east of true north (from the declination diagram lower right), implying that we are east of the central meridian for zone 15, and that UTM eastings should be greater than 500,000.

the spatial extent is perfectly aligned and north is the same on both maps, but different projections and other differences make this a rare occurrence. Also, we should be sure to use precision and numbers that make sense. Can we really measure distances over entire states down to the micrometer level or below? And even if we can, is this efficient to use in the GIS? In contrast, there is also a tendency to throw away precision needlessly.

Finally, while coordinates are the way that a GIS records information about location, location is just one of the many facets of geographic data. In the following section, we take a look at the full set of properties. Many of these become important to understand as we move into analysis and description of geographic features using a GIS.

2.4 GEOGRAPHIC INFORMATION

The purpose of geocoding is to encode the fundamental characteristics of geographic data in a digital form recognized by the GIS. Obviously, the most basic geographic characteristic is location. In a GIS, location is described by coordinates as numbers—and occasionally as letters. Obviously, just as a map contains many features, a map inside a GIS must contain a complete digital description of the features as coordinates. This means that a typical GIS database, especially the map component, is very large, especially if the coverage is detailed and the area large. Fortunately, the cost of storing data has decreased dramatically. Even on small computers, new storage methods have increased available memory from kilobytes to gigabytes within only a decade. The rapid growth of GIS has been heavily dependent on computer storage systems getting bigger and bigger over time.

Another basic characteristic of geographic data is *dimensionality*. Traditionally, cartography has divided data into *points*, *lines*, and *areas*. Basic to understanding how a GIS structures information is the idea that complex map features can be built up from simple ones. So a line can be constructed connect-the-dots style from a set of points.

Then an area or region can be made up of connected lines. This truly important concept is illustrated in Figure 2.19.

The attributes associated with a geographic feature are also important geographic information and can be categorized by their *level of measurement*. Levels of measurement are divided into *nominal*, *ordinal*, *interval*, and *ratio*. *Nominal* data are those that simply assign a label or class to a feature, such as a mine shaft or a ski resort. *Ordinal* features have a rank assigned to them, such as the sequence used for highways on maps of Jeep trails, unsurfaced roads, single-lane roads, two-way roads, state highways, and interstate highways. *Interval* values are those measured on a relative scale, such as elevations on a datum (based on mean sea level, an arbitrary zero point) or survey locations measured by pace and compass without a geodetic control. *Ratio* values are measured on an absolute scale, such as coordinates on a standard system or computed measures such as total precipitation. This division allows us to group features into classes, point-nominal, for example, or area-ratio. We return to this grouping system as a way of deciding what type of map to use in Chapter 7.

Another major characteristic of geographic information is *continuity*. Some types of maps, such as contour maps, assume a continuous distribution sometimes called a field, while others, such as choropleth maps, assume a discontinuous distribution. This distinction is made in detail in Chapter 7. Continuity is an important geographical property. The best example of a continuous variable is probably surface elevation. As we walk around on the earth's upper surface, we always have an elevation; at no point is elevation undetermined.

Continuity does not always apply to statistical distributions. For example, tax rates are a discontinuous geographic variable. A resident of New York has to pay the state personal income tax, but by living just 1 meter inside Connecticut, a person will not pay the New York tax. Geographic continuity is an important property. GIS coverage must be exhaustive for continuity; that is, there should be no holes or unclassified areas. Continuous variables in GIS, often called *field variables*, are most appropriately manipulated with raster (grid-based) GIS systems.

Once geographic features are built up from points, lines, and areas, their collective description can consist of measurements of the feature's size, distribution, pattern, orientation, neighborhood, contiguity, shape, and scale. Each of these properties defines

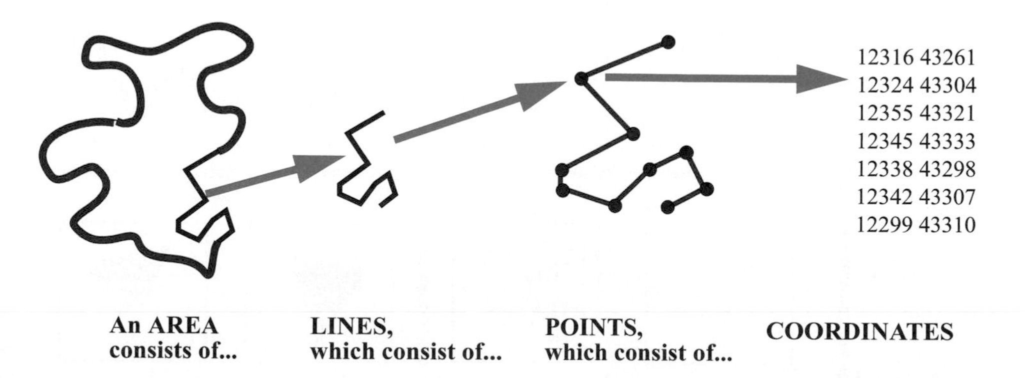

FIGURE 2.19: Geographic information has dimension. Areas are two-dimensional and consist of lines, which are one-dimensional and consist of points, which are zero-dimensional and consist of a coordinate pair.

the character of a feature and can usually be measured by the tools within a GIS and used for analysis. Usually, these higher-level descriptions are what prompted the use of a GIS in the first place. For example, we can measure the areas (size) of land parcels, or the orientation of highways, or the distribution of flora and fauna in a state park. The basic properties are summarized in Figure 2.20. Even though the GIS will directly hold only the coordinates and some additional information such as contiguity, information about every one of these properties will be available by using the tools within the GIS for higher-level analysis. Part of what the GIS user does is to coax descriptions of these properties out of the data that are available in the particular GIS in use. How well you can do this will depend on your skills as an intelligent GIS user.

Our digression on the cartographic roots of GIS is now complete. As we have seen, there are numerous important considerations to bear in mind that are directly related to the geometry of the map and the geometry of the features that we will store in the GIS. We are now ready to move on and begin covering GIS concepts proper. Step one is to cover how the map is structured as sets of digits inside the computer. Step two is to examine how to get the data from the map into the computer. As we will find out, this is another basic but overridingly important step in getting started with GIS.

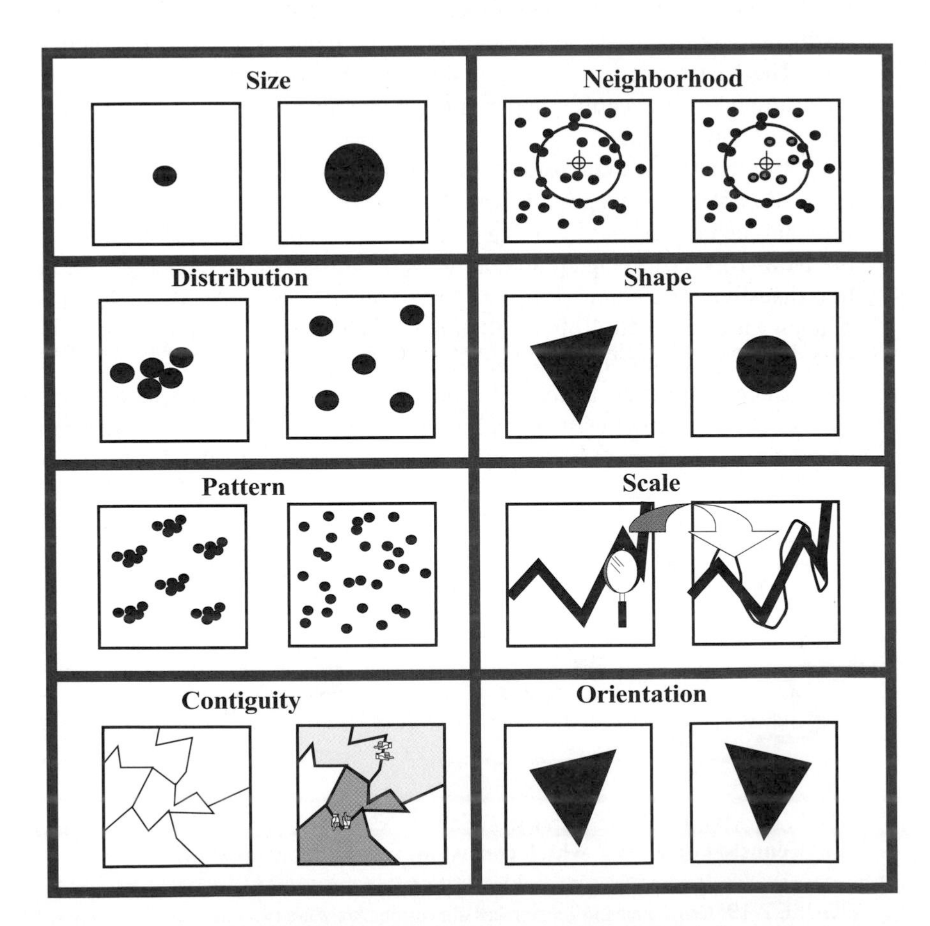

FIGURE 2.20: Basic properties of geographic features.

2.5 STUDY GUIDE

2.5.1 Summary

CHAPTER 2: GIS's Roots in Cartography

Map and Attribute Information (2.1)

- Information can be organized as lists, numbers, tables, text, pictures, maps, or indexes.
- Clusters of information called data can be stored together as a database.
- A database is stored in a computer as files.
- In a database, we store attributes as column headers and records as rows.
- The contents of an attribute for one record is a value. A value can be numerical or text.
- Data in a GIS must contain a geographic reference to a map, such as latitude and longitude.
- The GIS cross-references the attribute data with the map data, allowing searches based on either or both. The cross-reference is a link.
- Understanding the way maps are encoded to be used in GIS requires knowledge of cartography.
- Cartography is the science that deals with the construction, use, and principles behind maps.

Map Scale and Projections (2.2)

- The earth can be modeled as a sphere, an oblate ellipsoid, or a geoid.
  - The sphere is about 40 million meters in circumference.
  - An ellipsoid is an ellipse rotated in three dimensions about its shorter axis.
  - The earth's ellipsoid is only about 1/297 off from a sphere.
  - Many ellipsoids have been measured, and maps have been based on each. Examples are WGS83 and GRS80.
  - An ellipsoid gives the base elevation for mapping, called a datum. Examples are NAD27 and NAD83.
  - The geoid is a figure that adjusts the best ellipsoid and the variation of gravity locally. It is the most accurate, and is used more in geodesy than in GIS and cartography.
- Map scale is based on the representative fraction, the ratio of a distance on the map to the same distance on the ground.
- Most maps in GIS fall between 1 : 1,000,000 and 1 : 1,000.
- A GIS is scaleless because maps can be enlarged and reduced and plotted at many scales other than that of the original data.
- To compare or edge-match maps in a GIS, both maps MUST be at the same scale and have the same extent.
- The metric system is far easier to use for GIS work.
- Geographic coordinates are the earth's latitude and longitude system, ranging from 90 degrees south to 90 degrees north in latitude and 180 degrees west to 180 degrees east in longitude.
- A line with a constant latitude running east to west is called a parallel.

- A line with constant longitude running from the north pole to the south pole is called a meridian. The zero-longitude meridian is called the prime meridian and passes through Greenwich, England.
- A grid of parallels and meridians shown as lines on a map is called a graticule.
- A transformation of the spherical or ellipsoidal earth onto a flat map is called a map projection.
- The map projection can be projected onto a flat surface or a surface that can be made flat by cutting, such as a cylinder or a cone.
- If the globe, after scaling, cuts the surface, the projection is called secant. Lines where the cuts take place or where the surface touches the globe have no projection distortion.
- Projections can be based on axes parallel to the earth's rotation axis (equatorial), at 90 degrees to it (transverse), or at any other angle (oblique).
- A projection that preserves the shape of features across the map is called conformal.
- A projection that preserves the area of a feature across the map is called equal area or equivalent.
- No flat map can be both equivalent and conformal. Most fall between the two as compromises.
- To compare or edge-match maps in a GIS, both maps MUST be in the same projection.

Coordinate Systems (2.3)

- A coordinate system is a standardized method for assigning codes to locations so that locations can be found using the codes alone.
- Standardized coordinate systems use absolute locations.
- A map captured in the units of the paper sheet on which it is printed is based on relative locations or map millimeters.
- In a coordinate system, the x-direction value is the easting and the y-direction value is the northing. Most systems make both values positive.
- Some standard coordinate systems used in the United States are geographic coordinates, the universal transverse Mercator system, the military grid, and the state plane system.
- To compare or edge-match maps in a GIS, both maps MUST be in the same coordinate system
- A GIS package should be able to move between map projections, coordinate systems, datums, and ellipsoids.

Geographic Information (2.4)

- Geographic information has the characteristics of volume, dimensionality, and continuity.
- Simple geographic features can be used to build more complex ones. Areas are made up of lines, which are made up of points represented by their coordinates.
- Geographic features collectively have the properties of size, distribution, pattern, contiguity, neighborhood, shape, scale, and orientation.

- **Much of GIS analysis and description consists of investigating the properties of geographic features and determining the relationships between them.**

2.5.2 Study Questions

Map and Attribute Information

Define the following: *data*, *attribute*, *record*, *value*, and *database*. Using as an example the Yellow Pages part of a telephone directory for your town, discuss how you would build a database to hold the list information. What parts of the Yellow Pages attributes are geographic references?

Map Scale and Projections

Using an atlas, make a list of as many map projections as you can find. Are any of the atlas maps not annotated with their projection? Make a table listing the properties of each of the projections, plus any other information you can find out—for example, whether the projection is secant, transverse, based on an ellipsoid, conformal, and so on. In a final column, state what properties are distorted on the map; for example, "Map distorts area increasingly as one moves north and south."

Consult one of the sources listed in the References and try to find the sizes of as many ellipsoids as possible. Are any of them more or less suitable for foreign countries? Select a country, for example, Egypt or Australia, and research what ellipsoids have been used and whether any particular projection is favored for that country. Why would one projection be better than another?

Find the regulation sizes of either a baseball diamond or a soccer field. Draw maps of the fields at the following scales: 1 : 1000, 1 : 24,000, 1 : 100,000, and 1 : 1,000,000. What problems do you run into? What would be the effect of mapping both a winding river and an irregular patch of forest at these scales?

Coordinate Systems

The chapter covers several different coordinate systems. Determine which coordinate systems are shown on a map of your local area.

For a single location, such as your house or school, try to find the coordinates of the position in as many coordinate systems as possible. How might you make sure that your result from the map is correct?

Geographic Information

Make a table of levels of measurement versus dimension. In each cell of the table, write in as many types of geographic data or features as you can think of.

Using the example of a lake, write out sample measurements that might describe each of the major geographic properties covered in Figure 2.20. For example, the size of the lake is its area in square meters. For which properties is it most difficult to think of representative numbers? (*Hint*: Can a single number describe the shape?)

FIGURE 2.21: Mercator projection of North America secant at 40°N overlain with a Lambert conformal conic with standard parallels at 30°N and 45°N.

2.6 EXERCISES

1. *Use the manual that came with your GIS software package to see what map projections the software can support and whether it is possible to transform between map projections and from geographic coordinates into a map projection. For any digital map, ideally at a smaller scale such as 1 : 1 million, plot the same map in two map projections one on top of the other. What magnitude of error do you see?*
2. *Again, using your GIS package, draw a map of a small feature such as a lake. Manipulate the scale of the feature on the plot so that the same feature is plotted at a number of different map scales. How might the GIS deal with the small scales when the feature is too detailed? How might the GIS deal with the feature when it is enlarged beyond its ability to "look" like a map feature?*
3. *Using the database component of your GIS package and an existing database such as the tutorial data set that came with the software, list all the attributes for one or more records. Which are numerical attributes? What are the ranges and legal values for the attributes values? How would the GIS detect a value that was out of bounds? (Try it!) How does the GIS software allow you to change the value for an attribute for one or more records? (Hint: You will have to dig into the documentation for this.)*
4. *Use your GIS package to print out the location of a feature in its raw coordinates. What is the coordinate system in use? What is the map projection? What ellipsoid and datum were used? Does the software (or documentation) supply any information about the accuracy of the data? Using the printed coordinates, give a numerical estimate of the precision of the data. How might the accuracy and precision be improved?*
5. *Using a GIS map showing a polygonal feature such as a lake, a city boundary, or a soil or land-cover polygon, think of three different ways to quantify the orientation of the feature. (Hint: What "line" best represents the polygon?)*

2.7 BIBLIOGRAPHY

Campbell, J. (1993) *Map Use and Analysis*. 2nd ed. Dubuque, IA: William C. Brown.

Clarke, K. C. (1995) *Analytical and Computer Cartography*. 2nd ed. Upper Saddle River, NJ: Prentice Hall.

Department of the Army (1973) *Universal Transverse Mercator Grid*, TM 5-241-8, Headquarters, Department of the Army. Washington, DC: U.S. Government Printing Office.

Snyder, J. P. (1987) *Map Projections—A Working Manual*. U.S. Geological Survey Professional Paper 1396. Washington, DC: U.S. Government Printing Office.

2.8 KEY TERMS AND DEFINITIONS

absolute location: A location in geographic space given with respect to a known origin and standard measurement system, such as a coordinate system.

accuracy: The validity of data measured with respect to an independent source of higher reliability and precision.

attribute: A numerical entry that reflects a measurement or value for a feature. Attributes can be labels, categories, or numbers; they can be dates, standardized values, or field or other measurements. An item for which data are collected and organized. A column in a table or data file.

azimuthal: A map projection in which the globe is projected directly on a flat surface. Only one "side" of the globe can be shown at a time.

cartography: The science that deals with the principles, construction, and use of maps.

compromise: A map projection that is neither area preserving nor shape preserving. An example is the Robinson projection.

conformal: A type of map projection that preserves the local shape of features on maps. On a conformal projection, lines on the graticule meet at right angles, as they do on a globe.

conic: A type of map projection involving projecting part of the earth onto a cone-shaped surface that is then cut and unrolled to make it flat.

continuity: The geographic property of features or measurements that gives measurements at all locations in space. Topography and air pressure are examples.

coordinate pair: An easting and northing in any coordinate system, absolute or relative. Together these two values, usually termed (x, y), describe a location in two-dimensional geographic space.

coordinate system: A system with all the necessary components to locate a position in two- or three-dimensional space: that is, an origin, a type of unit distance, and axes.

cylindrical: A type of map projection involving projecting part of the earth onto a cylinder-shaped surface that is then cut and unrolled to make it flat.

data: A set of measurements or other values, such as text for at least one attribute and at least one record.

database: A collection of data organized in a systematic way to provide access on demand.

datum: A base reference level for the third dimension of elevation for the earth's surface. A datum can depend on the ellipsoid, the earth model, and the definition of sea level.

dimensionality: The property of geographic features by which they are capable of being broken down into elements made up of points, lines, and areas. This corresponds to features being zero-, one-, and two-dimensional. A drill hole is a point, a stream is a line, and a forest is an area, for example.

distortion: The space distortion of a map projection, consisting of warping of direction, area, and scale across the extent of the map.

easting: The distance of a point in the units of the coordinate system east of the origin for that system.

edge matching: The GIS or digital map equivalent of matching paper maps along their edges. Features that continue over the edge must be "zipped" together and the edge dissolved. To edge-match, maps must be on the same projection, datum, ellipsoid, and scales and show features captured at the same equivalent scale.

equal area: A type of map projection that preserves the area of features on maps. On an equal-area projection, a small circle on the map would have the same area as on a globe with the same representative fraction. See also **equivalent**.

equatorial radius: The distance from the geometric center of the earth to the surface, usually averaged to a single value for a sphere.

equirectangular: A map projection that maps angles directly to eastings and northings. A cylindrical projection, made secant by scaling the height-to-width ratio. The nonsecant or equatorial version is called the Plate Carree. Credited to Marinus of Tyre, about A.D. 100.

equivalent: A type of map projection that preserves the area of features on maps. On an equal area projection, a small circle on the map would have the same area as on a globe with the same representative fraction. See also **equal area**.

field variable: A geographic value that is continuous over space.

file: Data logically stored together at one location on the storage mechanism of a computer.

flattening (of an ellipsoid): The ratio of the length of half the short axis of the ellipse to half the long axis of the ellipse, subtracted from 1. The earth's flattening is about 1/300.

geocode: A location in geographic space converted into computer-readable form. This usually means making a digital record of the point's coordinates.

geodesy: The science of measuring the size and shape of the earth and its gravitational and magnetic fields.

geographic coordinates: The latitude and longitude coordinate system.

geographic property: A characteristic of a feature on earth, usually describable from a map of the feature, such as location, area, shape, distribution, orientation, adjacency, and so on.

geography: (1) A field of study based on understanding the phenomena capable of being described and analyzed with a GIS. (2) The underlying geometry and properties of the earth's features as represented in a GIS.

geoid: A complex earth model used more in geodesy than cartography or GIS that accounts for discrepancies over the earth from the reference ellipsoid and other variations due to gravity, and so on.

globe: A three-dimensional model of the earth made by reducing the representative fraction to less than 1 : 1.

GPS (Global Positioning System): An operational, U.S. Air Force–funded system of satellites in orbits that allow their use by a receiver to decode time signals and convert the signals from several satellites to a position on the earth's surface.

graticule: The latitude and longitude grid drawn on a map or globe. The angle at which the graticule meets is the best first indicator of what projection has been used for the map.

GRS80 (Geodetic Reference System of 1980): Adopted by the International Union of Geodesy and Geophysics in 1979 as a standard set of measurements for the earth's size and shape. The length of the semimajor axis is 6,378,137 meters. Flattening is 1/298.257.

information: The part of a message placed there by a sender and not known by the receiver.

interval: Data measured on a relative scale but with numerical values based on an arbitrary origin. Examples are elevations based on mean sea level, or coordinates.

latitude: The angle made between the equator, the earth's geometric center, and a point on or above the surface. The south pole has latitude −90 degrees, the north +90 degrees.

level of measurement: The degree of subjectivity associated with a measurement. Measurements can be nominal, ordinal, interval, or ratio.

link: The part or structure of a database that physically connects geographic information with attribute information for the same features. Such a link is a defining component of a GIS.

location: A position on the earth's surface or in geographic space definable by coordinates or some other referencing system, such as a street address or space indexing system.

longitude: The angle formed between a position on or above the earth, the earth's geometric center, and the meridian passing through the center of the observing instrument in Greenwich, England, as projected down onto the plane of the earth's equator or viewed from above the pole. Longitudes range from −180 (180 degrees West) to +180 (180 degrees East).

map: A depiction of all or part of the earth or other geographic phenomenon as a set of symbols and at a scale whose representative fraction is less than 1 : 1. A digital map has had the symbols geocoded and stored as a data structure within the map database.

map millimeters: A coordinate system based on the dimensions of the map rather than those of the features represented on the earth itself, in metric units.

map projection: A depiction of the earth's three-dimensional structure on a flat map.

mean sea level: A local datum based on repeated measurements of sea level throughout all of its normal cycles, such as tides and seasonal change. The basis for elevations on a map.

meridian: A line of constant longitude. All meridians are of equal length on the globe.

metric system: A system of weights and measures accepted as an international standard as the Systeme International d'Unites (SI) in 1960. The metre (meter in the United States) is the unit of length.

military grid: A coordinate system based on the transverse Mercator projection, adopted by the U.S. Army in 1947 and used extensively for world mapping.

mosaicing: The GIS or digital map equivalent of matching multiple paper maps along their edges. Features that continue over the edge must be "zipped" together and the edge dissolved. A new geographic extent for the map usually has to be cut or clipped out of the mosaic. To permit mosaicing, maps must be on the same projection, datum, ellipsoid, and scale, and show features captured at the same equivalent scale.

NAD27 (North American Datum of 1927): The datum used in the early part of the national mapping of the United States. The Clarke 1866 ellipsoid was used, and locations and elevations were referenced to a single point at Meade's Ranch in Kansas.

nominal: A level of measurement at which only subjective information is available about a feature. For a point, for example, the name of the place.

northing: The distance of a point in the units of the coordinate system north of the origin for that system.

oblate ellipsoid: A three-dimensional shape traced out by rotating an ellipse about its shorter axis.

oblique: A map projection in which the centerline of the map is not at right angles to the earth's geographic coordinates, following neither a single parallel nor a meridian.

ordinal: A level of measurement at which only relative information is available about a feature, such as a ranking. For a highway, for example, the line is coded to show a Jeep trail, a dirt road, a paved road, a state highway, or an interstate highway, in ascending rank.

origin: A location within a coordinate system where the eastings and northings are exactly equal to zero.

parallel: A line of constant latitude. Parallels get shorter toward the poles, becoming a point at the pole itself.

perfect sphere: A three-dimensional figure traced out by all possible positions of an arc of a fixed radius about a point. A good approximation of the shape of the earth.

polar radius: The distance between the earth's geometric center and either pole.

precision: The number of digits used to record a measurement or which a measuring device is capable of providing.

prime meridian: The line traced out by longitude zero and passing through Greenwich, England. The prime meridian forms the origin for the longitude part of the geographic coordinates and divides the eastern and western hemispheres.

ratio: A level of measurement at which numerical information is available about a feature, based on an absolute origin. For land parcels, for example, the assessed value in dollars would be an example, the value zero having real meaning.

record: A set of values for all attributes in a database. Equivalent to a row of a data table.

relative location: A position described solely with reference to another location.

representative fraction: The ratio of a distance as represented on a map to the equivalent distance measured on the ground. Typical representative fractions are 1 : 1,000,000, 1 : 100,000, and 1 : 50,000.

scale: The geographic property of being reduced by the representative fraction. Scale is usually depicted on a map or can be calculated from features of known size.

scaleless: The characteristic of digital map data in abstract form of being usable and displayable at any scale, regardless of the scale of the map used to geocode the data.

secant: A map projection in which the surface used for the map "cuts" the globe at the map's representative fraction. Along this line there is distortion-free mapping of the geographic space. Multiple cuts are possible, for example, on a conic projection.

standard parallel: A parallel on a map projection that is secant and therefore distortion-free.

state plane: A coordinate system common in utility and surveying applications in the lower 48 United States and based on zones drawn state by state on transverse Mercator and Lambert conformal conic projections.

transverse: A map projection in which the axis of the map is aligned from pole to pole rather than along the equator.

UTM (Universal Transverse Mercator): A standardized coordinate system based on the metric system and a division of the earth into sixty 6-degree-wide zones. Each zone is projected onto a transverse Mercator projection, and the coordinate origins are located systematically. Both civilian and military versions exist.

value: The content of an attribute for a single record within a database. Values can be text, numerical, or codes.

WGS84 (World Geodetic Reference System of 1984): A higher precision version of the GRS80 used by the U.S. Defense Mapping Agency in world mapping. A common datum and reference ellipsoid for hand-held GPS receivers.

zone (of a coordinate system): The region over which the coordinates relate with respect to a single origin. Usually, some part of the earth or a state.

CHAPTER 3

Maps as Numbers

3.1 REPRESENTING MAPS AS NUMBERS
3.2 STRUCTURING ATTRIBUTES
3.3 STRUCTURING MAPS
3.4 WHY TOPOLOGY MATTERS
3.5 FORMATS FOR GIS DATA
3.6 EXCHANGING DATA
3.7 STUDY GUIDE
3.8 EXERCISES
3.9 REFERENCES
3.10 KEY TERMS AND DEFINITIONS

Yes raster is faster, but raster is vaster, and vector just seems more corrector.
—*C. Dana Tomlin*, Geographic Information Systems and Cartographic Modeling *(1990), p. 44*

Your lucky number is 3552664958674928. Watch for it everywhere.
—*Anonymous.* *Berkeley Unix "Fortunes" file*

...the Map must go ever forward, no matter what obstacles and hazards its servants may encounter.
—*Showell Styles.* The Forbidden Frontiers: The Survey of India from 1765 to 1949 *(1970), p. 78*

"Prepare to perish, carbon-based Earth-scum! (GAK!!) OK guys, who fouled up on the scale conversion for the invasion map?"

3.1 REPRESENTING MAPS AS NUMBERS

In this chapter we look at the various ways that maps can be represented using numbers. All GISs have to store digital maps somehow. As we will see, there are some critical differences in how the various types of GIS navigate on this ocean of geographic numbers. The organization of the map into digits has a major impact on how we capture, store, and use the map data in a GIS. In Chapter 4 we will see that an important first stage in working with GIS is just getting the map in the right form of numbers into the computer.

Obviously, there are many ways that the conversion of a visual or printed map to a set of digits can be done. Over the years, the designers of GIS and computer mapping packages have devised an amazing number of ways that maps can be converted into numbers. The difference between the ways is not trivial, not only because different types of files and codes are needed, but because the entire way that we think about the data in a GIS is affected. The link between how we imagine the features that we are working with in the GIS and the actual files of bytes and bits inside the computer is a critical one. To the computer, the data are stored in a physical structure that is quite tangible, at least to the GIS software. The physical structure is not only how computer memory such as disk and RAM is used, but also how the files and directories store and access the map and attribute information.

On the physical level, the map, just like the attributes, is eventually broken down into a sequence of numbers, and these numbers are stored in the computer's files. In general, two alternative ways exist of storing the numbers. In the first, each number is saved in the file encoded into binary digits or bits. A number in base 10, decimal, can be converted into base 2, binary, and the binary representation stored as on or off, 1 or 0, in the files on the computer. Eight bits in a row are termed a *byte*, and one byte can hold all numerical values from 0000 0000 to 1111 1111 binary or 0 to 255 decimal. Many computer programmers use shorthand to represent a byte, since in base 16 (hexadecimal) one byte can hold two hexadecimal digits. Hexadecimal runs out of counting digits at 9, so it fills the gaps with letters. Counting in hexadecimal then goes in the sequence `0,1,2,3,4,5,6,7,8,9,A,B,C,D,E,F`. To a computer programmer, the range of numbers that can be stored in one byte goes from hexadecimal `00` to `FF`. Typically, as a GIS user you will never see hexadecimal values unless you crash the system or try to look at files stored in this binary format rather than the alternative.

The second way of encoding numbers into files is to treat each number the way that humans do—one decimal digit at a time. Coincidentally, this is the same way that we deal with text, punctuation, and so on, so this format is often called *text* or *ASCII files*. ASCII stands for the American Standard Code for Information Interchange, and the codes are 256 standard meanings for the values that fall into one byte. Some of the ASCII codes are numbers, some lowercase and some uppercase letters, some special characters such as "$" or ">" and some are actually keys to operations (such as the escape codes or tabs). Each fits into one byte, and we store in the file the numbers just as they come, even using indentation and spacing. These files are usually suitable for use with an editor, and they can be printed and read without a program.

It is the logical structure of the data that requires us to have a mental "model" of how the physical data represent a geographic feature, just as a sheet map is a flat paper "symbol" model of the landscape it covers. Traditionally in GIS and computer cartography, there were two basic types of data model for map data, and only one for attribute data. Map data could be structured in *raster* or *vector* format, and attributes as flat files.

A raster data model uses a grid, such as the grid formed on a map by the coordinate system, as its model or structure to hold the map data. Each grid cell in the grid is one map unit, often chosen so that each grid cell shows on the GIS map as one screen display point or *pixel*, or on the ground as a whole-number increment in the coordinate system. A pixel is the smallest unit displayable on a computer monitor. If you get a magnifying glass and look at a monitor or a television set, you will see that the picture is made from thousands of these tiny pixels, each made up of a triangle of three phosphor dots, one dot for red, one for green, and one for blue (if your screen displays color). When we capture a map into the raster data model, we have to assign a value to every cell in the grid. The value we assign can be the actual number from the map, such as the terrain elevation in a digital elevation model (DEM), or more usually, it is an index value standing for an attribute that is stored separately in the attribute database.

Key elements in raster data are shown in Figure 3.1. First, the cell size determines the resolution of the data, and the cell size has both a map and a ground expression. We often talk about 30-meter Landsat data, for example, meaning that each cell in the data is 30 meters by 30 meters on the ground. On the map, we may use several pixels to display the grid cell, or on paper we may use a dot of a certain size in a given color. Second, the grid has an *extent*, often rectangular since a grid has columns and rows, and even if we do not wish to store data in the GIS for grid cells outside our region (such as a state), we still have to place something (usually a code for "outside") in the grid cells. Third, when we map features onto the grid there is sometimes an imperfect fit. Lines have uneven widths, points must be moved to the center or the intersection points of the grid, and areas may need to have their edges coded separately. We sometimes have to determine in advance what connections within the grid are legal. For example, taking a single cell, we can allow connections only north, south, east, and west, like the way a rook moves in chess, or we can allow diagonal connections as well. Which we choose can mean a great deal as to how the GIS works at storing and using the features. Fourth, when we deal with a grid, each grid cell can usually only be "owned" by one feature, that is, the one whose attribute it holds. In many cases, map data are not so simple. Soils, for example, are often listed by their percentage of sand, silt, and clay at every point. Finally, when we have a grid, every cell in the grid has to be made big enough to hold the largest value of the attribute or index to be stored in the grid. You may have had the experience of using a spreadsheet or table to store people's names. Even when we store "JaneDoe" with only eight characters, we still have to allow for the occasional very long name. Every grid cell pays the storage penalty of the extra space, and with the total number of cells being the product of the numbers of rows and columns, the amount of space needed

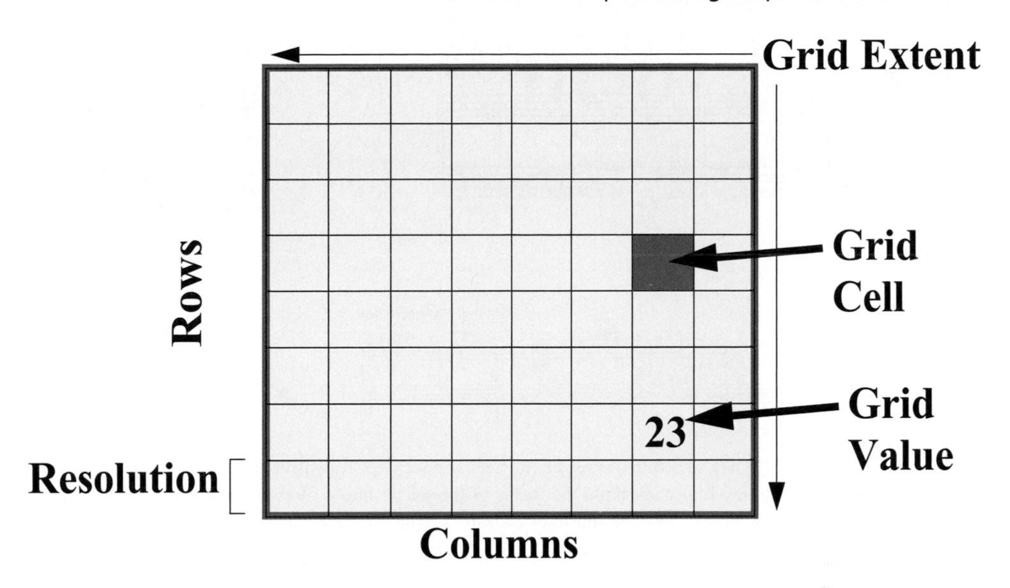

FIGURE 3.1: Generic structure for a grid.

can add up (or rather multiply up) quickly. Storage sizes for grids often increase by powers of 2 as more and more "bytes" of 8 bits are needed to store larger and larger values.

Nevertheless, raster grids have many advantages. They are easy to understand, capable of rapid retrieval and analysis, and are easy to draw on the screen and on computer devices that display pixels. Mark Bosworth imagines raster grids as being like the music of Mozart (Figure 3.2): detailed, repetitive, highly structured, and elegant. Slowly, dainty step by tiny step, they build the theme of the music into a glorious single structure, even though it may have "too many notes."

The other major type of data model for map data is the vector. The vector is composed of points, each one represented by an exact spatial coordinate. For a point or a set of points, vectors just use a list of coordinates. For a line, we use a sequence of coordinates; that is, the sequence of points in the list is the order in which they must be drawn on the map or used in calculations. Note that this gives lines a "direction" in which their points should be read. Areas in the vector model are the space enclosed by a surrounding ring of lines, either one or several of them.

Vectors are obviously very good at representing features that are shown on maps as lines, such as rivers, highways, and boundaries. Unlike the raster grid, where we have to store a grid cell's attribute whether we need it or not, we need only place points precisely where we need them. A square can be four lines connecting four points, for example. Even wiggly lines can be captured quite well in this way, by using more points for the bends and fewer when the line is straight. Using vectors, we can draw an outline map with only a few thousand points, far fewer than the number of grid cells that would be required. To complete the musical analogy, vectors are more like the music of Beethoven.

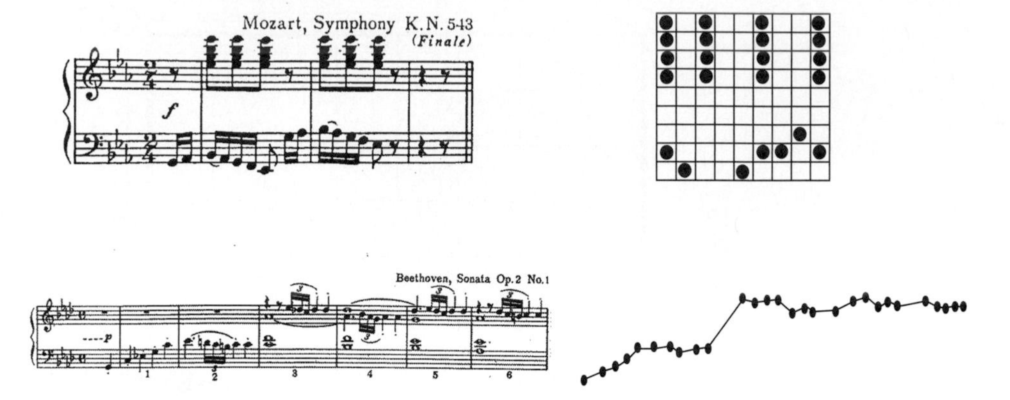

FIGURE 3.2: Data structures as music. Rasters are detailed, repetitive, highly structured, and elegant; slowly, dainty step by tiny step, they build the theme of the music into a glorious single structure, even though it may have "too many notes," like the music of Mozart. Vectors are bold, leaping streaks that go from place to place with rapidity and efficiency, like the music of Beethoven.

Vectors use bold, leaping streaks that go from place to place with rapidity and efficiency. There is little repetition, and the vectors get straight to the guts of the geographic features they represent.

Vectors have the advantage of accuracy, since they can follow features very closely. They are efficient at storing features. They are very suitable for plotting devices that draw with pens or points of light, as the features can be drawn one at a time in their completeness. They can also be adjusted to store information about connectivity to other features, as we discuss under topology in Section 3.4. On the down side, the vector is not very good at representing continuous field variables such as topography, except in a way we consider later called the *triangulated irregular network*. Vectors are also not a good structure to use if the maps to be generated involve filling areas with shades or color.

We have said little yet about the attribute data other than that its model is as a *flat file*. A flat file is how numbers are stored in tables or in a spreadsheet. The model is also a sort of grid, with rows for records and columns for attributes (Figure 3.3). Just like for the raster grids, we have to store values in the cells of the table. As we have already noted, these values must somehow link the data in the flat file to the data in the map. For a raster grid, we could store index numbers in the grid and any number of attributes for the index numbers in the flat file. For example, on a land use map, 1 could stand for forest, 2 for farmland, and 3 for urban. For vector data, we need a little more complexity. Point data are simple; we can even put the coordinates in the flat file itself. Lines and areas, however, have variable numbers of points. We again need to number off or "identify" the lines and store an attribute for the whole line in the flat file. We can do the same for an area, except that we need the line flat file as well to refer to in the polygon or area file. If we called the lines *arcs*, for example, we might need both a polygon attribute table file, and a file of arcs by polygon.

So far we have touched only briefly on the data models that GISs use. In the sections that follow, we first delve more closely into the way that attribute data are stored in files and how the tricks of data storage have improved over the last 30 years.

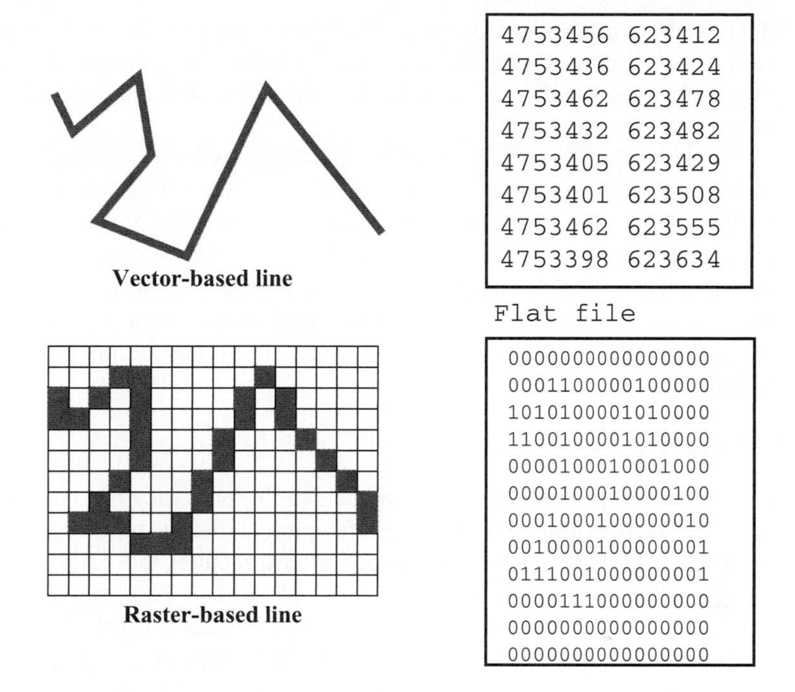

FIGURE 3.3: Both vectors and rasters can be stored in flat files, if they are simple.

Then, in the following sections we go into more detail and cover different logical and physical data models that GIS systems use to store the map data. These too have evolved over time and have improved significantly over the years. We will look at some of the formats that are used by GIS data providers and finish the chapter by raising some of the technical problems that have faced GIS users who need to move data between formats and between systems. The latter aspect has immense importance for the future of GIS and will be revisited in Chapter 10.

3.2 STRUCTURING ATTRIBUTES

In Chapter 5, we consider in more detail the logical way that attribute data are stored in files. For now, a simple way to imagine this is to recall the flat file we covered in this chapter. A flat file is a table. Columns store attributes, and rows store records. We know in advance what sort of information is stored in each attribute, whether it is text or numbers, how large the numbers are and so on. We can then write a sequence into the file. For each record we can write the ASCII codes for the values in each attribute (in database terms often called a *field*) in a consistent way. At the end of each record we could start a new line. The file then would be a sort of table or matrix with rows and columns.

It is now easy to see what some of the database operations actually do. For example, if we wished to sort the data we could renumber the lines in the file. If we wanted a particular record, we could search line by line until we found the correct one and then print it. These operations would be much faster if we could encode the numbers in binary or sort them in the file so that the most commonly referenced records were first in the file.

Most database management systems (DBMSs) do exactly this, and some use very clever ways of placing records into files. If you use a database manager, or even a spreadsheet program on a personal computer (PC), you have probably noticed these files, which hold your records in one place.

Another important part of structuring attributes is that the *database dictionary*, the list of all the attributes along with all their characteristics, must also be written into the file. This is sometimes done by using a separate file, but usually the dictionary is written into the top or *header* of the file, before the data begin. So the attribute database part of a GIS is fairly simple. At the most basic it consists of a file. At the most complex, it might be several files in a directory. From a data management point of view, manipulating the attributes is a piece of cake. Unfortunately, the maps are a little harder to deal with.

3.3 STRUCTURING MAPS

Maps have at least two dimensions; in the earth's space they have latitude and longitude, and in the map's space they have the left-right (x) and the up-down (y) directions. They are also scaled-down representations of features, features that can be points, lines, areas, or even volumes. Point features are very simple to deal with, and you could easily argue that you don't really even need a GIS for point features other than to draw them. This is because x and y can be stored just as regular attributes in a standard database. Line and area features are more complicated because they can be different shapes and sizes. A stream and a road would be captured with different numbers of points, and these would not fit easily into the attribute database.

Nothing says that we have to capture points, however. In Section 3.1, we met the two basic models for map data, the vector and the raster. While a GIS can usually deal with data from either format, only one structure can be used for retrieval and analysis of the map. We have already seen how the structure we choose can influence many GIS operations. The data structure also affects the type and amount of error involved in the process and the type of map we can use for display. It is worthwhile, then, to consider in turn each of the ways in which data can be structured.

3.3.1 Vector Data Structures

Vector data structures were the first to be used for computer cartography and GIS because they were simply derived from digitizing tablets, because they are more exact in representing complex features such as land parcels, and because they are easily drawn on pen-type output devices such as plotters. Surprisingly, few people in the early days thought of standardizing how digitizing was to take place, and since there were different technologies, many different formats evolved. The earliest included ASCII files of (x, y) coordinates, but these soon became very unwieldy in size, so binary files rapidly took over.

The first generation of vector files were simply lines, with arbitrary starting and ending points, which duplicated the way a cartographer would draw a map. Obviously the pen would be lifted from the paper to start a new line, but it could be lifted anywhere else. The file could consist of a few long lines, many short lines, or even a mix of the two. Typically, the files were written in binary or ASCII and used a flag or code coordinate to signify the end of a line.

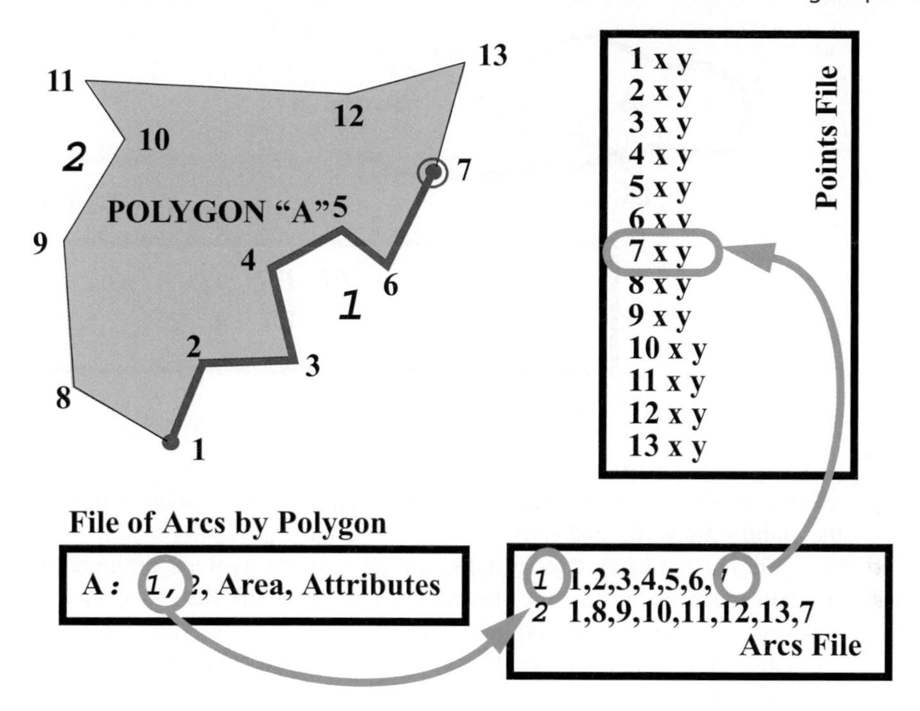

FIGURE 3.4: Arc/node map data structure with files.

As a computer programmer, having to follow the line from one place to another in the file was compared by early programmer Nick Chrisman to following the path of a single strand of spaghetti through a pile of spaghetti on a plate. The name stuck, and to this day unstructured vector data are called *cartographic spaghetti*. A surprising number of systems still use this basic structure, however, and the structure survives in many data formats, such as the Defense Mapping Agency's standard linear format. Most systems allow users to import data in this structure, but almost all now convert it to topological data after entry.

Just as the hierarchical system had caught on as a way of organizing attribute databases, starting in the 1960s a hierarchy for spatial data was worked out that became the arc/node model. Many first-generation systems, including POLYVRT, GIRAS, and ODYSSEY, used this system. This data structure uses the fact that each type of feature—point, line, and area—consists of features with the next fewer dimensions. So area features consist of connected lines, and lines consist of connected points. The most important advantage that this buys us is that we can have separate files for areas, lines, and points. The price is that we need to keep track of links between the files in a fairly arbitrary way. For example, in Figure 3.4, a single polygon consists of two lines or arcs, each with a set of points. The points are ordered from node to node in a sequence.

At the least, we need a file containing the attributes for the polygon, a file listing the arcs within the polygon, and, finally, a file of coordinates that are referenced by the arc file. Figure 3.4 shows that we need to store in each of these files a set of references between the files. For example, an entry in the arcs files states that to get the points for

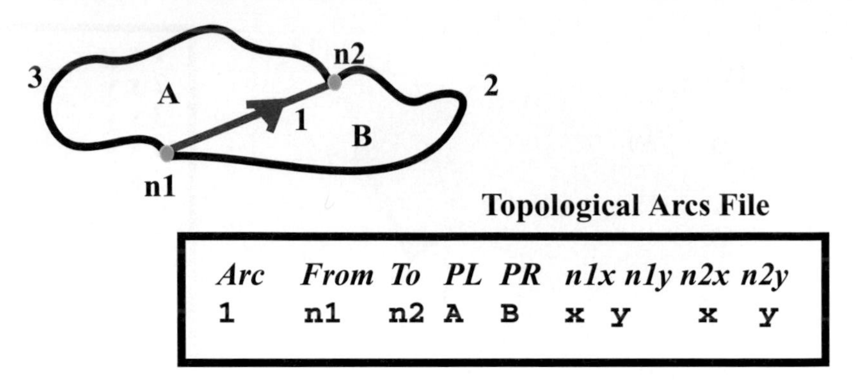

FIGURE 3.5: A topological structure for the arcs.

arc 2 from the points file they begin at the first coordinate point in the file, followed by the eighth, the ninth, and so on.

During the early days of GIS, several systems evolved different versions of this structure. Obviously, to save space we could write the files in binary. There are few ways, however, to store point, line, and area data that are as efficient. As long as the data are valid, this is a very powerful way to store data for map features. When the system breaks down, however, is when data contain errors, which is virtually always.

A new generation of arc/node data structures arrived after the First International Advanced Study Symposium on Topological Data Structures for Geographic Information Systems, held in 1979. This elegant new structure used the arc as the basis for data storage and relied on reconstructing a polygon when it was needed. The way that this was accomplished turned out to have other practical benefits, as we discuss in Section 3.4. The system stored point data as before, but included in the file of arcs linked to the points file was an abbreviated "skeleton" of the arc (Figure 3.5). This consisted of just the first and last points in the arc, called *end nodes*, and information that related not to this particular arc but to its neighbors in geographic space. This included the arc number of the next connecting arc, and the polygon number of which polygon lay to the left and right of the arc. If the line was just a river or a road, this information was not essential. If, however, the arc was part of a network that formed enclosed areas or polygons, the polygon identifier number became the key to polygon construction.

The way that a polygon could be built was by extracting all of the arcs that a specific polygon had as a neighbor. If the polygon is the right-hand neighbor of each of the arcs, the end nodes can be tested against each other to see which sequence they should be drawn in. The use of the arc as the basic unit meant that when a map was digitized, the user only needed to trace each arc once, instead of twice if each area was traced around the edge. As exactly the same arc was used in both cases, the type of error known as a *sliver* was avoided completely.

One problem with the vector data structure was that it did not really deal very well with geographical surfaces such as topography or air temperature. This was corrected by a research team that devised a new data structure called the *triangulated irregular network* or TIN (Figure 3.6). The TIN is really just a list of points with their coordinates; stored with the points is a file containing information about the topology of a network.

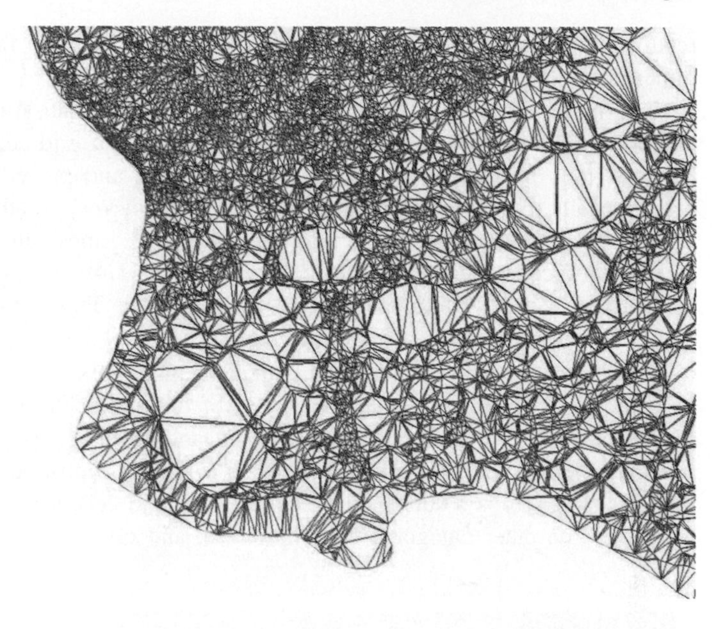

FIGURE 3.6: A triangulated irregular network (TIN) covering metropolitan Lisbon in Portugal. (Courtesy of Elisabete de Silva.)

The network is a set of triangles, constructed by connecting the points in a network of triangles called a *Delaunay triangulation*. This way of drawing triangles is optimal, because changing any one triangle makes the angles within the triangle less similar to each other.

Two sorts of TIN can be built, one with a file containing information about the arcs that connect points, and one containing all the data about one triangle. The TIN became popular as a way of storing topography or land elevation data for visualization and engineering. With a TIN it is easy to draw contours, make a three-dimensional view of an area, estimate how water would run downhill over a digital landscape, or calculate how much material would have to be moved in a construction project. Many GIS programs that work with computer-aided drafting (CAD) systems or with surveying software use TIN as their data structure. The TIN has proven to be both efficient in storing data and versatile in finding new uses within GIS.

3.3.2 Raster Data Structures

Raster or grid data structures have formed the basis for many GIS packages. The grid is a surprisingly versatile way of storing data. The data form an array or matrix of rows and columns. Each pixel or grid cell contains either a data value for an attribute, or an index number that points to a reference in the attribute database. So a pixel containing the number 42, for example, could correspond with the number 42 or "deciduous forest" in the Anderson Level II system, or just the 42nd record in the attribute file.

To write the numbers to a file, we can just start the file with any necessary attribute codes, perhaps the number of rows and columns and the maximum size of one value, and then write the data into the file in binary across all columns for all rows, one long stream of data with a start and an end, like an unraveled sweater. When

reading the data back in, we just place the data back into a raster grid of the correct dimensions.

A major advantage of the raster system is that the data form their own map in the computer's memory. An operation such as comparing a grid cell with its neighbors can be performed by looking at the values in the next and preceding row and column of the grid cells in question. However, the raster is not very good at representing lines or points, since each becomes a set of cells in the grid. Lines can become disconnected or "fat" if they cross the grid at too shallow an angle. However, variables that needed the TIN in vector data structures fit easily into the raster. This is particularly suited to data that come from remote sensing or scanning.

One major problem with raster data is the mixed pixel problem. Figure 3.7 provides an example. The photographs show a part of the outline of a lake in an oblique view. In the top picture, there is only one type of land cover, "grass," so all pixels belong in one class. In the bottom picture, there are two types of attributes, water and grass. Even though a wet foot is a sure indicator of a water grid cell, it is hard to assign each pixel to one or the other category. A compromise, and one often used in GIS, is to assign

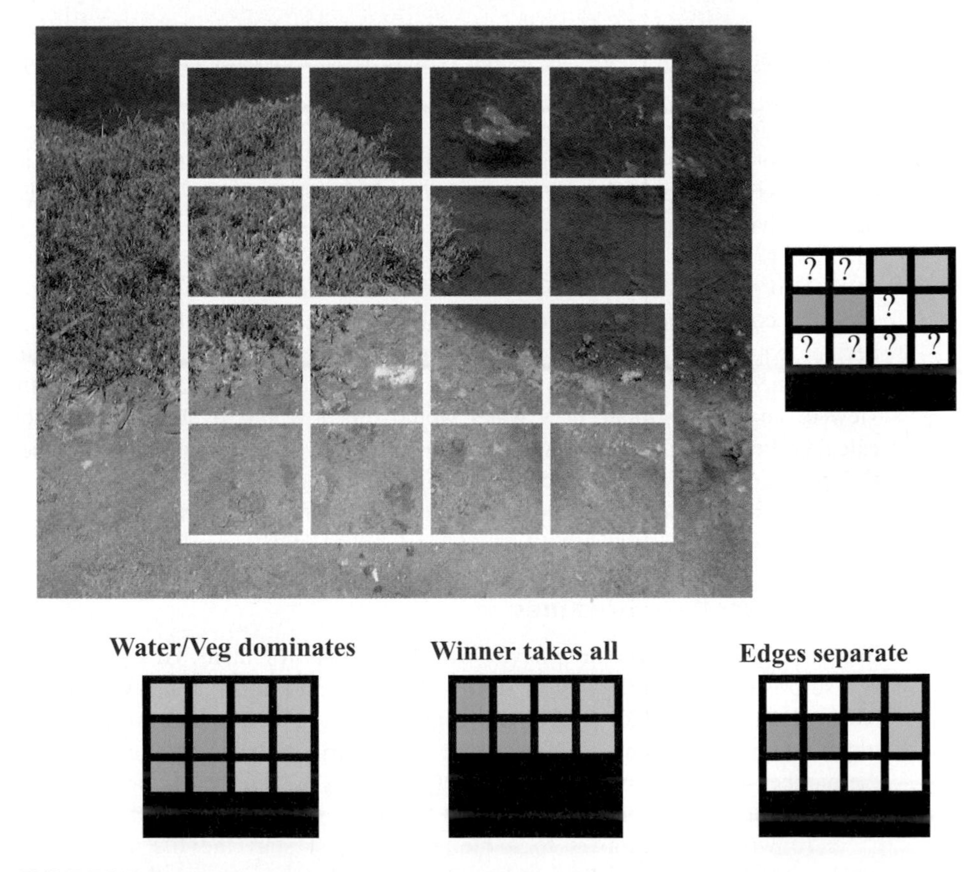

FIGURE 3.7: The mixed pixel problem. Domination can be applied in sequence, the area covering most of the pixel assigned the pixel, or edges can be assigned a separate value. Any of the three options is acceptable, or any other, as long as the rule is applied consistently. (Water = Cyan, Vegetation = Green, Bare Ground = Black).

FIGURE 3.8: March 2002 Landsat 7 image of the Santa Barbara channel, CA. This data layer is a grid with a large section of "missing data," in this case, the zeros in the ocean, and the unregistered image edges.

edge pixels, those that are not exclusively in one class or another. Finally, when several classes are involved, we either have to make up rules for assignment, such as assigning a mixed pixel to the class that occupies the most area within it, or we have to put up with edges and mixed pixels. Even when the boundary is absolutely clear, a vector data representation may be better than the raster.

At least two ways have been devised within GISs to deal with the problem that a grid often contains redundant or missing data (Figure 3.8). The first of these is a compression mechanism called *run-length encoding*. Along each row, only changes between attributes and the numbers of pixels of that same attribute are stored. If a whole row is all one class, it is stored as the class and the number of pixels only, quite a saving in space. As the raster becomes more and more varied, however, fewer and fewer savings can be made over the original grid. Many GIS packages and many industry-standard image formats use run-length encoding.

Another way to save space is to use a data structure called a *quad tree*. A quad tree works by dividing a grid into four quadrants, saving a reference to the quadrant of the grid only if it contains data. Then the quadrant is split into four half-size quadrants, and so on until the individual pixel is reached. If we split a quadrant and all the pixels in it have the same attribute, then obviously we would not need to store anything else for that entire quadrant. Each attribute becomes a list of quadrants needed to get to the

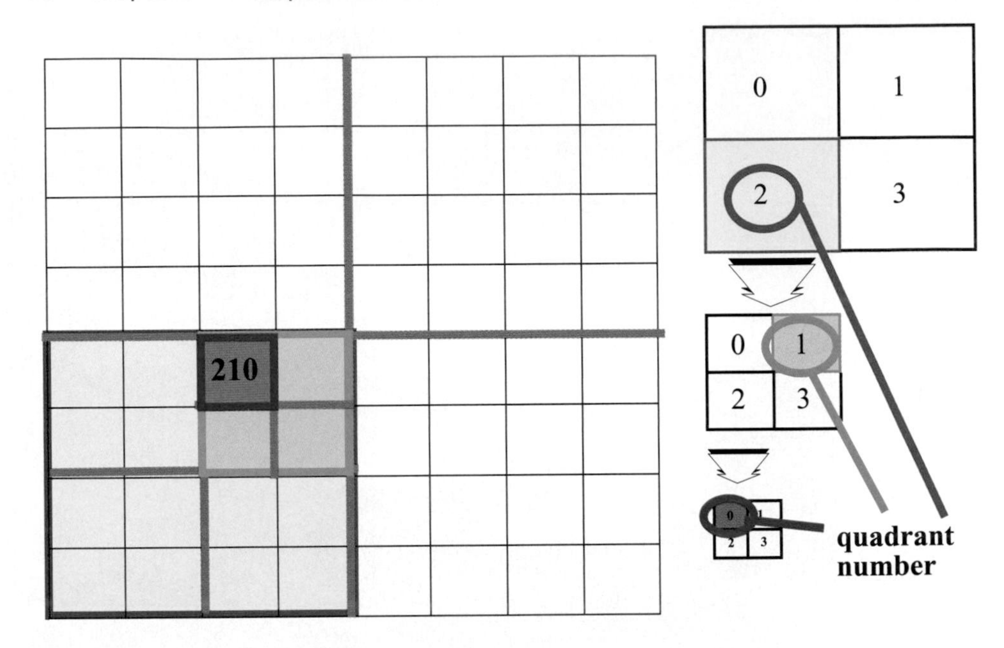

FIGURE 3.9: The quad-tree structure. Reference to code 210.

area, like a coordinate reference (Figure 3.9). Quad trees have been used more in image processing than in GIS, but at least one GIS package, SPANS, uses them as its primary data structure.

3.4 WHY TOPOLOGY MATTERS

When topological data structures became widespread in GIS, some significant benefits resulted, enough that today the vector arc/node data structure with topology probably is the most widespread for GIS data. Typically, a GIS maintains the arc as the basic unit, storing with it the polygon left and right, the forward and reverse arc linkages, and the arc end nodes for testing. This means that each line is stored only once and that the only duplication is the endpoints. The disadvantage is that whenever areas or polygons are to be used, some recomputing is necessary. Most programs save the result, however, such as the computed polygon areas, so that recalculation is unnecessary.

Topology allowed GIS for the first time to do error detection. If a set of polygons is fully connected, and there are no gaps at nodes or breaks in the lines defining the areas, the set of areas is called *topologically clean.* When maps are first digitized, however, this is rarely the case. The topology can be used to check the polygons. Polygon interiors are usually identified by digitizing a point inside a polygon, a label point, and by keeping track of the arcs as they are entered. A polygon gets the label from the label point when the point is found to be inside the polygon. A GIS will have the ability to build the topology from the unconnected arcs. First, each endpoint is examined to see if it is "close" to another. If it is, the points are "snapped" together; that is, their (x, y) coordinates are averaged and each is replaced with exactly the same values (Figure 3.10).

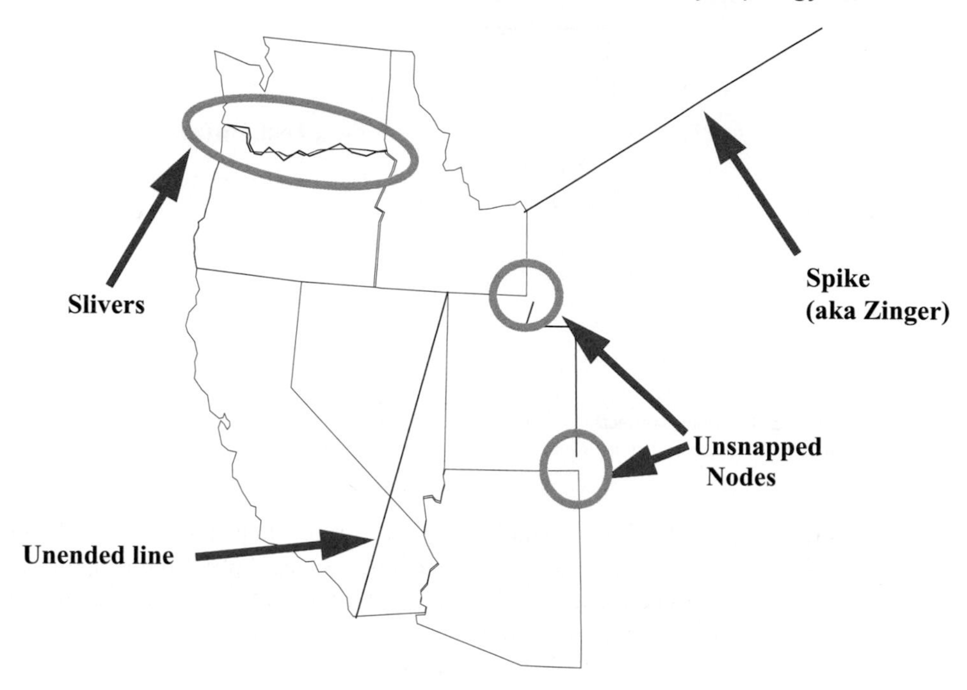

FIGURE 3.10: Example of slivers (unmatched nodes along two lines), unsnapped nodes (endpoints of two lines that should be the same point), spikes (erroneous coordinates), and unended lines.

In addition, arcs that connect the same nodes are tested to see if they are duplicates, and the user is asked which to delete. Any small areas, probably the result of errors called *slivers,* caused by double digitizing, are also eliminated. So, an error automatically detected can become an error automatically eliminated. Obviously, the separation between nodes required before they are treated separately and the size that a polygon must reach before it is retained are critical values for the map. Sometimes called *fuzzy tolerances*, these values should be handled with caution, because they allow the map's features to move around. Short lines, small polygons, or precisely measured point locations have critical significance and should not be deleted by automatic testing for topological completeness. For example, a map of Europe should include Andorra, Monaco, and Liechtenstein.

The primary advantage of having a topologically consistent map is that when two or more maps must be overlain, much of the initial preparation work has been done. What still has to be established are where new points must be added along lines to become nodes, and how to deal with any small or sliver polygons that are created (Figure 3.10). The latter can be a real problem. Many borders between regions, states, counties, and so on match along lines such as rivers, which are generalized differently at different map scales. Although the line should be the same, in fact it is not. Some packages allow the extraction of a line from one map to be "frozen" for use on another. This seemingly small difference can be very significant, especially if areas or densities are being calculated.

The final advantage to topology is that many operations of retrieval and analysis can be conducted without having to continuously deal with the (x, y) data. In some cases,

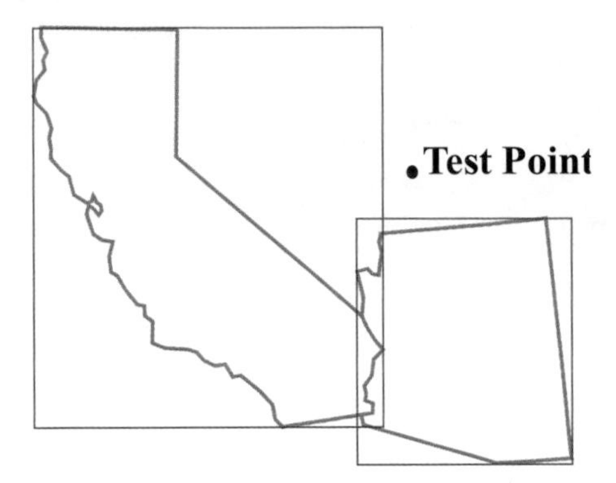

FIGURE 3.11: Bounding rectangles: rectangles that contain a polygon completely. If a test point lies wholly outside the bounding rectangle, it must be outside the enclosed polygon.

pretesting can be done. For example, say that a point is to be tested to see if it falls inside an area. If the point falls outside the bounding rectangle of all the endpoints, the point is most likely outside the region (Figure 3.11). So useful are these bounding rectangles, the highest and lowest x and y values along an arc, that they are often computed once and saved with the topological information in the arcs file.

3.5 FORMATS FOR GIS DATA

With a 30-year history and with so many alternative ways to structure map and attribute data, it is hardly surprising that most GISs use radically different approaches to handling their content. The data structures used are often invisible as far as the GIS user is concerned. We might not even need to understand exactly what is happening when two maps are overlain. However, if we are to be objective, scientific GIS users, at the very least we must have a full understanding of the errors and transformations involved. Regardless of how a GIS structures its maps as numbers, it must be able to import data from other GIS packages and from the most common data sources, as well as scanned and digitized data, and to convert the result into its own internal format. In some cases this is an open process. Some GIS companies have published and documented their internal or exchange data formats, including Intergraph and Autodesk. Others protect their internal data as a trade secret, in the hope of being able to sell data and data converters as well as their GIS.

The most common data formats for GIS data have been used by so many GIS operations and for so much existing data that a GIS ignores them at its peril. Some are so common that utility programs and even operating systems read, process, and display these formats automatically. These formats include some that have arisen because they are a common data format, such as TIGER and DLG. Others are industry-standard formats, proprietary formats that have been used so much that they are documented and published, although their use may have restrictions.

In the GIS world, a small subset of these formats has become commonplace, and we cover them here for completeness. We then finish the chapter by discussing some of the issues of data exchange between GIS systems and take a look at the accepted national standard, the Spatial Data Transfer Standard (SDTS).

3.5.1 Vector Data Formats

A general distinction between industry and commonly used standards for GIS data is that between formats that preserve and use the actual ground coordinates of the data and those that use an alternative *page coordinate* description of the map. The latter are the coordinates used when a map is being drafted for display in a computer mapping program or in the data display module of a GIS (Figure 3.12).

HPGL

```
IN;IP0 0 8636 11176;
SC-4317 4317 -5586 5586;
SP1;
SC-4249 4249 -5498 5498;

SP1;
PU-2743 847;PD -2743 3132 608 3132
608 847 -2743 847;
```

PostScript

```
%%BeginSetup
11.4737 setmiterlimit
1.00 setflat
/$fst 128 def

%%EndSetup
@sv
/$ctm matrix currentmatrix def
@sv
%%Note: Object
108.58 456.98 349.85 621.50 @E
0 J 0 j [] 0 d 0 R 0 @G
0.00 0.00 0.00 1.00 K
0 0.22 0.22 0.00 @w
 0 0 0 @g
0.00 0.00 0.00 0.00 k
%%RECT 241.272 -164.520 0.000
108.58 621.50 m
349.85 621.50 L
349.85 456.98 L
108.58 456.98 L
108.58 621.50 L
@c
B
@rs
@rs
%%Trailer
end
```

AutoCAD DXF

```
POLYLINE
  8
7
  6
CONTINUOUS
 66
  1
  0
VERTEX
  8
7
 10
-2.742
 20
3.132
  0
VERTEX
  8
7
 10
0.608
 20
3.132
  0
VERTEX
  8
7
 10
0.608
 20
0.848
  0
VERTEX
  8
7
 10
-2.742
 20
0.848
  0
VERTEX
  8
7
 10
-2.742
 20
3.132
  0
SEQEND
  0
ENDSEC
  0
EOF
```

FIGURE 3.12: Some alternative industry-standard vector formats. Headers have been removed. Graphic is the same four-point rectangle in each case.

The Hewlett-Packard Graphics Language (HPGL) is a page description language designed for use with plotters and printers. The format is simple and the files are plain ASCII text. Each line of the file contains one move command, so a line segment connects two successive lines or points. The format works with a minimum of header information, so that files can be written or edited easily. However, the header can be manipulated to change the scaling, size, colors, and so on. The HPGL is an unstructured format and does not store or use topology.

Another industry-standard format is the PostScript page definition language, developed by the Adobe Corporation for use in its desktop and professional publishing products and now in widespread use. So common is this format that most laser-quality printers use it as the printer device control format. PostScript, at least in its vector mode, is a page description language. This means that its coordinates are given with respect to a printed page, say an 8-1/2-by-11-inch sheet.

PostScript uses ASCII files but has particularly complex headers controlling a very large number of functions, such as fonts, patterns, and scaling. PostScript is really a sort of programming language, and to be viewed it must be interpreted. There are many commercial and shareware packages for viewing PostScript files, and many word processors and graphics packages will both read and write the format, although many more write it than read it. In GIS, PostScript is usually used to export or print a finished map rather than data as such.

The popular CAD package AutoCAD by Autodesk has made commonplace the AutoCAD digital exchange format (DXF) for drawing data. The AutoCAD Map GIS software also makes use of these formats. While Autocad uses its own internal format for data storage, it uses an external format called DXF for transfer of the files between computers and between packages. Again, these are simple ASCII files (although there is a binary mode), but in the DXF case there is a very large and mandatory file header containing significant amounts of metadata and file default information.

Although DXF does not support topology, it does allow the user to maintain information in separate layers, a familiar GIS concept. There is considerable support for details of drawings, line widths and styles, colors, and text, for example. DXF is importable by almost all GIS packages other than those that use raster formats. Some GIS packages work directly with the Autocad or other CAD software and can manage these files internally.

Two formats have become widespread largely because so many important data have been made available using them: the DLG and the TIGER formats. The digital line graph (DLG) format of the U.S. Geological Survey's (USGS) National Mapping Division are available at two scales, the scales of the map series from which they were captured (Figure 3.13). These scales are 1 : 100,000, for which almost all of the country has some data available, and 1 : 24,000, with only a small portion of the country but in extreme detail.

The formats of the data are documented formally, and the files are ASCII. They use the ground coordinates in UTM, truncated to the nearest 10 meters to reflect their locational precision and to save space (Figure 3.14). Features are handled in separate files—for example, hydrology, hypsography (contours and topographic features), transportation, and political. Many GIS packages will import these files, but often some extra data manipulation is necessary, such as making the records of some fixed line length in bytes.

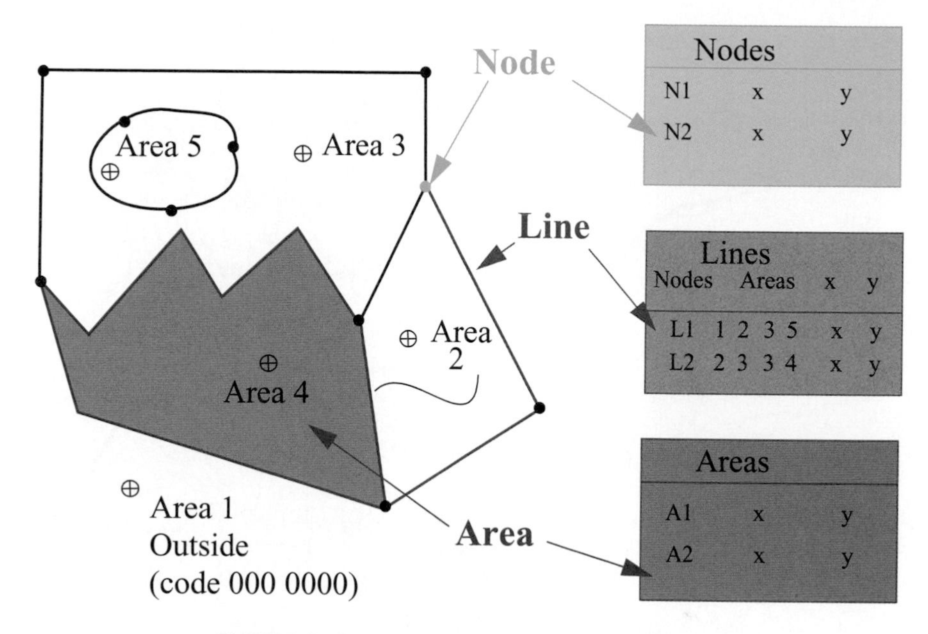

FIGURE 3.13: Sample digital line graph coding format.

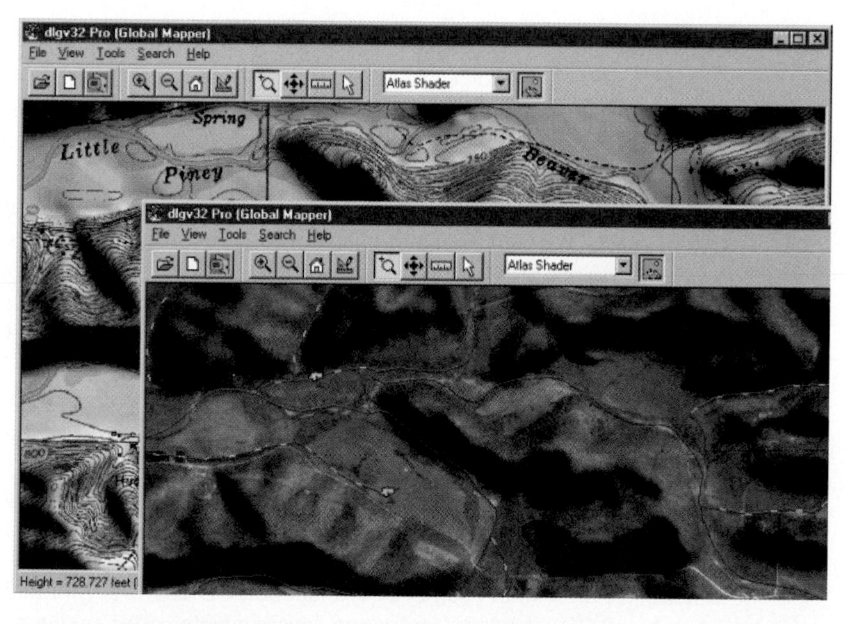

FIGURE 3.14: Sample DLG obtained from the USGS and displayed with the dlgv32 DLG viewer software.

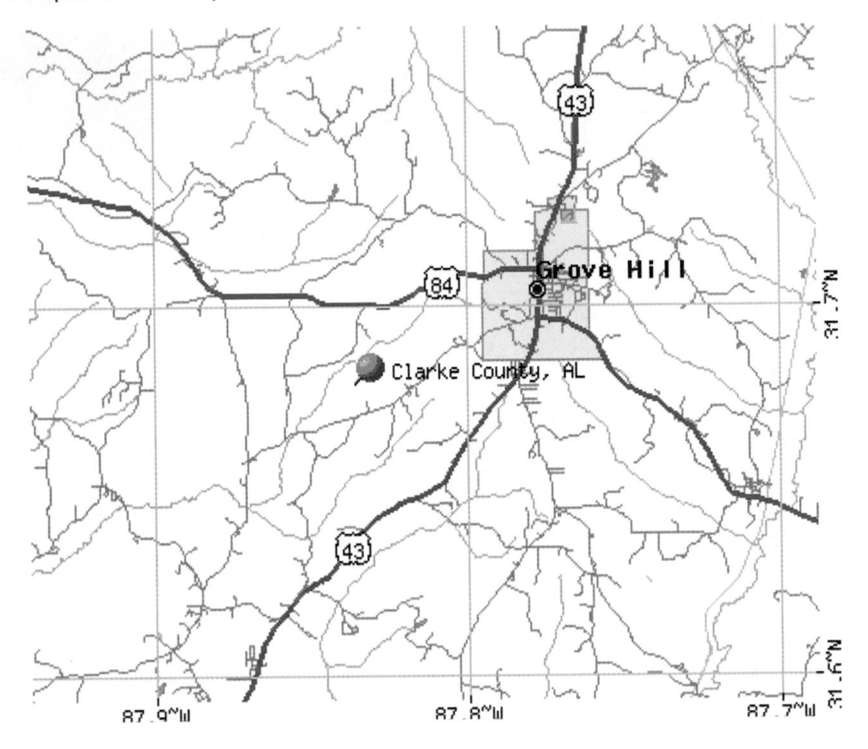

FIGURE 3.15: 2000 Census TIGER files plotted for part of Clarke County, Alabama.

The TIGER formats are from the enumeration maps of the U.S. Census Bureau, as used for the decennial census (Figure 3.15). They are vector files and contain topology; in fact, their predecessor the GBF/DIME files were highly instrumental in popularizing topological data structures. They consist of an arc/node type arrangement, with separate files for points, lines, and areas linked together by cross references. The TIGER terminology calls points *zero cells*, lines *one cells*, and areas *two cells*. The cross indexing means that some features can be encoded as landmarks, and these include rivers, roads, permanent buildings and so forth, which allow GIS layers to be tied together. The TIGER files exist for the entire United States, including Puerto Rico, the Virgin Islands, and Guam. They are block level maps of every village, town, and city, and include geocoded block faces with address ranges of street numbers (Figure 3.16). This means that the address matching function is possible, and a large proportion of GIS use depends highly upon this capability. The data are also obviously referenced to the U.S. Census, so that many population, ethnicity, housing, economic, and other data can also be used with TIGER.

Furthermore, the whole country is available on CD-ROM at minimal cost and over the Internet. Most GIS vendors, and some independents, offer updated and enhanced TIGER files as their own products. Although topologically correct, TIGER has been criticized as not being particularly geographically correct. More recent GIS functions have made the addition of higher levels of geographic accuracy possible, and many data suppliers have enhanced the TIGER files.

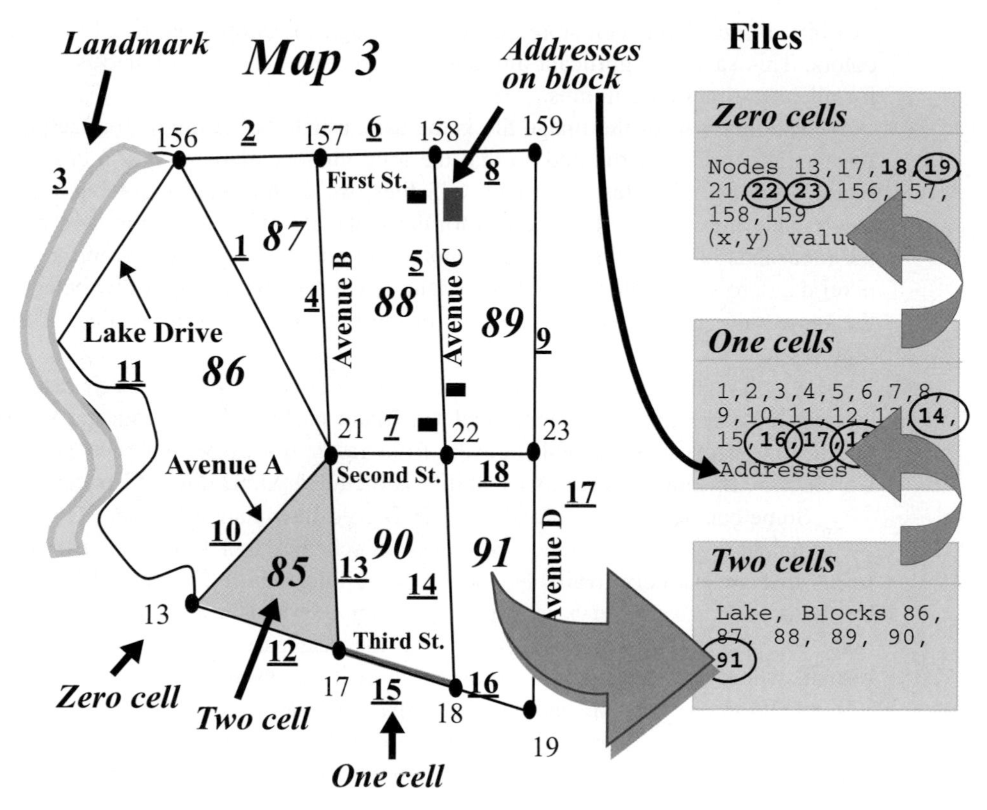

FIGURE 3.16: The U.S. Census Bureau's TIGER data structure.

3.5.2 Raster Data Formats

Raster data formats have been much more widely used, especially since the arrival of networking, because many of them are the same formats that are used to store digital images and pictures. Image formats are particularly simple to create, and as a result there are many. Some of the formats have been optimized for passing the images through networks, and it is these formats that are now most common. Raster formats are mostly similar in how the files are arranged. Few of the formats use ASCII; most use binary. Usually the file structure is a header with a fixed length and a keyword or "magic number" to identify the format. Included in the header is at least the length of one record in bits (called the *image depth*), the number of rows and the number of columns in the file.

Optionally, the file contains a color table. The color table allows the data file to consist of indices, say the numbers 1, 2, 3, 4 and so forth, and each index to correspond to a value, usually three values in each of three bytes as intensities between 0 and 255 in the red, green, and blue, (RGB) respectively. For example, if an image file were to consist of only four colors, red, green, blue, and white, there is no need to store values in each pixel like (255,255,255) for white's RGB value. Instead, the color table could assign white to 0, red to 1, green to 2, and blue to 3. Now no pixel need be more than two bits in size (2 bits can store decimal 0 through 3), as opposed to 24 bits for the full colors. This saves a significant amount of space, since the record size is multiplied up for all rows times all columns.

The final part of the image file is the data, usually all columns for each row. Some formats are padded at the end so that the total number of bytes is a multiple of a key factor, such as 512 bytes. For many GIS files, we do not want color but want to store single-band files, using all of the bits for data. Elevations, for example, which may be in the thousands of feet or meters, need more than one byte per pixel. Alternatively, if the pixel data are simple attributes such as land cover type with only a few categories, then the same value can be assigned to each of the red, green, and blue values if the format requires it.

A surprising number of utility programs exist to convert between raster formats. Among them are Image Alchemy and xv. Many packages also read and write a huge number of formats. Some will convert from raster to vector and vice versa, such as CorelDraw! In some cases, this capability is included within the GIS package.

Some common raster formats are the Tagged Interchange Format (TIF), which can use run length and other image compression schemes and has a number of different forms that are publicly available; the Graphics Interchange Format (GIF), popularized by the online network services, especially CompuServe (the developers), which uses a quite sophisticated compression scheme on the data part of the image; and the JPEG format, which uses a variable resolution compression system offering both partial and full resolution recovery depending on the space available.

Finally, the PostScript industry standard includes a specification for images called encapsulated PostScript. This is a very simple image format indeed, which literally encodes the hexadecimal values to be placed into the image inside an image macro within a regular PostScript file. Many PostScript devices and programs will not handle this format, but those with higher amounts of memory and more advanced capabilities can. This is far better for storing images than maps for GIS use, and it is rarely used to store data; rather, it is used to drive printers and plotters to generate GIS maps.

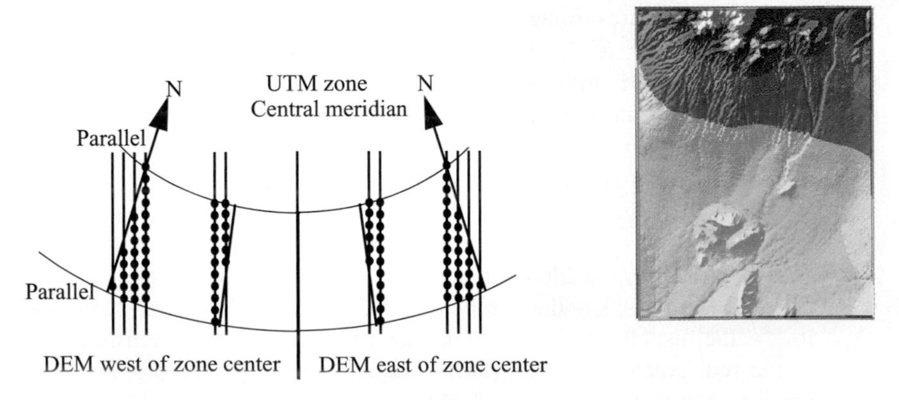

FIGURE 3.17: The 1 : 24,000, 30-meter DEM format from the USGS. Example on right is the Jackass Flats, Nevada quadrangle, hillshaded. Note the edge effect of the UTM grid angle.

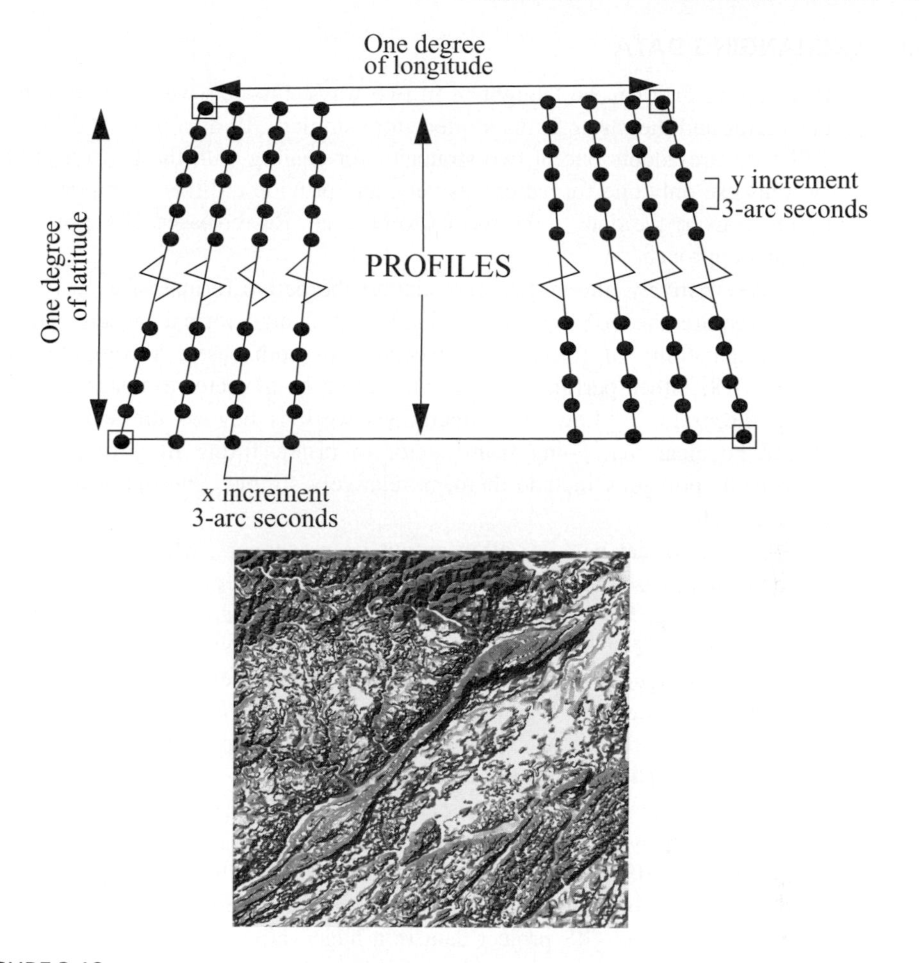

FIGURE 3.18: (Top) The 1 : 250,000, 3-arc second DEM format from the USGS. (Bottom) The Scranton, PA East 1 : 250,000 digital elevation model hill shaded to show topography. Each two-by-one degree quadrangle forms two 1201 by 1201 DEMs.

One raster format has gained widespread acceptance and is often read directly by GIS packages or stand-alone utilities that come with the software. This is the Digital Elevation Model format of the USGS. This format is one in which two types of data are distributed, the 30-meter elevation data from the 1 : 24,000, 7.5-minute quadrangle maps, and the 1 : 250,000, 3-arc second digital terrain data originally supplied by the Defense Mapping Agency but now distributed by the USGS. These are documented formats and are somewhat complex because of the map projections involved. The 30-meter data, for example, are sectioned by quadrangles of latitude and longitude, so in the UTM coordinate base they are stored in have blank sections at the edges where no data are stored. The 3-arc second data are less of a problem to read, but must be projected by the GIS for further use. Some examples of the two data sets and their formats are shown in Figures 3.17 and 3.18.

3.6 EXCHANGING DATA

Exchanging data can be thought of in two ways. First, as we have seen in this chapter, the vector and the raster formats often store similar GIS data in very different ways. The GIS software adopts one of two strategies for dealing with the two types of data. Some systems use only one format exclusively, and provide utilities or import options to bring in and convert the data to the format to be used. Raster-based GIS programs especially use this approach.

Alternatively, the system can support the native format of each type of data, and it can require the GIS operator to explicitly change formats when operations requiring compatability of formats are executed. In both cases, a computer program, part of the GIS, either performs a raster-to-vector or a vector-to-raster conversion. While a full discussion of how these operations work is beyond the scope of this book, it should be clear that going from vector to raster, filling in grid cells as lines cross them or as polygons include them, is relatively simple. The opposite is quite complex (Clarke, 1995).

Raster format data are often output from scanners, when the GIS requires vector data. In many cases, a special suite of software or even a special-purpose computer is used. When done by the GIS, especially for a large data set and for fine resolutions, the process can be very time-consuming. The program must try to follow each line from pixel to pixel, figure out where the end nodes are, and generate a vector equivalent of the line. Often the lines are jagged from the raster effect, and must be smoothed. If the lines are too fat, they sometimes must be thinned first, and this can generate false connections and loops in the lines (Figure 3.19).

The second way to envision data exchange is to consider the issue of transferring data not between formats, but between entirely different computer systems potentially using different GIS packages. This situation is quite ordinary. Different local authorities may have developed their GIS operations around different software. Different projects may have delivered GIS project data in a huge variety of formats. There is also a real need for data exchange. At state or local boundaries, for example, data should be able to be matched for continuity across the borders, just as the data should match from map sheet to map sheet.

The history of GIS has ensured that this commonality and sharing rarely have taken place. Even state and city GIS efforts have often had contradictory or even competing GIS data, and more than a few projects have found it easier to start digitizing and data assembly all over again rather than convert GIS data from an exchange-unfriendly data

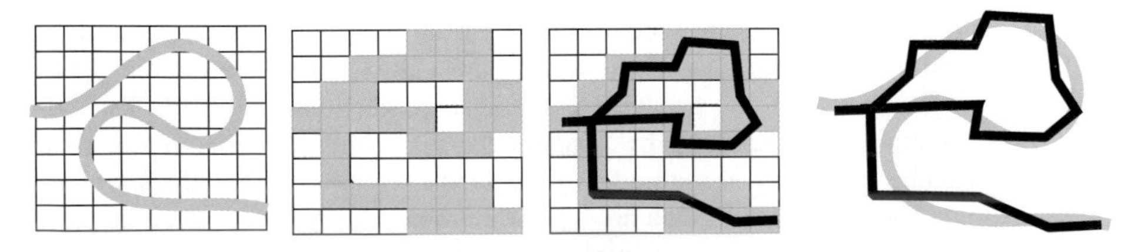

FIGURE 3.19: Errors caused by exchanging data between raster and vector formats. The original (cyan) river after raster-to-vector conversion appears to connect the loop back.

set. Added to this has been the marketing philosophy of many GIS vendors, which has left data formats as proprietary in nature, undocumented and therefore unable to be used as import/export modules for other packages. In the past, data exchange could only be characterized as haphazard and chaotic.

Not exchanging or reusing data between projects, especially within a single organization, is a good example of duplication and waste. There is a real advantage in starting from a single, standard data set for all development or enhancement, especially when GIS operations will eventually bring the data back together for analysis and display. Some industry standards have been quite useful for data exchange, as we saw in Section 3.5. However, two aspects remain a problem. First, none of the industry standards exchange topology with the data, transferring instead only the graphic information. Second, with many different formats, each package has to include a large number of format translators.

A parallel exists between GIS data formats and spoken languages (Figure 3.20). We can get by in isolation by knowing English alone, or perhaps we learn a little French.

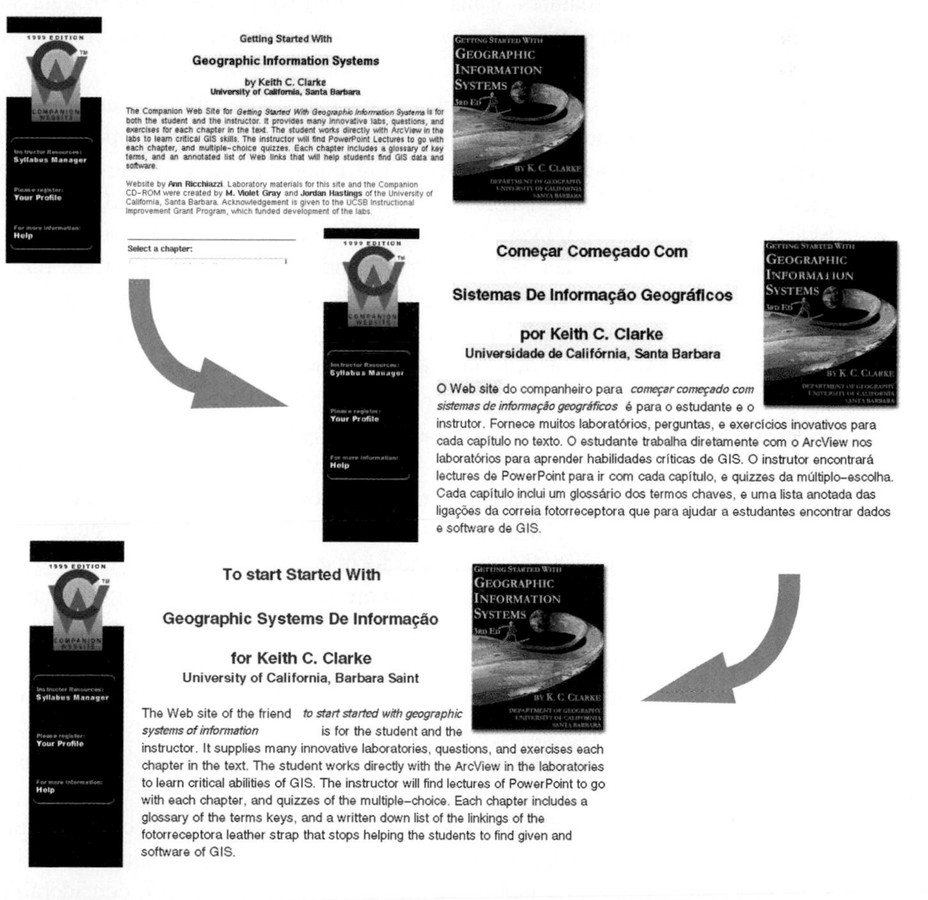

FIGURE 3.20: The Multiple Translators Problem. The web page for this book translated into Portuguese by Altavista's Babelfish (`http://babelfish.altavista.com`), then back into English. Unless a data translation or spatial operation is error free, the same thing happens with GIS data as they move through operations and across systems.

If we wish to speak to someone who speaks only Russian, however, we need someone who speaks either English and Russian or French and Russian. In the latter case, as I speak my words in French, they arrive at the destination having moved twice between languages. Anyone who has played the children's game "telephone" knows the outcome of this process. From a GIS context, we finish with data of unknown accuracy, source, projection, and with unspecified or imperfectly matched attributes. On one country map, perhaps, all streams and rivers are shown, on the other only the major ones. One might think that the climate was wetter, but in fact the difference is one of interpretation and lack of standardization.

The GIS industry in the United States began a standardization effort in the mid-1980s, which led to a federal standard approved in 1992. This standard, the Federal Information Processing Standard 173, called the Spatial Data Transfer Standard (SDTS), had to be quite broad because of the huge degree of complexity involved. Not only did the standard have to produce a bibliography, a terminology, and a complete list of geographic and map features, it also had to address the problems of data accuracy and the broader metadata issues for data description. The terminology created sets of terms for features and data structures that have become commonplace.

Best of all, the standard included a mechanism for file exchange. As a result, two implementations of the standard, called *profiles*, have been developed, for vector, raster, and point data. Already, data sets are being made available in the vector profile of the standard for DLG and for TIGER data. These have been termed DLG-SDTS and TIGER-SDTS. At the same time, many GIS vendors have incorporated input and output utilities that read and write the data in SDTS format and to SDTS specifications. Back to the spoken language analogy, I have convinced the Russian to learn English, and with luck most of the rest of the world too! The cost was many hours spent debating exactly what was meant by each and every word that was to be used in discussion in everyone's language.

The United States civilian mapping agencies have not been alone in seeking standardization of GIS information for data exchange (Figure 3.21). The U.S. military has drafted standards for use among the Army, Navy, and Air Force called the Tri-Service Spatial Data Standards. Within the members of NATO, an exchange standard called DIGEST was developed, with a vector data profile called the Vector Product Format (VPF). This format is best known as the format in which the Digital Chart of the World was released on CD-ROM. Similar efforts are under way to standardize data exchange in Germany, Australia, South Africa, the European Union, and for worldwide nautical chart data by the International Hydrographic Organization (DX-90). As data become used for global instead of local and national projects, the ability to exchange data internationally will increase in importance. International peacekeeping and disaster relief efforts, for example, involve cooperation and therefore exchange of GIS data among many quite different countries and organizations.

We will return to this issue in Chapter 10, when we discuss the future of GIS. Obviously, the open exchange of data can help those developing and using GIS considerably. The SDTS took many years to develop and met with considerable resistance. Nevertheless, the advantages for a GIS future when data can be imported and exported at will promises a more effective use of GIS, and allows GIS users to concentrate on the science of data analysis and the effectiveness of common-sense information use, rather than the politics of data acquisition.

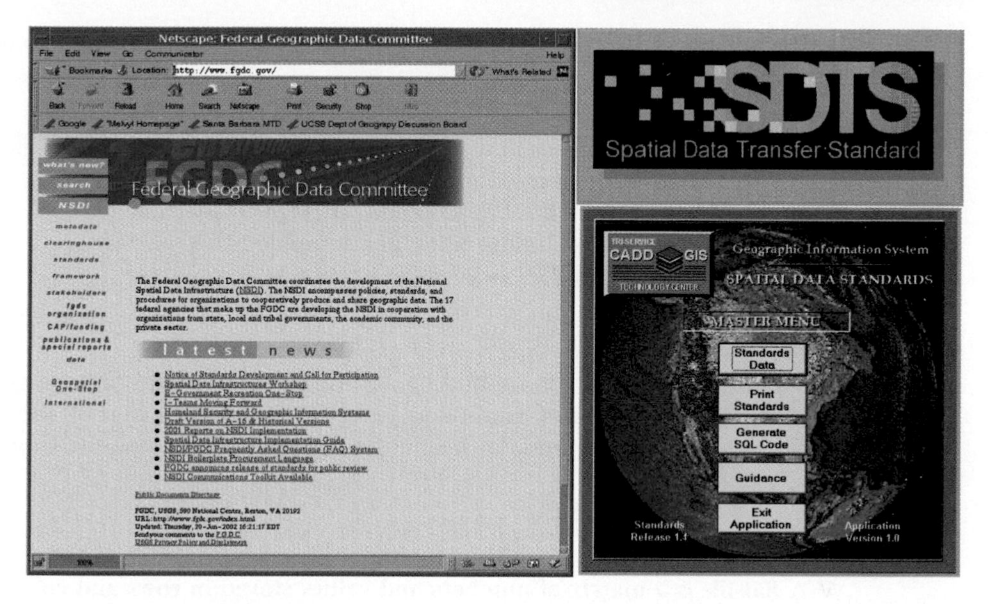

FIGURE 3.21: Some Spatial Data Exchange Standards documents.

3.7 STUDY GUIDE

3.7.1 Summary

CHAPTER 3: Maps as Numbers

Representing Maps as Numbers (3.1)

- GIS requires that both data and maps be represented as numbers.
- The GIS places data into the computer's memory in a physical data structure (i.e., files and directories).
- Files can be written in binary or as ASCII text.
- Binary is faster to read and smaller; ASCII can be read and edited by humans but uses more space.
- Programmers use hexadecimal as shorthand for binary, since two hexadecimal digits correspond to 8 bits (a byte).
- A logical data model is how data are organized for use by the GIS.
- GISs have traditionally used the logical data models of either raster or vector, and for attributes the flat file.
- A raster data model uses a grid.
  - One grid cell is one unit or holds one attribute.
  - Every cell has a value, even if it is "missing."
  - A cell can hold a number or an index value standing for an attribute.
  - A cell has a resolution, given as the cell size in ground units.
  - Points and lines in raster format have to move to a cell center.
  - Lines can become fat. Areas may need separately coded edges.
  - Each cell can only be owned by one feature.
  - As data, all cells must be able to hold the maximum cell value.

- **Rasters are easy to understand, easy to read and write, and easy to draw on the screen.**
- **A vector data model uses points stored by their real coordinates.**
  - **Lines and areas are built from sequences of points in order.**
  - **Lines have a direction to the ordering of the points.**
  - **Polygons can be built from points or lines.**
  - **Vectors can store information about topology.**
  - **The model uses TIN to represent volumes.**
- **Vectors can represent points, lines, and area features very accurately.**
- **Vectors are far more efficient than grids.**
- **Vectors work well with pen and light plotting devices, and tablet digitizers.**
- **Vectors are not good at continuous coverages or plotters that fill areas.**
- **Vectors are like the music of Beethoven, rasters are like Mozart's music.**

Structuring Attributes (3.2)

- **Attribute data are stored logically in flat files.**
- **A flat file is a matrix of numbers and values stored in rows and columns, like a spreadsheet.**
- **Both logical and physical data models have evolved over time.**
- **DBMSs use many different methods to store and manage flat files in physical files.**

Structuring Maps (3.3)

- **A GIS map is a scaled-down digital representation of point, line, area, and volume features.**
- **While most GIS systems can handle raster and vector, only one is used for the internal organization of spatial data.**
- **VECTOR**
- **At first, GISs used vector data and cartographic spaghetti structures.**
- **Vector data evolved the arc/node model in the 1960s.**
- **In the arc/node model, an area consists of lines and a line consists of points.**
- **Points, lines, and areas can each be stored in their own files, with links between them.**
- **The topological vector model uses the line (arc) as a basic unit. Areas (polygons) are built up from arcs.**
- **The endpoint of a line (arc) is called a node. Arc junctions are only at nodes.**
- **Stored with the arc is the topology (i.e., the connecting arcs and left and right polygons).**
- **Volumes (surfaces) are structured with the TIN model, including edge or triangle topology.**
- **TINs use an optimal Delaunay triangulation of a set of irregularly distributed points.**
- **TINs are popular in CAD and surveying packages.**
- **RASTER**
- **A grid or raster maps directly onto a programming computer memory structure called an array.**

- **Grids are poor at representing points, lines, and areas, but good at surfaces.**
- **Grids are good only at very localized topology, and weak otherwise.**
- **Grids are a natural for scanned or remotely sensed data.**
- **Grids suffer from the mixed pixel problem.**
- **Grids must often include redundant or missing data.**
- **Grid compression methods used in GIS include run-length encoding and quad trees.**

Why Topology Matters (3.4)

- **Topological data structures dominate GIS software.**
- **Topology allows automated error detection and elimination.**
- **Rarely are maps topologically clean when digitized or imported.**
- **A GIS has to be able to build topology from unconnected arcs.**
- **Nodes that are close together are snapped.**
- **Slivers due to double digitizing and overlay are eliminated.**
- **The tolerances controlling snapping, elimination, and merging must be carefully considered, because they can move features.**
- **Complete topology makes map overlay feasible.**
- **Topology allows many GIS operations to be done without accessing the point files.**

Formats for GIS Data (3.5)

- **Most GIS systems can import different data formats, or use utility programs to convert them.**
- **Data formats can be industry standard, commonly accepted, or standard.**
- **Vector formats are either page definition languages or preserve ground coordinates.**
- **Page languages are HPGL, PostScript, and AutoCAD DXF.**
- **True vector GIS data formats are DLG and TIGER, which has topology.**
- **Most raster formats are digital image formats.**
- **Most GISs accept TIF, GIF, JPEG, or encapsulated PostScript, which are not georeferenced.**
- **DEMs are true raster data formats.**

Exchanging Data (3.6)

- **Most GISs use many formats and one data structure.**
- **If a GIS supports many data structures, changing structures becomes the user's responsibility.**
- **Changing vector to raster is easy; raster to vector is hard.**
- **Data also are often exchanged, or transferred between different GIS packages and computer systems.**
- **The history of GIS data exchange is chaotic, and has been wasteful.**
- **Data exchange by translation (export and import) can lead to significant errors in attributes and in geometry.**
- **In the United States, the SDTS was evolved to facilitate data transfer.**
- **SDTS became a Federal Standard (FIPS 173) in 1992.**

- **SDTS contains a terminology, a set of references, a list of features, a transfer mechanism, and an accuracy standard.**
- **Both DLG and TIGER data are available in SDTS format.**
- **Other standards efforts are DIGEST, DX-90, the Tri-Service Spatial Data Standards, and many other international standards.**
- **Efficient data exchange is important for the future of GIS.**

3.7.2 Study Questions

Representing Maps as Numbers

Define bit, byte, file, attribute, record, hexadecimal, binary. Now define data model, data structure, logical and physical data structure, vector, and raster. Place each word on a "word line" with the computer hardware at one end and the GIS user at the other end.

Make a list of the reasons why different GIS packages might have different data structures.

Structuring Attributes

Make up a small attribute database, say with six records for the locations of six local businesses. Color in or highlight the attributes that contain spatial data, and think about how you would place the six locations on a street map by hand. If your table were to be resorted using one of the attributes, would the spatial indexing still work?

Structuring Maps

Make a table of advantages and disadvantages of vector and raster data for GIS. What sorts of applications would each be most suited to?

Explain Dana Tomlin's statement about raster and vector at the beginning of the chapter.

Vector Data Structures

Define the following: cartographic spaghetti, point file, arc, polygon, topology, forward link, polygon left, TIN.

Raster Data Structures

Define the following: pixel, resolution, grid extent, fat line, mixed pixel, polygon boundary, array.

Why might the attribute in a particular pixel be wrong as soon as it is geocoded?

Why Topology Matters

Write your own definition of topology. Make a simple diagram of one or two polygons, connected by arcs. Label the polygons A, B, C, and so on. Label the arcs 1, 2, 3, and so on. Now create the arcs file as a table. Make the first arc 1, second 2, and so on. Add extra columns to the table for forward and reverse links, and polygon left and right. How do you deal with the "outside"? How might you deal with a "hole" inside a polygon?

Formats for GIS Data

List three characteristics of each of the following GIS data formats: TIGER, DLG, DEM, TIF, GIF, JPEG, HPGL, DXF, PostScript.

Exchanging Data

Make a list of the advantages and disadvantages of sharing GIS data. What obstacles to data sharing exist at the level of one company, a municipality, a state, or between nations?

3.8 EXERCISES

1. *Carefully examine the documentation for your GIS, or find details of a GIS from one of the sources listed in Chapter 1. Make a table with entries for the following: primary data structure, import formats, export formats, attribute data structure, data model, files and directories used, data transfer method. Look up the advantages and disadvantages of those that your GIS supports. Now list three applications your GIS would be best suited to and three it would be worst suited to.*
2. *Using a word processor, editing tool, or your GIS create a file in one of the formats in this chapter, such as HPGL or PostScript. Use an editor or word processor to look at the file if it is in ASCII format. Try to identify the various parts of the file such as the header, the data, and how images are stored. How big is the file? What coordinates are used inside the file?*
3. *Use a library or the Internet to find out as much as possible about data transfer standards. How many different countries have or are working on standards? Use the Internet to download a file in SDTS format, such as a TIGER or DLG file, and try to import it into your GIS. Explain the irony in the statement "the nice thing about standards is that there are so many to choose from."*
4. *By using your GIS, make an assessment of how much about the data structure is "hidden" from the GIS user by performing three different GIS operations and noting how much information the GIS requires you to supply about the data. Make a list of the advantages and disadvantages of hiding the data structure from the GIS user.*

3.9 REFERENCES

3.9.1 Chapter References

Burrough, P. A. and R. A. McDonnell (1998) *Principles of Geographical Information Systems.* Oxford: Oxford University Press.

Clarke, K. C. (1995) *Analytical and Computer Cartography* 2nd ed. Upper Saddle River, NJ: Prentice Hall.

Dutton, G., ed. (1979) *Harvard Papers on Geographic Information Systems.* First International Advanced Study Symposium on Topological Data Structures for Geographic Information Systems. Reading, MA: Addison-Wesley.

Peucker, T. K. and N. Chrisman (1975) "Cartographic data structures." *American Cartographer,* vol. 2, no. 1, pp. 55–69.

Peucker, T. K., R. J. Fowler, J. J. Little, and D. M. Mark (1976) *Digital Representation of Three-dimensional Surfaces by Triangulated Irregular Networks (TIN).* Technical Report No. 10, U.S. Office of Naval Research, Geography Programs.

Samet, H. (1990) *Design and Analysis of Spatial Data Structures.* Reading, MA: Addison-Wesley.

Tomlin, D. (1990) *Geographic Information Systems and Cartographic Modelling.* Upper Saddle River, NJ: Prentice Hall.

3.10 KEY TERMS AND DEFINITIONS

address range: The range from the highest to the lowest street number on one side of a street, on one block.

arc: A line that begins and ends at a topologically significant location, represented as a set of sequential points.

arc-node: Early name for the vector GIS data structure.

area: A two-dimensional (area) feature represented by a line that closes on itself to form a boundary.

array: A physical data structure for grids. Arrays are part of most computer programming languages, and can be used for storing and manipulating raster data.

ASCII: The American Standard Code for Information Interchange. A standard that maps commonly used characters such as the alphabet onto one-byte-long sequences of bits.

attribute: An attribute is a characteristic of a feature that contains a measurement or value for the feature. Attributes can be labels, categories, or numbers. Attributes can be dates, standardized values, or field or other measurements. Item for which data are collected and organized. A column in a table or data file.

Autocad: A leading CAD program by Autodesk, often interfaced with GIS packages and used for digitizing, especially floor plans and engineering graphics.

block face: One side of a street on one block, which is between two street intersections.

bounding rectangle: The rectangular region defined by the maximum extent of a map feature in the x and y directions. All parts of the feature must lie within or on the edge of the bounding rectangle.

bit: The smallest storable unit within a computer's memory with only an on and an off state, codable with one binary digit.

byte: Eight consecutive bits.

CAD: Computer aided design. Computer software used in producing technical and design-type drawings.

cartographic spaghetti: A loose data structure for vector data, with only order as an identifying property to the features.

color table: Part of the header record in a digital image file that stores specifications of colors based on simple index values, which are then stored in the data part of the image file.

computer memory: A sequence of nonrandom bytes that are recoverable after a computer has been turned off and on again.

data analysis: The process of using organized data to test scientific hypotheses.

database: Any collection of data accessible by computer.

data dictionary: The part of a database containing information about the files, records, and attributes rather than just the data.

data exchange: The exchange of data between similar GIS packages but groups with a common interest.

data format: A specification of a physical data structure for a feature or record.

data model: The logical means of organization of data for use in an information system.

data retrieval: The ability of a database management system to get back from computer memory records that were previously stored there.

data structure: The logical and physical means by which a map feature or an attribute is digitally encoded.

data transfer: The exchange of data between noncommunicating computer systems and different GIS software packages.

DBMS: Database management system. Part of a GIS, the set of tools that allow the manipulation and use of files containing attribute data.

decennial census: The effort required by the U.S. constitution that every 10 years all people be counted and their residences located.

decimal: The counting system when people have 10 fingers.

Delaunay triangulation: An optimal partitioning of the space around a set of irregular points into nonoverlapping triangles and their edges.

DEM: Digital elevation model. A raster format gridded array of elevations.

DIGEST: The NATO transfer standard for spatial data.

DIME: Dual Independent Map Encoding. The data model used for the Census Bureau's Geographic Base Files, the predecessor of TIGER.

digital elevation model: A data format for digital topography, containing an array of terrain elevation measurements.

DLG: A vector format used by the USGS for encoding lines on large-scale digital maps.

double digitized: The same feature captured by digitizing twice.

DXF: Autocad's digital file exchange format, a vector mode industry-standard format for graphic file exchange.

editor: A computer program for the viewing and modification of files.

elevation: The vertical height above a datum, in units such as meters or feet.

encapsulated PostScript: A version of the PostScript language that allows digital images to be included and stored for later display.

end node: The last point in an arc that connects to another arc.

enumeration map: Map designed to show one census enumerator the geographic extent and address ranges within one's district.

export: The capability of a GIS to write data out into an external file and into a nonnative format for use outside the GIS, or in another GIS.

fat line: Raster representation of a line that is more than one pixel wide.

feature: A single entity that composes part of a landscape.

field: The contents of one attribute for one record, as written in a file.

file: A collection of bytes stored on a computer's storage device.

file header: The first part of a file that contains metadata rather than data.

FIPS 173: The Federal Information Processing Standard maintained by the USGS and the National Institute of Standards and Technology that specified a standard organization and mechanism for the transfer of GIS data between dissimilar computer systems. FIPS 173 specifies terminology, features types, and accuracy specifications, as well as a formal file transfer method.

forward/reverse left: Moving along an arc, the identifier for the arc connected in the direction/opposite direction of the arc to the immediate left.

forward/reverse right: Moving along an arc, the identifier for the arc connected in the direction/opposite direction of the arc to the immediate right.

fully connected: A set of arcs in which forward and reverse linkages have identically matching begin and end nodes.

GBF: Geographic Base File. A database of DIME records.

geographical surface: The spatial distribution traced out by a continuously measurable geographical phenomenon, as depicted on a map.

GIF: An industry standard raster graphic or image format.
grid cell: A single cell in a rectangular grid.
grid extent: The ground or map extent of the area corresponding to a grid.
hexadecimal: The counting system that would be used if people had 16 fingers.
hierarchical: System based on sets of fully enclosed subsets and many layers.
HPGL: Hewlett Packard Graphics Language. A device-specific but industry-standard language for defining vector graphics in page coordinates.
image depth: The numbers of bits stored for each pixel in a digital image.
import: The capability of a GIS to bring data in an external file and in a nonnative format for use within the GIS.
industry standard format: A commonly accepted way of organizing data, usually advanced by a private organization.
internal format: A GIS data format used by the software to store the data within the program, and in a manner unsuitable for use by other means.
label point: A point digitized within a polygon and assigned its label or identifier for use in topological reconstruction of the polygon.
landmark: TIGER term for a geographic feature not a part of the census features.
layer: A set of digital map features collectively (points, lines, and areas) with a common theme in coregistration with other layers. A feature of GIS and most CAD packages.
line: A one-dimensional (length) map feature represented by a string of connected coordinates.
logical structure: The conceptual design used to encrypt data into a physical structure.
magic number: Any number that has a specific value for a specialized need.
matrix: A table of numbers with a given number of rows and columns.
metadata: Data about data, usually for search and reference purposes.
missing data: Elements where no data is available for a feature or a record.
mixed pixel: A pixel containing multiple attributes for a single ground extent of a grid cell. Common along the edges of features or where features are ill defined.
node: The end of an arc. At first, any significant point in a map data structure. Later, only those points with topological significance, such as the ends of lines.
page coordinates: The set of coordinate reference values used to place the map elements on the map, and within the map's own geometry rather than the geometry of the ground that the map represents. Often page coordinates are in inches or millimeters from the lower left corner of a standard size sheet of paper, such as A4 or 8 1/2 by 11 inches.
physical structure: The mechanical mapping of a section of computer memory onto a set of files or storage devices.
pixel: The smallest unit of resolution on a display, often used to display one grid cell at the highest display resolution.
point: A zero-dimensional map feature, such as a single elevation mark as specified by at least two coordinates.
polygon: A many-sided area feature consisting of a ring and an interior. An example is a lake on a map.
polygon interior: The space contained by a ring, considered part of a polygon.
polygon left: Moving along an arc, the identifier for the polygon adjacent to the left.
polygon right: Moving along an arc, the identifier for the polygon adjacent to the right.
PostScript: Adobe Corp.'s page definition language. An interpreted language for page layout designed for printers but also an industry standard for vector graphics.

quad tree: A way of compressing raster data based on eliminating redundancy for attributes within quadrants of a grid.

RAM: The part of a computer's memory designed for rapid access and computation.

raster: A data structure for maps based on grid cells.

ring: A line that closes upon itself to define an area.

run-length encoding: A way of compressing raster data based on eliminating redundancy for attributes along rows of a grid.

SLF: An early Defense Mapping Agency data format.

sliver: Very small and narrow polygon caused by data capture or overlay error that does not exist on the map.

snap: Forcing two or more points within a given radius of each other to be the same point, often by averaging their coordinate.

Spatial Data Transfer Standard: The formal standard specifying the organization and mechanism for the transfer of GIS data between dissimilar computer systems. Adopted as FIPS 173 in 1992, SDTS specifies terminology, features types, and accuracy specifications, as well as a formal file transfer method for any generic geographic data. Subsets for the standard for specific types of data, vector, and raster, for example, are called profiles.

spreadsheet: A computer program that allows the user to enter numbers and text into a table with rows and columns, and then maintain and manipulate those numbers using the table structure.

table: Any kind of organization by placement of records into rows and columns.

TIF: An industry-standard raster graphic or image format.

TIGER: A map data format based on zero-, one-, and two-cells, used by the U.S. Census Bureau in the street-level mapping of the United States.

TIN: A vector topological data structure designed to store the attributes of volumes, usually geographic surfaces.

tolerance: The distance within which features are assumed to be erroneously located different versions of the same thing.

topologically clean: The status of a digital vector map when all arcs that should be connected are connected at nodes with identical coordinates, and the polygons formed by connected arcs have no duplicate, disconnected, or missing arcs.

topology: The property that describes adjacency and connectivity of features. A topological data structure encodes topology with the geocoded features.

USGS: The United States Geological Survey, part of the Department of the Interior and a major provider of digital map data for the United States.

vector: A map data structure using the point or node and the connecting segment as the basic building block for representing geographic features.

volume: A three-dimensional (volume) feature represented by a set of areas enclosing part of a surface, in GIS usually the top only.

VPF: Vector product format, a data transfer standard within DIGEST for vector data.

zero/one/two cell: TIGER terminology for point, line, and area, respectively.

CHAPTER 4

Getting the Map into the Computer

4.1 ANALOG-TO-DIGITAL MAPS
4.2 FINDING EXISTING MAP DATA
4.3 DIGITIZING AND SCANNING
4.4 FIELD AND IMAGE DATA
4.5 DATA ENTRY
4.6 EDITING AND VALIDATION
4.7 STUDY GUIDE
4.8 EXERCISES
4.9 REFERENCES
4.10 KEY TERMS AND DEFINITIONS

"Roll up that map; it will not be wanted these ten years."

—*(William Pitt, 1805)*

They say "to err is human but to really foul things up requires a computer."
But who, when lost, has not been heard to say:
"My map shows clearly that a road is here;
the land is wrong, why ask that trooper?"

4.1 ANALOG-TO-DIGITAL MAPS

Most people think of maps as drawings on paper. Maps hang on walls, lie in map drawers, and fill the pages of books, atlases, street guides, newspapers, and magazines. Maps roll off the nation's printing presses in the millions each year, and they fill the spaces in every car's glove compartment, neatly folded or not! The traditional paper maps of our everyday world can be called *real maps*, because they are touchable. We can hold them in our hands, fold them up, and carry them around. The computer, in contrast, has forced us to reconsider this simple definition of a map. In the digital era, and especially within GISs, maps can be both real and *virtual*.

A *virtual map* is a map waiting to be drawn. It is an arrangement of information inside the computer in such a way that we can use the GIS to generate the map however and whenever we need it. We may have stored map information about roads, rivers, and forests, for example, but may decide that only the forests and rivers need be shown on any map that the GIS produces. Every real map is simply a conversion of the virtual map into a medium, the form that the map will take. In most cases, the medium we use is paper.

Using maps within GISs means that somehow they have already been turned from real into virtual maps. Another way to say this is that a paper map has gone through a conversion, from a paper or analog form into a digital or number form. We start with paper, or sometimes film, Mylar, or some other medium, and we end up with a set of numbers inside files in the computer. This conversion process is called *geocoding*, which we can define as the conversion of spatial information into computer-readable form. Some GIS vendors would be pleased to help you acquire the data you need, but at an immense price. Studies have shown that finding the right maps, and converting these maps from real to virtual form by geocoding, takes up anywhere between 60% and 90% of both the time and money spent on a typical GIS project. Fortunately, this is a once-only cost. As soon as we have the map in digital form, we can use it in a GIS over and over again for different uses and projects unless it needs an update.

Digital map data for use in GIS really falls into two categories. Either the data already exist and all we have to do is find or buy them, or they don't exist and we have to geocode paper maps or maps on some other medium. A third case is that the maps don't even exist, and here we often turn to remote sensing, aerial photography, or field data collection by surveyors or the global positioning system (GPS), to get our first map of a new location. Also, sometimes the maps we need already exist, but whoever geocoded them is not interested in sharing the data with you, even for a price! Even when we can get the maps we need in digital form, they may not suit our particular type of GIS, or may be out of date or not show the features we want. The bottom line is that sooner or later, and usually sooner, we end up geocoding at least some of our own maps.

Before we cover the ways that maps can be converted into numbers, scanning and digitizing, we will take a look at how we might go about finding digital map data that already exist. If we are successful, with a little effort, some conversion programs, and a knowledge of GIS data formats, we can reuse one of the many excellent maps already available for us. Many of these maps can be read straight into a GIS, sometimes without any need to research the way the files and numbers are structured. In this chapter we take a guided tour of the various flavors of data, their formats, and the way the information in the maps has been structured during geocoding.

These days, very few GIS projects have to start with no data at all. The vast amount of data collected and made available by the various branches of government is an excellent base on which to start building. The trick is knowing where to look, what to do when you find what you want, and how to get the data into your GIS.

4.2 FINDING EXISTING MAP DATA

The search for paper maps is often started in a map library. The libraries most likely to carry maps and to support cartographic research are the research libraries in the largest cities and those attached to major universities. Map librarians make use of computer networks to share information and conduct searches, and they are increasingly making census and other digital maps available both in libraries and via computer networks.

Another place to look for map information is in books. An excellent starting point is the *Map Catalog* by John Makower (1986). *Maps for America*, by M. M. Thompson (1987) of the USGS, is a good survey of existing published maps for the United States. Another information source, especially internationally, is the *Inventory of World Topographic Mapping* (Bohme, 1993). The appendices in John Campbell's book *Map Use and Analysis* (Campbell, 1993) show how to use map series and their indices, and many other information sources are listed, especially Chapter 21, "U.S. and Canadian Map Producers and Information Sources."

In many cases, state and local governments keep collections of paper maps. A local planning or building permit office can often find maps of your property or of parks and business properties. Make sure to call ahead. How good the service of providing maps to the public is depends a great deal on the office and its policies and services. Some larger agencies have their own map division. A state highway authority, park service, or industrial development organization may have its own maps available, sometimes free or at little cost.

Commercial companies sell cartographic data and some will conduct map data searches. Imagery from most commercial vendors can be searched and browsed using an online database. Many commercial services offer not only packaged existing data for your use, but will digitize or scan data and even write the data in GIS format for you at a cost. Two companies offering such services are ETAK and Geographic Data Technology.

Obviously, each company has its own strengths and types of map for sale. Commercial companies are not, however, for the novice. They are primarily used by large corporations, governments, the real estate industry, and so on. For a first cut, the usually free public data are the best starting point, and in many cases enough, even many times more, than you will ever need to work with your GIS.

Digital map data by public agencies have been dominated by data from the federal government. In the United States, digital map data created at the federal level for its own use are the property of the American people, with the obvious exception of sensitive data of use in national security—although recently even spy satellite data have been made available. The Freedom of Information Act guarantees every American the right to get copies of digital map data used by the federal government, subject to a distribution or copying cost that may not exceed a reasonable marginal cost of providing the data.

Not all data has to be extracted from the government using the act, however. Government agencies have made it their mission to make map data as freely accessible as possible to any interested party. Computer networks have made this not only accessible to almost any computer user but have also made it more flexible.

4.2.1 Finding Data on the Networks

An excellent way to begin a data search is to use a computer network. Several computer packages allow you to do this over the various network access methods, such as America

Online and CompuServe. The most sophisticated tools, however, are those available on the Internet. Among the various tools, such as Archie, Veronica, WAIS, and Gopher, is a computer program called Mosaic, from the National Supercomputing Center at the University of Illinois. Mosaic allows you to search the World Wide Web (WWW), an interlinked set of computers and servers, or data repositories on the Internet. Similar and more widely used programs are Netscape Navigator and Microsoft Explorer. Each major agency has a World Wide Web server, or *gateway*, through which data can be searched and downloaded. Simply enormous amounts of data are available through this simple mechanism.

While many U.S. government agencies create and distribute digital maps, data from three agencies, each one with its own different types, have been most used in GISs. The agencies are the U.S. Geological Survey (USGS), part of the Department of the Interior; the U.S. Bureau of the Census; and the National Oceanic and Atmospheric Administration (NOAA), both part of the Department of Commerce. Data they supply cover the land and its features, the population, and the weather, atmosphere, and oceans across the United States.

Each of these agencies is worth covering here, although there are many others. Finding information in any of them has been made much easier by several public information service and computer network services, especially over the Internet. The Internet is a network of computer networks and is accessible to all users through a computer that is attached to the system. More information can be found on the World Wide Web site for this book, given in Chapter 1.

4.2.2 U.S. Geological Survey

Digital cartographic data from the USGS are distributed by the Earth Science Information Centers as part of the National Mapping Program. Information is available by calling 1-888-ASK-USGS in the United States and Canada or by writing to the addresses listed in the study guide for this chapter. The USGS digital data fall into six categories: *digital line graphs* (DLGs), *digital elevation models* (DEMs), *land-use and land-cover digital data*, *digital cartographic text* (Geographic Names Information System, GNIS), *digital orthophotoquads* (DOQ), and *digital raster graphics* (DRG). The USGS Web portal is shown in Figure 4.1.

The USGS continues to improve coverage of the United States and distributes the map data products on computer tape, floppy disk, and CD-ROM. These data formats are covered in this chapter, and many GIS packages support them directly. Arc/Info, for example, will recognize a digital line graph file and read the data accordingly.

Many of the USGS data sets are now available directly through the Internet by the file transfer protocol (FTP), a tool for moving files over computer networks. Data sets are maintained on USGS servers, usually listed alphabetically by the name of the map quadrangle concerned. This means that to retrieve the data set, you should both know what data are needed and what format the data are to be found in.

The USGS also now distributes data on land cover derived from classifications of NOAA's AVHRR (*advanced very high resolution radiometer*) measurements. These data are distributed by the EROS Data Center on CD-ROM and on the Internet, with a ground resolution of 1 kilometer. Biweekly composites showing a vegetation index for North

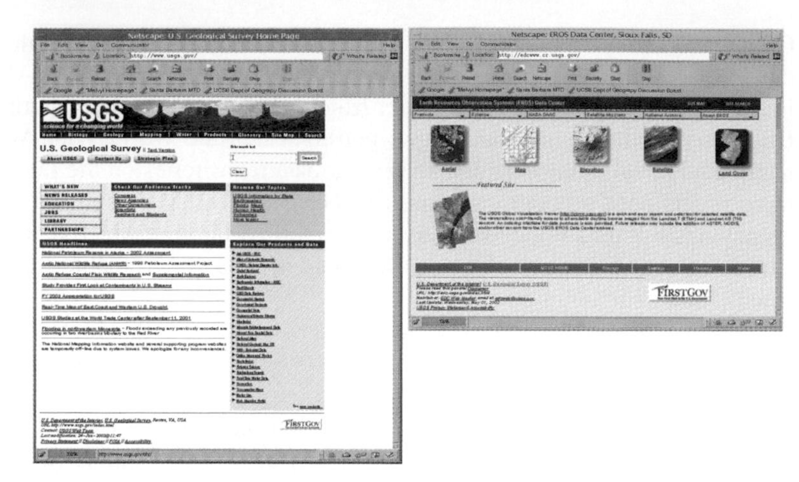

FIGURE 4.1: World Wide Web gateway for the U.S. Geological Survey. Left: The main home page `http://www.usgs.gov`. Right: The EROS data center home page at `http://edcwww.cr.usgs.gov`. (Used with permission of the U.S. Geological Survey.)

America are available, and efforts are under way to release a global AVHRR data set at this resolution.

An important global data set is the *Digital Chart of the World* (DCW). This digital map contains coverage of the entire world, including major cities, rivers, lakes, coastlines, contours, vegetation, and transportation routes. The DCW is a digital version of the Defense Mapping Agency's operational navigation chart (ONC) and jet navigation chart (JNC) at a scale of 1 : 1,000,000. The data are stored and distributed on four CD-ROMs, which include software for viewing the data files on IBM-PC compatible microcomputers. Distribution of the data is via the USGS's Earth Science Information Centers.

4.2.3 National Oceanic and Atmospheric Administration

The NOAA concentrates on marine and aeronautical navigation systems that electronically integrate digital charts, global positioning system-based locations, and real-time environmental information. Examples are the daily weather map, satellite and radar images, and maps used by pilots and air traffic control. The NOAA charts must be carried aboard all large ships in U.S. waters. The National Geophysical Data Center, part of NOAA, has released numerous digital map data sets on CD-ROM, most recently including detailed bathymetry of the ocean and land-surface topography as well as geodetic and magnetic data for the earth's surface (Figure 4.2).

4.2.4 U.S. Bureau of the Census

The mapping of the U.S. Census Bureau is to support the decennial census by generating street-level address maps for use by the thousands of census enumerators. For the 1990 census, the Census Bureau developed a system called TIGER (*topologically integrated geographic encoding and referencing*). The TIGER system (Figure 4.3) uses the block face or street segment as a geographic building block and recognizes cartographic objects of different dimensions, points (nodes), lines (segments), and areas (blocks, census tracts,

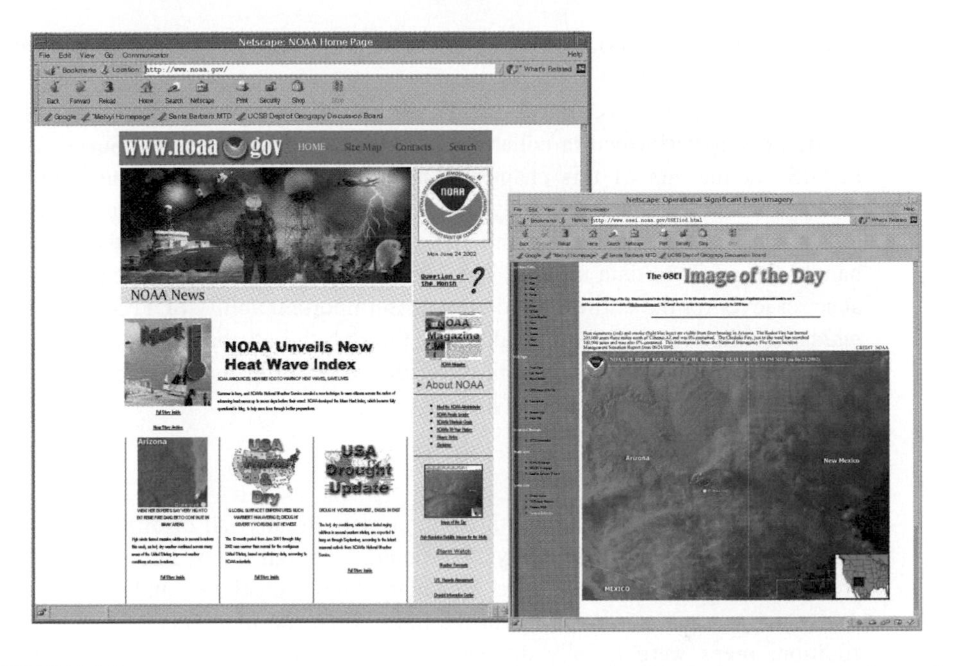

FIGURE 4.2: World Wide Web gateway for NOAA, and a sample satellite image map extracted over the Internet at `http://www.noaa.gov`. (Used with permission.)

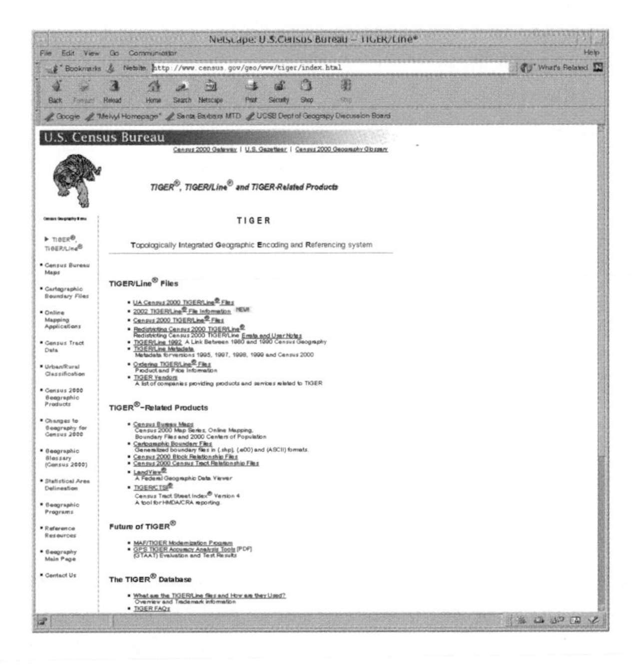

FIGURE 4.3: World Wide Web gateway for the census TIGER data at the U.S. Census Bureau. The TIGER logo is a copyright of the U.S. Census Bureau. (Used with Permission.)

or enumeration districts). In TIGER terminology, points are *zero cells*, lines are *one cells*, and areas are *two cells*.

A large-scale cooperative effort prepared these maps for the 1990 census. Map digitizing was performed in collaboration with the U.S. Geological Survey (see "People in GIS" at the end of this chapter). The maps are distributed along with the census data on computer tape, separately on CD-ROM, and over the Internet. Virtually every GIS allows TIGER files to be imported directly into the system, although not all GISs handle the attribute data as well. The TIGER was the first comprehensive GIS database at street level for the entire United States. An important ability of TIGER is to do *address matching*: the search for street addresses through the attribute files to match a block or census tract in the TIGER graphic files; that is, finding its geographic location on the map solely from a street address listing.

4.2.5 Creating New Data

Wonderful as it is to find an existing digital map, the myriad of different data formats is usually the least of the GIS analyst's problems. Digital maps, like their analog sources, are specific to a given map scale. Boundary lines, coastlines, and so on all reflect the degree of generalization applied to the lines when the map was originally digitized. In addition, maps were usually digitized with different levels of precision, from source maps that were out of date, or that have become out of date since they were captured by computer, or sometimes even have errors or problems with their accuracy. Two different maps of the same area rarely agree over every detail, yet the computer is unable to resolve the differences in the same way that the human mind can reason about the reliability of information, its timeliness, and so on.

In summary, like it or not, sooner or later if you are involved with GIS you will find yourself digitizing a map. Although this is a tedious, time-consuming, and potentially frustrating exercise, the learning process involved will greatly increase your awareness of the limitations of digital maps for GIS use. It is far better to persevere and learn, than to make a million errors and misjudgments for the lack of a little hands-on experience. Time, then, to get a little digital (or at least virtual) "mud" on our boots!

4.3 DIGITIZING AND SCANNING

Historically, many different means have been used to geocode. At first, some very early GIS packages required maps to be encoded and entered by hand. The hours of monotonous work required for this task made errors common and their correction difficult. Since special-purpose digitizing hardware became available, and especially since the cost of this hardware fell substantially, virtually all geocoding has been performed by computer.

Two technologies have evolved to get maps into the computer. Digitizing mimics the way maps were drafted by hand and involves tracing the map over using a cursor while it is taped down onto a sensitized digitizing tablet. The second method involves having the computer "sense" the map by scanning it. Both approaches work and have their advantages and disadvantages. Most important, the method of geocoding stamps its form onto the data in such a way that many other GIS operations are affected afterwards.

4.3.1 Digitizing

Geocoding by tracing over a map with a cursor is sometimes called *semiautomated digitizing*. This is because in addition to using a mechanical device, it involves a human operator. Digitizing means the use of a digitizer or digitizing tablet (Figure 4.4). This technology has developed as computer mapping and computer-aided design have grown and placed new demands on computer hardware.

The digitizing tablet is a digital and electronic equivalent of the drafting table. The major components are a flat surface, to which a map is usually taped, and a stylus or cursor, with the capability of signaling to a computer that a point has been selected (Figure 4.4). The mechanism to capture the location of the point can differ. Many systems have connected arms, but most have embedded active wires in the tablet surface that receive an electrical impulse sent by a coil in the cursor. In some rare cases, the cursor transmits a sound, which is picked up and recorded by an array of microphones.

The actual process of digitizing a map proceeds as follows (Figure 4.5). First, the paper map is tailored or preprocessed. If the map is multiple sheets, the separate

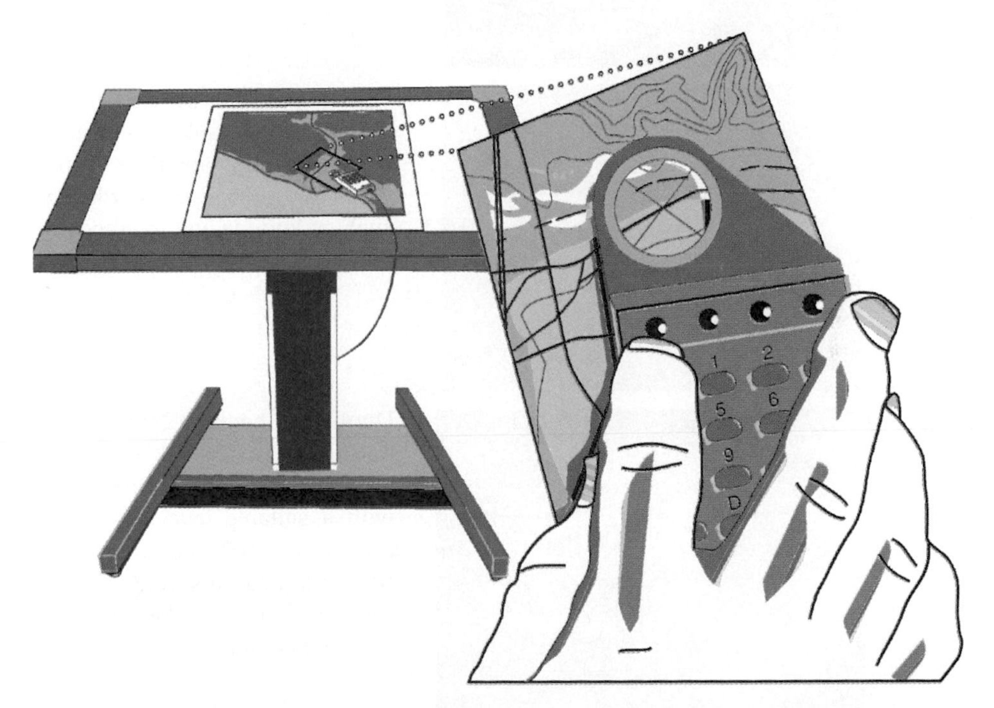

1. Digitizer cursor transmits a pulse from an electromagnetic coil under the view lens.
2. Pulse is picked up by nearest grid wires under tablet surface.
3. Result is sent to computer after conversion to x and y units.

FIGURE 4.4: Digitizing tablet system of operation.

Stable base copy of map is prepared by choosing digitizing control points at known locations. Any features to be selected should be marked in advance.

Map is firmly taped or fixed down to tablet. No movement of the map should be possible. Surface should be flat and free of folds, bubbles, and so forth. Double tape over edges that will be rubbed by elbows and forearms.

Control points to be used in registering the map to the tablet are entered one at a time, along with their map coordinates in geocoded coordinate space such as latitude and longitude or UTM meters.

Digitizing begins. Map features are traced out using the cursor. Care is taken to capture features accurately, with a suitable level of detail. Points can be selected one at a time or in a stream turned on and off from the cursor. Attributes can be entered as features are completed.

FIGURE 4.5: The digitizing process.

sheets should be digitized independently and digitally merged (zipped) later. Unless annotations have to be made onto the map to assist the geocoding, the next major step is to derive a coordinate system for the map. Most applications use Universal Transverse Mercator coordinates (UTM), or latitude and longitude (see Chapter 2), but many people use hardware coordinates or map inches or millimeters. Map units are sometimes used when precise matching between the digitized map and its source is required. The map is then transformed into geographic coordinates when both the editing and the proofing are complete.

As a minimum, the coordinate locations of three points are required, usually the upper right easting and northing, the lower left easting and northing, and at least one other corner. From these points, with their map coordinates and their raw digitizer coordinates, all the parameters can be computed for converting the data into the map's coordinates. Many GIS map entry and digitizing software packages require four of these control points for computing the map geometry, and it is advisable to repeatedly digitize control points and to average the coordinates to achieve higher accuracy.

The beginning of the digitizing sequence involves selecting the control points and interactively entering their world coordinates. This is a very important step, because an error at this stage would lead to an error in every pair of coordinates. After the map is taped to the tablet, it should not be moved without entering the control points again, and it is preferable to perform this step only once per map. Ideally, the entire digitizing process should be finished at one sitting.

Tape should be placed at each map corner after smoothing the map, and care should be taken to deal with folds and the crinkles that develop with certain papers during periods of high humidity. A stable base product such as Mylar or film is preferable for digitizing. The lower edge, which will have the cursor and your right sleeve (if you are right-handed) dragged over it many times, should be taped over its entire length. You should always permanently record the x and y values of the map control points, ideally digitally and with the geocoded data set. This may allow later recovery of lost resolution or systematic errors.

Digitizing then proceeds with the selection of points. The cursor may have multiple buttons and may be capable of entering text and data without using the keyboard. Voice data entry and commands are also sometimes used. On specialized workstations, there may even be a second tablet with its own mouse or cursor for commands. Errors can be reduced during this process by reading the documentation in advance and by occasionally stopping to review the actual data being generated on the screen.

Points are usually entered one at a time, with a pause after each to enter attributes such as labels or elevations. Lines are entered as strings of points and must be terminated with an end-of-chain signal to determine which point forms the node at the end of the chain. This signal must come from the cursor in some way, either by digitizing a point on a preset menu area or by hitting a preset key.

Areas such as lakes or states are usually digitized as lines. Sometimes an automatic closure for the last point (snapping) can be performed. Finally, the points should be checked and edited. The digitizing software or GIS may contain editing features, such as delete and add a line or move and snap a point. The software may also support multiple collection modes. *Point mode* simply digitizes one point each time the button on the cursor is pressed. *Stream mode* generates points automatically as the cursor is moved, either one point per unit of time or distance. This mode can easily generate very large data volumes and should be avoided in most cases. Error correction is especially difficult in this mode. *Point select mode* allows switching between point mode and stream mode. This mode is sometimes used when lines are both geometric and natural, such as when following a straight road and then a river.

At this point the data are ready for direct integration into the GIS. Usually, a separate module of the GIS is used for digitizing and editing, and the map can now be passed on for use. Digitizing should be approached with caution and a desire for accuracy. Errors in digitizing can usually be eliminated using some very simple procedures and rules.

4.3.2 Scanning

The second digitizing process is *automated digitizing* or more usually, just *scanning*. The scanner you may have seen at a computer store or in an advertisement, or perhaps the one you use for scanning documents, is a *desktop scanner*. The *drum scanner* is most commonly used for maps. This type of scanner receives an entire sheet map, usually clamped to a rotating drum, and scans the map with very fine increments of distance, measuring the amount of light reflected by the map when it is illuminated, with either a spot light source or a laser (Figure 4.6). The finer the resolution, the higher the cost and the larger the data sets. A major difference with this type of digitizing is that lines, features, text, and so on, are scanned at their actual width and must be preprocessed for the computer to recognize specific cartographic objects. Some plotters can double as scanners, and vice versa.

For scanning, maps should be clean and free of folds and marks. Usually, the scanned maps are not the paper products but the film negatives, Mylar separations, or the scribed materials that were used in the map production. An alternative scanner is

FIGURE 4.6: Digitizing by the map scanning process. (Photo by Susan Baumgart/Matt Eimers.)

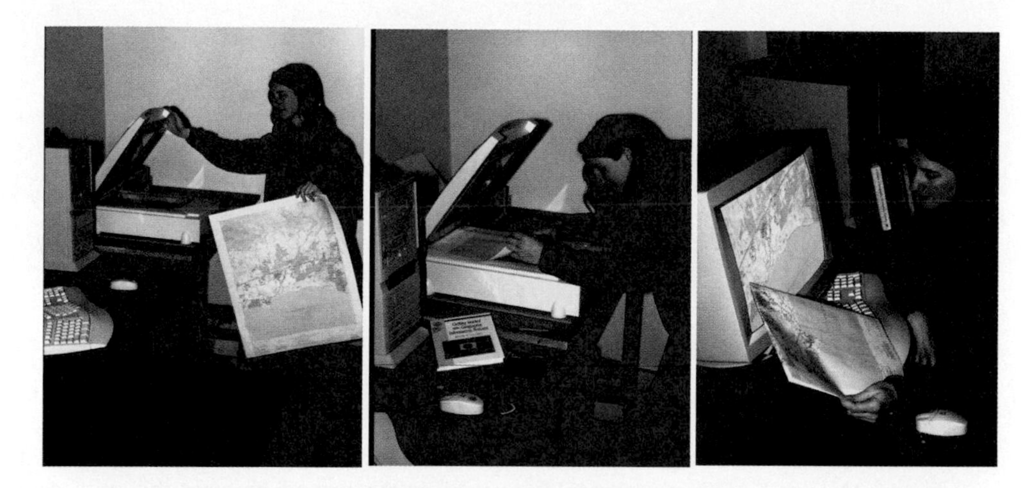

Left: Map is cut or folded to frame the section which is to be scanned. Center: Map is placed in scanner, aligned to edge as upright. Top is closed to clamp map and prevent movement. Right: Software is used to frame area to be scanned, to set resolution and mode (color, monochrome, etc.). When all is set, map is scanned, and saved on hard disk. Scan is checked against original.

FIGURE 4.7: Scanning a map with a desktop scanner. (Images courtesy of David Lawson and Jeannette Candau.)

the *automatic line follower*, a scanner that is manually moved to a line and then left to follow the line automatically. Automatic line followers are used primarily for continuous lines, such as contours. These and other scanners are very useful in CADD (computer-aided drafting and design) systems, where input from engineering drawings and sketches is common.

Simple desktop scanners are becoming important geocoding devices as their resolutions improve and their prices fall. The process of scanning usually begins with preparing the section of map, which obviously needs to be as clean and with as solid and crisp lines as possible. Next, the map is placed on the desktop scanner. The software is told which window to scan, the scan is previewed, and the scan is then saved to the resultant scan file (Figure 4.7). The process can be very quick; nevertheless, care and attention can save considerable work later on. Desktop or low-resolution scanning is rarely adequate for GIS purposes but can be used to put a rough sketch into a graphic editing system for reworking. In this way, a field sketch can be used as the primary source of information for developing the final map for the GIS.

It is important to have a clear concept of *scale* and *resolution* when scanning. In Figure 4.8, the same map, part of the Goleta, California, 7.5-minute USGS quadrangle map, was scanned at two different resolutions. The two larger squares outlined on the map are 100 millimeters on a side. At 1 : 24,000, this distance is $24{,}000 \times 100 = 2{,}400{,}000$ millimeters, or 2400 meters.

Although Figure 4.8 shows a scanned 100-mm-by-100-mm segment of the map, the pixel density of the scan is given per inch. A scan at 400 dots per inch (DPI) translates to 15.748 pixels per millimeter, making the scanned map square about 1575 by 1575 pixels. The same area was also scanned at 100 DPI, or 3.937 pixels per millimeter, for an image

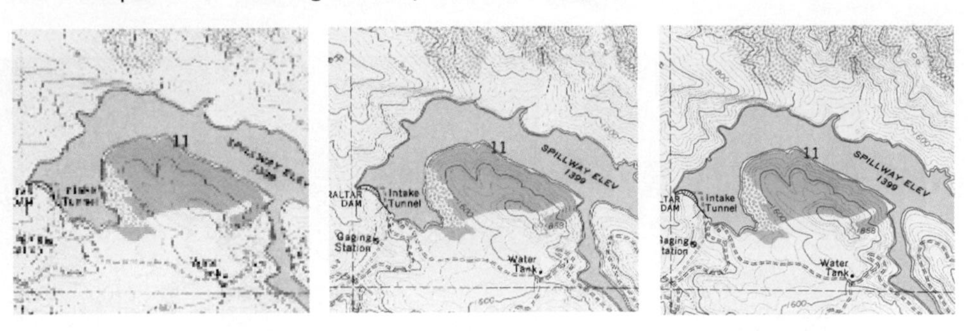

FIGURE 4.8: Drum scanner images of two sections of the Little Pine Mountain, California topographic quadrangle at 1 : 24,000. Left at 50 dpi, center at 100 dpi, and right at 200 dpi.

that is 394 by 394 pixels. On the two scans at their ground equivalent, one pixel is 1.524 meters on the first but 6.096 meters on the second. It is not this print density but the equivalent scale that is important for the map's accuracy. The width of a very thin line on the map, such as a small stream or a contour, is about 0.1 millimeter.

At 1 : 24,000, this means that the contour line would be 2.4 meters wide if it were painted on the ground, more than the pixel size on the 400 DPI scan but less than a single pixel of 6.1 meters on the 100 DPI scan. Most of the line would be skipped, and only occasionally would the pixel and the line coincide. This can be shown clearly in the insets in Figure 4.8. Losing features in this way is called *dropout*. Dropout can virtually eliminate a feature on the map, or at best make it seem like background "noise." Another fact to note about scanning is that folds in the map, pencil lines, coffee stains, paper discoloration, and, in particular, wrinkles all show up. This can lead to problems, as we will see in Section 4.6.

4.4 FIELD AND IMAGE DATA

4.4.1 Field Data Collection

An increasing amount of data for GIS projects comes from a combination of field data, global positioning system data, and imagery. Field data are collected using standard surveying methods, in which locations are established in the field as *control points* and then additional locations, tracing out features or covering terrain, for example, are traced out by large numbers of measurements using instruments designed to measure angles and distances. The highest accuracy instruments, called *total stations* are digital recorders as well as measurement instruments and use laser ranging to prism reflectors to calculate distance.

Less expensive instruments such as theodolites, engineers' transits, and levels often use a technique for measuring distances called *stadia*, which involves reading the numbers on a calibrated pole through the lens of the instrument. Data are recorded in notebooks, and usually the data are then entered into a computer program to turn the bearings, angles, and distances into eastings, northings, and elevations. The type of software used is called COGO, for "coordinate geometry," and many COGO packages either write data directly into GIS format or are capable of writing files that can transfer directly. Figure 4.9 shows a typical field mapping project, the mapping of a lake in upstate New York.

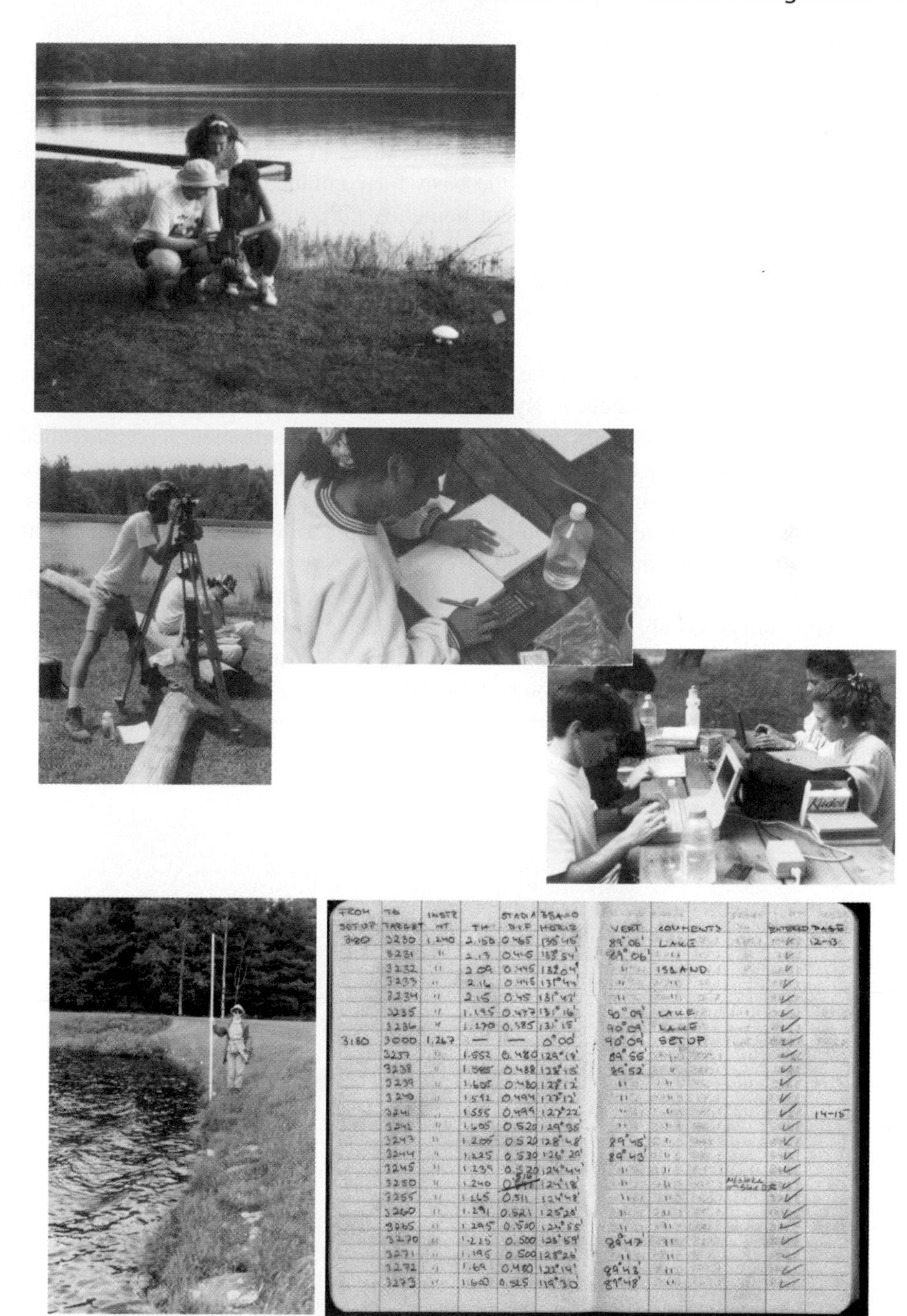

FIGURE 4.9: Getting field data into the GIS. Top left: Collecting base control data with a GPS receiver. Center left: Surveying detailed lake edge information using a theodolite to measure angles and (bottom left) stadia to measure distance. Bottom right: Field note-book data from the surveying. Center middle: Transcribing and averaging the field notes. Center Right: Entering the data into a surveying COGO package to translate distances and angles into coordinates. (Bowman Lake State Park, NY: Hunter College GTECH 350 students.)

4.4.2 GPS Data Collection

The first stage in the surveying process, that of setting the control, is usually done by locating a USGS control point (a bench mark) or by using GPS. Using two GPS receivers together in *differential mode* it is possible to locate control points to sub-meter accuracy. These points are then used as the basis for continued extension of the survey network going outward and between these points. A GPS unit being used for control surveying at extremely high precision is shown in Figure 4.10.

The GPS is a system of 24 orbiting satellites in medium earth orbits (about 20,000 km), each transmitting a time signal. At any given time, at least four of the satellites are above the local horizon at every location on earth 24 hours a day. When a GPS receiver is activated, the nearest satellites are located and the signals are received from each visible satellite. By decoding the time differences between the signals from each satellite, combined with data from the satellite itself about its orbit (called *ephemeris data*) it is possible to solve the three unknowns of latitude, longitude, and elevation. Many receivers can do direct conversion into any of several coordinate systems and datums, and most can download the data directly to a computer. Some GPS equipment can download directly in common GIS formats.

Prior to May 2000, the GPS signal was accurate in its coarse acquisition (C/A) code mode to only about 75 to 100 meters because the signal was deliberately degraded under a system called *selective availability*. The use of selective availability has now been determined to be no longer of national security interest and has consequently been turned off, so that accuracies of 10 to 25 meters are normally possible.

FIGURE 4.10: GPS control being established using a high-precision differential GPS receiver. (Photo courtesy of Magellan Systems Corporation, San Dimas, CA. Used with permission.)

By using two units, one at a known location and one "rover" unit, the degradation can be measured and eliminated, usually by processing the data from the two units on a computer after collecting the data. This is called *differential-mode use* of the GPS. It is possible by using a radio receiver or a cellular telephone link to receive real-time differential corrections in the field. The corrections are broadcast as an aid to navigation in the United States and elsewhere, and are also available from private services.

Many hand-held GPS receivers are now capable of downloading their data to computer software, either for post-processing for accuracy enhancement or for direct integration into GIS. In some cases, GPS receivers have elaborate map displays integrated into the portable units. In others, GPS units work in conjunction with software (including GIS) to display GPS locations directly onto maps. Several GPS vendors now offer software for portable digital assistants, and even cellular telephones, which will soon come equipped with their own GPS receivers inside.

4.4.3 Image and Remote Sensing Data

Imagery data are very common input layers to a GIS. They are most frequently air photos such as the USGS's orthophotos or satellite images. Air photos are usually black and white and are produced from scanning or accessed from CD-ROM. The National Airphoto Program makes photography at a variety of scales available in the United States, and private vendors also sell images. Digital orthophotos are at an equivalent scale of 1 : 12,000 and have a 1-meter ground resolution. The current program calls for national coverage soon after the turn of the century and a 5-year update after that. Examples of a digital orthophoto and satellite data from the Landsat program are shown in Figure 4.11.

The Landsat program has been generating imagery of many locations in the world since 1972. Two scanners, the multispectral scanner and the thematic mapper, image areas on the ground at 79 and 30 meters, respectively. The images are geometrically corrected into the space oblique Mercator projection and are available for use in GIS projects. Coverage can be quite discontinuous due to gaps in the program. Other satellites also generate imagery, including the French SPOT satellite series and the Canadian RADARSAT. In addition, GIS projects at small scales also often use the NOAA polar orbiting satellites carrying the AVHRR (advanced very high resolution radiometer) and the geostationary GOES weather satellite, the one seen every evening on the television weather report.

4.5 DATA ENTRY

Geocoding is the part of GIS data input that results in getting a map into the computer. It is not the entire story, however, for as yet we have not dealt with getting the attributes into the GIS. An attribute is a value, usually a number, containing information about the features contained in the GIS. If the feature we are geocoding is a road, for example, then capturing the route of the road from a map as it winds from intersection to intersection is pure geocoding. We also have to tell the computer what this long and winding line is: a road, and anything else that the GIS needs to know about it. Relevant attributes for a road might be its state route number, the year it was built, what the surface is made of, how many traffic lanes are on the road, if the road is one-way or two-way, how many bridges it goes over, how many cars travel along the road per hour, and so on. These

FIGURE 4.11: Imagery often used in GIS. Lower left: Digital orthophoto of Washington, DC. Ground resolution of 1 meter. Upper left: Landsat 7 thematic mapper image of New England. Ground resolution of 30 meters. Upper right: High resolution air photography matched with vector data. Sources: USGS EROS Data Center and county of Santa Barbara.

values are the road's *attributes*. They are the very meat and potatoes of GIS analysis. Somehow, we have to get them into the computer, too.

The simplest way to think of attributes is in a flat file. A flat file is really like a table of numbers. The columns of the table are the attributes, and the rows of the table are the features themselves. Each line in the table is a record, but the name used depends on who you talk to. A computer scientist would call a row a *tuple*, a statistician would call it a *case* or an *observation*. A programmer might call it an *instance of a geographic object*. They are all pretty much the same. *Record* sounds simpler.

Take a look at the flat file in Figure 4.12. The records and attributes relate to the example we discussed above, a road. The attribute table then consists of several parts. First, it has attributes with their names. Setting up the attributes means deciding what values are going to be associated with each of our features. At the time of setup, it is easy to anticipate something we may want to collect in the future and to leave a column in the table for it.

Second, there are records. A record usually has a value in every one of the columns. Software programs such as spreadsheets and some databases allow you to click into a

Features on Adams, NY Map

ID #	Feature	Name	Surface	Lanes	Traffic per hour
1	Road	US 11	tarmac	3	113
2	Road	I 81	concrete	4	432
3	Road	Lisk Bridge Road	tarmac	2	12

***value*, is the number or text associated with a record for an attribute**

***attribute* has a name and a value for each record**

***record*, all attributes for one feature**

FIGURE 4.12: An attribute table organized as a flat file.

cell in the table and put in a value. Nevertheless, setting up the table has to be a little more formal than that.

Each attribute has more than simply a name associated with it. For example, if we tried to put "US11" into the attribute column "Surface," something is obviously wrong. Each attribute should have several characteristics, all of which usually have to be known in advance. The following is a list of what has to be considered.

1. What is the *type* of the value? For example, values could be text, number, decimal value or units such as meters, vehicles per hour, and so on.
2. What is the legitimate *range* of the values? For example, percentages should be between 0 and 100. Are negative numbers allowed? For text values, what spellings or range of choices (known as *categories*) are allowed? For text, how many characters long is the longest string?
3. What happens when there is a missing cell in the table? For example, a record could be missing an attribute such as the traffic count in Figure 4.12 because nobody was available to make the count. Often a *missing value flag*, such as the value −999 or NULL, is used in these cases. We obviously would not want these to count if we summed or averaged the rows or columns.
4. Are duplicates allowed? What if we had two road entries for Interstate 81 on Figure 4.15, one for the northbound and one for the southbound lanes? The traffic counts, road surfaces, and so on may be different and worthy of their own record. In this case, the values entered in the attribute column under "Name" would be identical.
5. Which attribute is the *key*? The key is the link between the two databases. So in the example of Figure 4.15, the attribute "ID #" should match the tag that was placed on the road when it was digitized from the map. Otherwise, all our attributes would be "lost in space."

Many of these questions must be answered when we set up the database to begin with. The tool within the database manager that allows this attribute setup is called the *data definition module*. It often has its own menus, language, and so on, and may need a programmer rather than a typical GIS user to set it up. In some cases, just as with the digital map data, the attribute data will have been found from an existing source, such as the Census Bureau's data files that link to the TIGER files. In this case, the links will already be made between the attributes and the features on the map. If there are new data, however, or if we make our own database for our own purposes, we have to make the links and check them ourselves.

A complete listing of all of the above information is called a *data dictionary*. Having the data dictionary in advance allows the part of the GIS that handles data entry, or the spreadsheet or database program we choose to use, to check each value as we enter it. Sometimes we enter the numbers and values one by one into a special part of the database manager called the *data-entry module*.

Often, we import into our GIS data manager all the records in a preexisting setup. Some of the more common databases and spreadsheets support specific formats for data exchange to allow this. The simplest form is to write a file with each of the attributes and their labels written as text, one per line, sometimes separated by commas and quote marks so that blanks and other symbols can be included. For example, the data in Figure 4.12 could be "unraveled" into the file listed in Figure 4.13.

```
Attribute_labels = "ID #", "Feature",
"Name" , "Surface" , "Lanes", "Traffic" , "per hour"
"1",
"Road",
"US 11",
"tarmac",
"3",
"113"

"2",
"Road",
"I 81",
"concrete",
"4",
"432"

"3",
"Road",
"Lisk Bridge Road",
"tarmac",
"2",
"12",
"4"
```

FIGURE 4.13: An unraveled or ASCII text from version of the flat file in Figure 4.12.

For new GIS data, the process of entering attributes eventually comes down to someone (usually the lowest-paid person) entering the attribute values one by one into a database manager's data-entry module. The data usually come from a data form of some kind, onto which they had been recorded painstakingly by the person collecting the data.

Some data-entry systems are better than others. At the very least, the system should check the type and range of the value for each attribute at the time of entry. At best, it is helpful if the software allows things like copying a record but then changing it to reflect a new value, deletion, or changing of values that are wrongly entered at the time of entry, and if the software brings errors to your attention with beeps and messages so that correction can take place immediately. No software package should allow data to be lost if the computer crashes, the file fills up, or the user presses the wrong button.

Most GIS packages allow the use of almost any spreadsheet, such as Microsoft Excel, or database systems such as DBase IV or Borland. Some require that you use the database entry system that comes with the GIS and no other. Each is slightly different, although all share the items discussed in this section.

4.6 EDITING AND VALIDATION

Many early geocoding systems had only limited editing capabilities. They allowed data entry, but error detection was by after-the-fact processing, and correction was by deletion of records or even whole data sets and reentry. Anything we can do in the geocoding process that reduces errors, or that makes errors easily detectable, we should indeed do. As an absolute minimum, data for lines and areas can be processed automatically for consistency, and any unconnected lines or unclosed polygons can be detected and signaled to the user. The connection between lines, known bordering of areas, and inclusion of points in areas is called map *topology*. Topology really comes into its own during the map validation stage.

The easiest way to avoid errors in geocoding is to ensure that errors are detected as soon as possible and then to make their correction easy. Video display during digitizing and audio feedback for error messages is essential. GIS software should spell out exactly what will happen in the case of an error. A common geocoding error is to overflow a hard or floppy disk or perhaps a disk-size quota while digitizing. It helps also to be able to recognize errors when they appear and to be able to understand their origin.

Some easy-to-detect errors are *slivers*, *spikes*, *inversions*, *lines that are not ended*, and *unsnapped nodes*, which we discussed in Chapter 3. Scaling and inversion errors are when the map appears squashed, like the titles at the beginning of a wide-screen movie shown on TV, or flipped. These are usually due to an incorrect digitizer setup procedure; that is, they are systematic errors caused by incorrectly entering the control points for establishing the map geometry. *Spikes* are random hardware or software errors in which a zero or extremely large data value erroneously replaces the real value in one of the coordinates. Spikes are also sometimes known as *zingers*. Errors in topology, missing or duplicate lines, and unsnapped nodes are operator errors.

Plotting the data becomes a useful aid because unplottable data often have bad geocodes. Similarly, attempting to fill polygons with color often detects gaps and slivers not visible in busy polygon networks. The best check for positional accuracy is a check against an independent source map of higher accuracy.

The equivalent of a plot for the attribute data is a *data listing* or *report*. Most data management systems have the ability to generate a report, listing the attributes as a table, or formatting them neatly for printing and checking. You should go line by line, checking the attributes and their values. However, even when the attributes and the map are validated by checking, it is still likely that errors exist in the links. One New York City database had more than 20 spellings for a single street name, for example.

The GIS often allows check plots to be generated that simply plot the label or identification number of the key within a polygon or next to a line. These maps and the tedious process of checking them should never be skipped. Moving straight on to making elegant graphics or doing a GIS-based analysis with erroneous data can be anything from embarrassing to dangerous, or even life-threatening.

A data set that is correctly geocoded both positionally and with attributes is not necessarily logically consistent. Logical consistency can be checked most easily for topological data. Topologically, data can be checked to see that all chains intersect at nodes, that chains cycle correctly in a ring around polygons, and that inner rings are fully enclosed within their surrounding polygons. Otherwise, attributes can be checked to ensure that they fall within the correct range and that no feature has become too small to be represented accurately.

Everyone would like to say that the data in his or her GIS are accurate and correct. Obviously, this means several things. Accuracy of position means that the locations shown on the map are in their correct locations with respect to the real world. Of course, there may be a difference between the map that was geocoded and the "best possible" map. Positional error is sometimes tested or measured, and this is best done against another map of higher accuracy or against accurate field measurements such as GPS fixes. Another aspect of data is the accuracy of the attribute. A map may be perfect as far as appearance is concerned, but the roads and rivers could both be mislabeled as power lines. This type of error can be treated as a misclassification. Testing can also be conducted and can even be automated as GIS data are already in a database management system.

A final issue is that of scale and precision. A map used for geocoding has a particular scale, such as 1:24,000. If this is the case, while the GIS allows us to compare data from another scale, say 1:250,000, it may not be appropriate to do so, as attributes, generalization of the features, and other properties of the map may be different at the two scales. Also, all data in the GIS have a degree of precision associated with them. If a highly detailed line is geocoded only to the nearest 10 meters on the ground, comparison with more detailed data becomes problematical. Generally, we should apply the same concerns and considerations of limitations to digital maps as we do to paper maps. Unfortunately, many people treat digital maps as absolutely correct maps instead of the digital alternative form of the analog maps to which they owe their humble origins.

The intelligent GIS user should know and understand the amount and distribution of error in a GIS database. Many of the sources of error are due to the method and process of geocoding. Some of the errors multiply as we move through the stages of data management, storage, retrieval, GIS use, and analysis. An understanding of error is essential to working effectively with GIS.

4.7 STUDY GUIDE

4.7.1 Summary

CHAPTER 4: Getting the Map into the Computer

Analog-to-Digital Maps (4.1)

- A GIS depends on maps being available in digital rather than analog form.

Finding Existing Map Data (4.2)

- GIS data can be

  (1) Purchased
  (2) Found from existing sources in digital form
  (3) Captured from analog maps by geocoding

- Geocoding is the conversion of spatial information into digital form.
- Geocoding involves capturing the map, and also capturing the attributes.
- Existing map data can be found through a map library, via network searches, or on media such as CD-ROM and disk.
- Many major data providers make their data available via the World Wide Web, a network of file servers available over the Internet.
- Major federal agencies with WWW servers are the USGS, NOAA, and the Census Bureau.

Digitizing and Scanning (4.3)

- The method of geocoding can influence the structure and error associated with the spatial information that results.
- The two major geocoding methods for maps are digitizing and scanning.
- Digitizing on a tablet captures map data by tracing lines by hand, using a cursor and an electronically sensitive tablet, resulting in a string of points with (x, y) values.
- Scanning places a map on a glass plate and passes a light beam over it, measuring the reflected light intensity. The result is a grid of pixels. Image size and resolution are important to scanning. Small features on the map can drop out if the pixels are too big.

Field and Image Data (4.4)

- Much GIS data comes from field data collection, GPS data collection of positions, or from digital versions of air photos or satellite images.

Data Entry (4.5)

- Attribute data can be thought of as being contained in a flat file. This is a table of attributes by records, with entries called values.
- A database system contains a data definition module, which sets up the constraints on the attribute values, a data-entry module to enter and correct values, and a data management system for storage and retrieval.

- **The legal data definitions can be listed as a data dictionary. A database manager can check values with this dictionary, enforcing data validation.**

Editing and Validation (4.6)

- **Validation and checking for map data is usually done using topology.**
- **Map and attribute data errors are the data producer's responsibility, but the GIS user must understand error.**
- **Accuracy and precision of map and attribute data in a GIS affect all other operations, especially when maps are compared across scales.**

4.7.2 Study Questions

Analog-to-Digital Maps

Define the following: *digital, analog, real, virtual*, and *geocoding*. Give some examples of maps that can be brought into a GIS free, and maps that would have to be geocoded.

Finding Existing Map Data

What GIS data are available to a user without access to the Internet? Give three examples of agencies that provide data over the World Wide Web, and list the type of data that each provides. What GIS applications might need each type?

Digitizing and Scanning

Discuss the differences between digitizing and scanning. What hardware and software would each require? What sorts of geocoding errors can each method generate?

Data Entry

What is the difference between map data and attribute data? Give the stages of setting up a database from scratch. Why is the data definition stage so important?

Editing and Validation

What software tools might be used in data editing? Name some of the common errors in geocoding that can be corrected by editing. Why might the value of an attribute in a record be invalid? What part of the database manager allows data editing and validation?

4.8 EXERCISES

1. *If you have access to the Internet, use the tools you have available (perhaps Netscape Navigator or Microsoft Explorer) to search for information about GIS and for digital map data online covering your town or city. Are there any attribute data for the maps, perhaps in gazetteers, almanacs, or data books? You may be able to get these from Internet providers. If you do not have access to the Web, visit a library and use its facilities, or look in the reference section for information. If you live near a map library, perhaps at a university, see if you can use this facility in your search. After a few searches, make an inventory of what data you were successful in locating. Put the inventory in the form of a list. Add a column to show which agencies supplied the data you found and how recent the data were. How are the data made available to the public?*

2. *Find the most detailed map you can of a place you know well. You can try your local planning office, a town or college map library, the county records office, or the town hall. Make a fact sheet about your search for other people to use, showing the source of the information, how you got the data, what the map showed, its scale, the date(s) of the information, and what features are shown. If you were to digitize this map for use in a GIS, what sort of applications would be possible? What other maps would you need?*
3. *Buy a copy of the U.S. Geological Survey 7.5-minute quadrangle covering the area where you live. You can find the maps in local camping stores, a map store, or a book store; copy them at a library, or you can get the name of map dealers near you by calling the Earth Science Information Center at 1-800-ASK-USGS (in the United States). Box out the section centered on your house, say 3 kilometers by 5 kilometers. Figure 4.8 is a section of a USGS quadrangle map scanned on a desktop scanner. This section of map was scanned, resulting in a file in TIF format that was 266,818 bytes in size. This was a file of color intensities between 0 and 255 for red, green, and blue in each of three layers spaced on a grid 0.25 millimeter apart. How much data would be necessary to capture the features on your map as vectors? Would it be more or less than the grid (raster) file?*
4. *Carefully read the documentation for any GIS package to which you have access. If you don't have a GIS, see if you can borrow the documentation, or get it as an online file. How does this package allow you to enter (1) data from other sources, such as the network, (2) raster data from a scanner, (3) vector data that you want to digitize from a paper map, or (4) data from other GIS packages? Make a log with entries for each of these capabilities. Keep this log, and compare it with logs for other GIS packages.*
5. *How does your GIS allow you to enter attribute information? Can you enter the data into a spreadsheet and move it into the GIS? What sort of user interface is used? What capabilities does the software have for data set definition? For the capture of errors during data entry? For validation of the data? For editing the data?*
6. *What steps are necessary for effective geocoding? Where did you make most errors in geocoding your first map? What steps in the geocoding process enabled you to (1) notice, (2) find or locate, and (3) eliminate these errors? How could these errors have been avoided to begin with?*

4.9 REFERENCES

4.9.1 Books

Bohme, R. (1993) *Inventory of World Topographic Mapping*. New York: International Cartographic Association/Elsevier Applied Science Publishers.

Campbell, J. (1993) *Map Use and Analysis*, 2nd ed. Dubuque, IA: Wm. C. Brown.

Clarke, K. C. (1995) *Analytical and Computer Cartography*, 2nd ed. Upper Saddle River, NJ: Prentice Hall.

Makower, J. (ed.) (1986) *The Map Catalog*. New York: Vintage Books.

Thompson, M. M. (1987) *Maps for America*. 3rd ed. U.S. Geological Survey, Washington, DC: U.S. Government Printing Office.

4.9.2 Internet Addresses

U.S. Geological Survey `http://www.usgs.gov`

U.S. Census Bureau `http://www.census.gov/tiger/tiger.html`

NOAA `http://www.noaa.gov`

4.10 KEY TERMS AND DEFINITIONS

address matching: Using a street address such as *123 Main Street* in conjunction with a digital map to place a street address onto the map in a known location. Address matching a mailing list, for example, would convert the mailing list to a map and allow the mapping of characteristics of the places on the list.

analog: A representation where a feature or object is represented in another tangible medium. For example, a section of the earth can be represented in analog by a paper map, or atoms can be represented by Ping-Pong balls.

attribute: A characteristic of a feature that contains a measurement or value for the feature. Attributes can be labels, categories, or numbers; they can be dates, standardized values, or field or other measurements. An item for which data are collected and organized. A column in a table or data file.

data dictionary: A catalog of all the attributes for a data set, along with all the constraints placed on the attribute values during the data definition phase. Can include the range and type of values, category lists, legal and missing values, and the legal width of the field.

data entry: The process of entering numbers into a computer, usually attribute data. Although most data are entered by hand, or acquired through networks, from CD-ROMs, and so on, field data can come from a GPS receiver, from data loggers, and even by typing at the keyboard.

data-entry module: The part of a database manager that allows the user to enter or edit records in a database. The module will normally both allow entry and modification of values, and enforce the constraints placed on the data by the data definition.

digitizing: Also called semi-automated digitizing. The process in which geocoding takes place manually; a map is placed on a flat tablet, and a person traces out the map features using a cursor. The locations of features on the map are sent back to the computer every time the operator of the digitizing tablet presses a button.

digitizing tablet: A device for geocoding by semiautomated digitizing. A digitizing tablet looks like a drafting table but is sensitized so that as a map is traced with a cursor on the tablet and the locations are identified, converted to numbers, and sent to the computer.

drop-out: The loss of data due to scanning at coarser resolution than the map features to be captured. Features smaller than half the size of a pixel can disappear entirely.

drum scanner: A map input device in which the map is attached to a drum that is rotated under a scanner while illuminated by a light beam or laser. Reflected light from the map is then measured by the scanner and recorded as numbers.

editing: The modification and updating of both map and attribute data, generally using a software capability of the GIS.

flat file: A simple model for the organization of numbers. The numbers are organized as a table, with values for variables as entries, records as rows, and attributes as columns.

flatbed scanner: A map input device in which the map is placed on a glass surface, and the scanner moves over the map, converting the map into numbers.

FTP (File Transfer Protocol): A standardized way to move files between computers. It is a packet switching technique, so that errors in transmission are detected and corrected. FTP allows files, even large ones, to be moved between computers on the Internet or another compatible network.

gateway: A single entry point to all the servers and other computers associated with one project or organization. For example, the U.S. Geological Survey, though spread across the country and throughout dozens of computers, has a single entry point or gateway into these information sources.

geocoding: The conversion of analog maps into computer-readable form. The two usual methods of geocoding are scanning and digitizing.

Internet: A network of computer networks. Any computer connected to the Internet can share any of the computers accessible through the network. The Internet shares a common mechanism for communication, called a protocol. Searches for data, tools for browsing, and so forth ease the tasks of "surfing" the Internet.

medium: A map medium is the material chosen on which to produce a map; for example, paper, film, Mylar, CD-ROM, a computer screen, a TV image, and so on.

network: Two or more computers connected together so that they can exchange messages, files, or other means of communication. A network is part hardware, usually cables and communication devices such as modems, and part software.

NOAA (National Oceanic and Atmospheric Administration): An arm of the Department of Commerce that is a provider of digital and other maps for navigation, weather prediction, and physical features of the United States.

point mode: A method of geocoding in semiautomated digitizing, in which one press of the cursor button sends back to the computer only one (the current) tablet location.

real map: A map that has been designed and plotted onto a permanent medium such as paper or film. It has a tangible form and is a result of all of the design and compilation decisions made in constructing the map, such as choosing the scale, setting the legend, choosing the colors, and so on.

report: A listing of all the values of attributes for all records in a database. A report is often printed as a table for verification against source material, and for validation by examination.

scanning: A form of geocoding in which maps are placed on a surface and scanned by a light beam. Reflected light from every small dot or pixel on the surface is recorded and saved as a grid of digits. Scanners can work in black and white, in gray tones, or in color.

server: A computer connected to a network whose primary function is to act as a library of information that other users can share.

stream mode: A method of geocoding in semi-automated digitizing, in which a continuous stream of points follows a press of the cursor button. This mode is often used for digitizing long features such as streams and coastlines. It can generate data very quickly, so is often weeded out immediately by automated line generalization within the GIS.

TIGER: A map data format based on zero, one, and two cells; used by the U.S. Census Bureau in street-level mapping of the United States.

topology: The numerical description of the relationships between geographic features, as encoded by adjacency, linkage, inclusion, or proximity. Thus a point can be inside a region, a line can connect to others, and a region can have neighbors. The numbers describing topology can be stored as attributes in the GIS and used for validation and other stages of description and analysis.

U.S. Census Bureau: An agency of the Department of Commerce that provides maps in support of the decennial (every 10 years) census of the United States.

USGS (United States Geological Survey): A part of the Department of the Interior and a major provider of digital map data for the United States.

validation: A process by which entries placed in records in an attribute data file, and the map data captured during digitizing or scanning, are checked to ensure that their values fall within the bounds expected of them and that their distribution makes sense.

virtual map: A map that has yet to be realized as a tangible map; it exists as a set of possible maps. For example, the same digital base map and set of numbers can be entire series of possible virtual maps, yet only one may be chosen to be rendered as a real map on a permanent medium.

PEOPLE IN GIS

Susan Benjamin U.S. Geological Survey Physical Scientist

Susan Benjamin is a research physical scientist with the U.S. Geological Survey's National Mapping Division. She works at the EROS Data Center's field office at the NASA Ames Research Center at Moffett Field, California, near San Francisco. Raised and educated in the Bay Area, Susan has worked with remote sensing and GIS systems since graduating from San Jose State University with a B.A. and then an M.A. in geography. Susan has been involved in many research projects, including pioneering work on the new Digital Orthophoto Quadrangle images. She is the proud mother of a son and a daughter, ages 8 and 9.

KC: Susan, how did your GIS career begin?

SB: After college I went to work on the ILIAC IV supercomputer at NASA Ames. My high school had emphasized math, and it was my minor in college. At that time, GIS courses existed at only one or two universities. You couldn't even get a degree in computer science; that was considered a tool, not a discipline!

KC: You picked math because you knew you were going to work with GIS?

SB: Yes. I knew that the math would be important in remote sensing and GIS. I started out in remote sensing. I became interested in the use of digitized photographs to identify features for making topographic maps. The National Mapping Division then was involved in making and revising large-scale topographic maps at the Western Mapping Center in Menlo Park, California.

KC: Is that where you first heard about GIS?

SB: I encountered GIS there for the first time in about 1985. In 1987 I was hired as part of the GIS Lab staff at USGS's Western Mapping Center. It was very chaotic; the lab had just been formed and equipment from other projects was being reassigned. We had SCITEX workstations that were being used by the USGS to produce digital maps for the 1990 census. Another workstation was used mostly by the Geologic Division to produce geologic maps and I was interested and wanted to work on that. It was the SCITEX that lured me here. At the time they were just setting up their Prime systems, getting Arc/Info up and running, and learning how to integrate Arc/Info into the map production process.

KC: How did you define GIS back then?

SB: I thought of GIS as computer programs that would make my job easier. I always thought that GIS was more analysis and that digital cartography was more production and digitizing of attributes. I thought of GIS as looking at what the attributes were, and drawing conclusions from them.

KC: Did you train as a photogrammetrist?

SB: Yes, I was the last person to be trained at Menlo Park on the PG2 stereo plotters. I learned how to take stereo pairs and set models and to use the floating dot to follow the ground and compile a contour map.

KC: What is a floating dot?

SB: A stereoplotter takes a set of two aerial photographs that have overlap between them, and lets you assemble the views to see three dimensions. The machine provides you with a floating dot that you can set at different elevations in the image. You have the dot follow the ground surface in three dimensions, tracing out a contour from the image. The kind of contours that are in the digital line graph and the digital elevation model data are made like this, only now it's all digital.

KC: What sort of GIS software have you used over the years?

SB: GRASS and Arc/Info. I have been through SPANS training. For remote sensing I've used ERDAS Imagine, which is often considered a GIS, and LAS, EDCs image analysis software.

KC: What's the plaque on your office wall?

SB: Back in 1990 I put out a sampler CD-ROM of digital orthophotos of Dane County, Wisconsin, the first CD-ROM that the mapping division produced with nothing but the DOQ images on it. I was responsible for assembling the data sets, organizing them on the disk, and putting them together for GRASS. I got all the descriptive information for premastering the CD; I took the disk to the pressing factory and made sure they were pressed and delivered to customers. The plaque on my wall is an award from URISA from the 1991 conference for the most innovative project in the project show case. It was for a hypercard stack that accesses the Dane County DOQ and lets you select digital photos for display and further use.

KC: That is great, thank you very much.

(Used with permission.)

CHAPTER 5

What Is Where?

5.1 BASIC DATABASE MANAGEMENT
5.2 SEARCHES BY ATTRIBUTE
5.3 SEARCHES BY GEOGRAPHY
5.4 THE QUERY INTERFACE
5.5 STUDY GUIDE
5.6 EXERCISES
5.7 REFERENCES
5.8 KEY TERMS AND DEFINITIONS

If you don't know where you're going, you'll end up somewhere else.
—Yogi Berra.

Of all the gin joints in all the towns in all the world, she walks into mine.
—Julius J. Epstein, Casablanca *(1942 film)*

All military commanders ... should thoroughly store up [in their minds] the location of ways in and out of the terrain.... This is the constant value of maps.
—Di Tu, On Maps, *227 B.C., from the* Guanzi, Political, Economic and Philosophical Essays from Early China, *translation by W. Allyn Rickett.*

Philosophical GIS: Not the most respected subfield of geography.

5.1 BASIC DATABASE MANAGEMENT

A GIS can answer the two questions: "what?" and "where?" More important, a GIS answers the question "What is where?" The *where* component relates to the map behind all GIS activities. The *what* relates to the features, their size, geographical properties and, above all else, their attributes. Getting this information is what the toolbox definition of GIS in Chapter 1 meant by *retrieval.*

These are not trivial questions. Other forms of data organization often fall apart when dealing with "where." The telephone book, for example, a list organized alphabetically by last name, gives only relative locations (street addresses) and a house number. An entirely new directory is necessary for each new district, and to retrieve the telephone number of a friend in another town, perhaps just across the river, becomes a major problem because you require a different telephone directory than the one covering your own neighborhood.

The properties of geographic search, finding all the phone numbers of people on a single city block, for example, are not available easily to the user of a telephone directory. The secret to *data retrieval,* the ability to gain access to a record and its attributes on demand, is in data organization. In Chapter 4 we examined the various ways that the graphic part of a GIS can be structured for storage inside the computer. Which structure any given GIS uses, and how the map is encoded into that structure, fully determine how easy it is to find a record and extract its values for use. Once again, the attribute and the map data have different means of access. At the most simplistic level, the GIS is a computer program that accesses data stored in files. Obviously, with a great deal of data, the ability to access the files quickly is important if the user is to receive interactive control of the GIS.

At the logical level, access requires a *data model,* a theoretical construct that becomes the key for unlocking the data's door. In medieval times, when much information had to be memorized, monks would train their powers by spending hours committing to memory the parts of a specific cathedral, capturing a mental picture of the interior and exterior so that as they later learned text, such as the contents of the Bible, they could mentally "place" a chapter, a verse, a line, or even a word in a location that they had already memorized (Figure 5.1). Reciting that verse, then, became the memorization in reverse. The monk would mentally walk to the right place in the cathedral, and the words would be stored there. The cathedral, not in substance, but as a visual memory, became the data model. Without such a data model, data cannot be searched or extracted and therefore become worthless.

We can define a data model, then, as a logical construct for the storage and retrieval of information. It is the computer's way of memorizing all the GIS data that we need to use. This is different from the data structures we examined in Chapter 4, because these deal primarily with how the data are physically stored in files on the computer system. As we have seen, this means that a GIS must have at least two data models, and that the two must have a bridge or link between them to tie the attributes and the geography together. These are the *map data model* and the *attribute data model.* In Chapter 4 we illustrated some of the storage and organization aspects of map data models. In this chapter, we deal with the attribute models and then look closely at how the models help in locating and then extracting data from both map and attribute databases.

The database management system's (DBMS's) heritage is from within computer science, but the user community is as broad as that of GIS, literally millions of firms, accountants, colleges and universities, banks, and so forth that need to keep and organize records by computer. The earliest database management systems date from the efforts of the early 1970s, when large mainframe computers were used, data-entry was

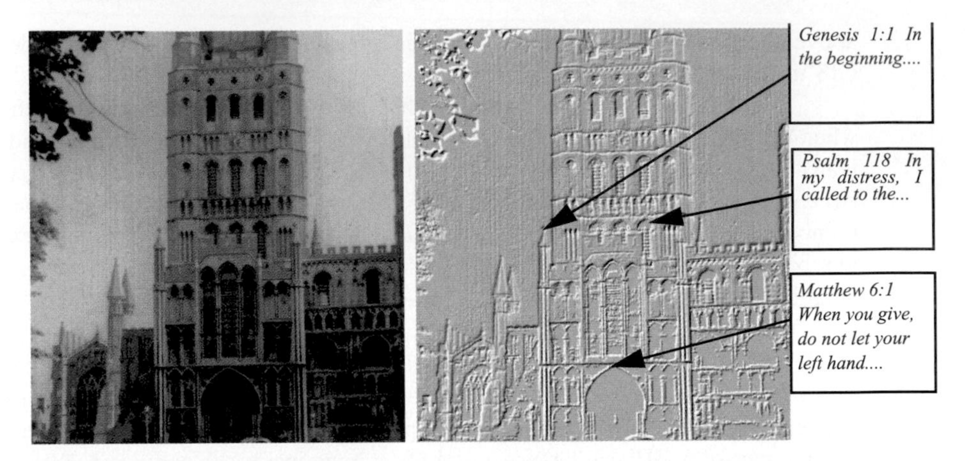

FIGURE 5.1: Medieval monks used the cathedral as a data model for storing documents such as the Bible. The data model in GIS is a logical construct for the storage and retrieval of spatial information. (Photo of Ely Cathedral, UK.)

by key punch and punched cards, and the technology was called *automatic data processing*.

Database management went through its own revolutions due to the technological trends we have already discussed from the GIS viewpoint: the microcomputer, the workstation, the network, low-cost bulk storage, interactive and graphical user interfaces, and so on. Database management, however, has also been significantly influenced by the intellectual breakthroughs that led to radical changes in the way that attribute data can be stored in files. The latest revolution, the object-oriented database system, is now under way and is discussed in Chapter 10.

The parts of a DBMS have remained fairly consistent over time, regardless of how the attribute data are actually placed into files. The *data definition language* is that part of the DBMS that allows the user to set up a new database, to specify how many attributes there will be, what the types and lengths or numerical ranges of each attribute will be, and how much editing the user is allowed to do. This establishes the *data dictionary*, a catalog of all of the attributes with their legal values and ranges. Every DBMS has the ability to examine the data dictionary, and the data dictionary itself is a critical piece of metadata (data about the data) that is often required when the database has to move between systems.

The most basic management function is *data entry*, and because most entry of attribute data is fairly monotonous and may be by transcription from paper records, the DBMS's data-entry system should be able to enforce the ranges and limits entered into the data dictionary by the data definition language. For example, if an attribute is to contain a percentage, and the data-entry person types a value of "110", the DBMS should refuse to accept the value and alert the person.

All data entry is subject to error, and the first step after entry should be *verification*. This is often done by producing a report or printed copy of the data in a standard form that can be checked against the original. Even if the data are error free, most databases must be *updated* to reflect change. Deletion, insertion, and modification of records, or sometimes changes to the data dictionary itself, must be made frequently. This is done by using the data maintenance part of the DBMS. Care must be taken when updates are made, because changes create a new updated version of the entire database. Sometimes modifications are done in batches and a new version of the database "released," reflecting a whole suite of changes, perhaps to reflect the calendar year.

With the preceding tasks complete, the DBMS can then be used to perform its more advanced functions. These are the *sorting*, *reordering*, *subsetting*, and *searching* functions. For example, a database of student records could be sorted by grade-point average to find all students below a "C" average. A database of students containing zip codes in their address could be reorganized by zip code to make mailing easier for the mailroom staff. Subsetting involves using the *query language*, that part of the DBMS that allows the user to interact with the data to perform these tasks, to create a new data set that meets certain search criteria, such as all students who have more than 100 credits toward their degree. Finally, a *search* for a specific record is often needed. At a public terminal, for example, students may be allowed to type in their Social Security numbers to allow them to examine the grades they got in a given semester. All of these functions are part of regular DBMS and are performed differently in each different system. Nevertheless, all of these functions are common to both DBMSs and GISs.

The first-generation DBMSs used a hierarchical structure for their file organization. For example, a university might contain a division or school, the school might contain a department for a single discipline such as Geography, and a department may have a group of students who are majors in that department (Figure 5.2). All students who are declared majors must be assigned to a department, each department belongs to one division, and the entire university consists of a group of divisions. The file structure of a hierarchical system can be organized in the same way. A top-level directory could contain a list of divisions and a set of divisional directories. Changing down to the next level, the divisional level, reveals a list of departments and departmental directories. Down in a departmental directory lie the files containing the student records, one student per record, with attributes such as the student's name and address, year of expected graduation, and grades in various classes. The first generation of database managers used exactly this type of organization of files and records as its data model.

Life is not as simple as the hierarchical model would like it to be, however. In many cases, relationships between records overlap. This is even more so for geographic data, as we will see below. For now, consider a simple hierarchy for the political administration of a state. The order could go state-county-city-district. For much of the country, this would work. In New York City, however, the city of New York consists of the five boroughs, each of them a county in its own right as far as the state of New York is concerned. So the simple hierarchy model already fails, because the hierarchy is literally stood on its head in one case. The data model would require that the county file contain records for cities, but the reality would be the opposite, that the city should contain the county.

Another complex case would be multiple membership. Consider a single house that falls into a fire district, a police district, a school district, a voting district, and a census tract. Five databases, each structured using a different hierarchy, would have five completely independent ways of getting to the attributes for the single house. Although

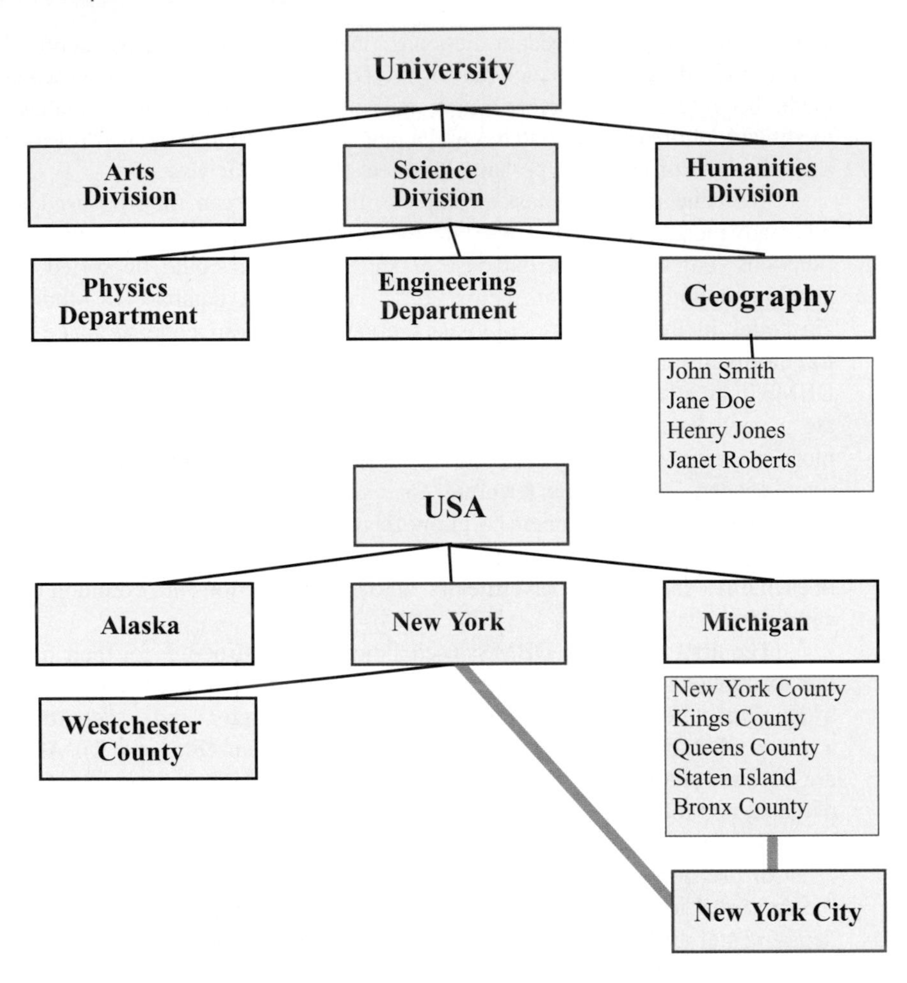

FIGURE 5.2: Hierarchical structures for information. The university example is a simple hierarchy. The nation/state/country/city example is an "inverted" hierarchy. The hierarchical file organization or data model would not work in this case.

each database might store different kinds of information, at some stage it might be useful to assemble together all of the data to be used. Under a hierarchical structure, this cannot be done.

The revolution in DBMS that untangled this database logjam was the use of relational database management systems. Beginning with a set of theoretical breakthroughs in the 1970s, relational DBMSs swept the field and replaced almost all existing systems during the 1980s. They remain the dominant form of DBMS today. The relational model is rather simple, and from the user's standpoint is an extension of the flat file model (Figure 5.3). The major difference is that a database can consist of several flat files, and each can contain different attributes associated with a record. Using the example of the house above, the house can now be a single record in several databases, none of which requires a hierarchy. If the hierarchy still makes sense, we can keep separate files by district or use codes such as zip codes to reveal districts.

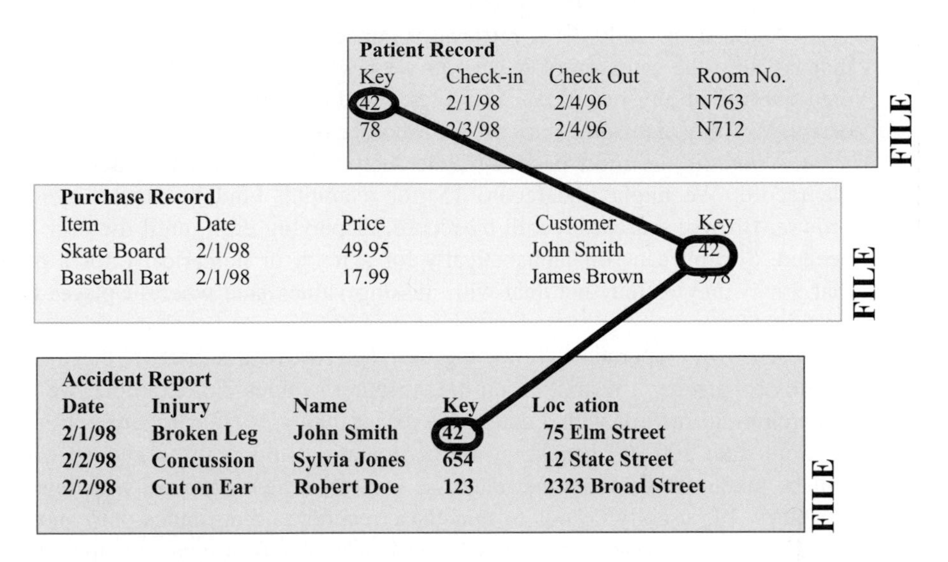

FIGURE 5.3: In a relational database, files with different structures are linked by keys. Records with a common key relate to one feature within the GIS.

Critical to each part of a relational database is a special attribute that serves as a marker rather than a regular attribute. We could assign to every record a unique identifier, for example, or time and date stamp a transaction record to make it different from all others. This "key" attribute can then serve as a link between the flat files. Because the key is unique it allows us to extract various attributes and records from one database and others from another as required. We can store the data in a group of linked files, each of which can be used, have data entered, and be edited and updated and searched separately and without affecting the others.

The relational database manager contained a new set of data management commands that allows the keys and links to be exploited. These typically include such actions as *relate*, to select from two flat files that have a common key attribute, and *join*, to take the relate output and merge them into a single database. So the relational data model, while permitting records and attributes to be separated into different files for storage and maintenance, also allows the user to assemble any combination of attributes and records, as long as they are linked by a key attribute. A *join* can create many unneeded subrecords for a single feature, such as multiple records with different dates, so care should be exercised when joining databases.

5.2 SEARCHES BY ATTRIBUTE

Most GIS systems include as part of the package a fairly basic relational database manager, or simply build on the existing capabilities of a database system. Searches by attribute then are controlled by the capability of the database manager. All DBMSs include functions for basic data display; that is, show all attributes in a database, show all records with their attributes, and show all existing databases. Most also allow records to be output in a standard form, with a particular page layout and style, called a *report generator*. If we need a paper copy of the database, perhaps for checking and verification, then the report generator is used.

As far as actually doing retrieval is concerned, the DBMS must support functions that fall into the category of *query*. As we have seen, a DBMS should allow sufficient data query that any record can be isolated and any subset required for mapping found easily. We may also sometimes wish to reorder or renumber an attribute.

A *find* is the most basic attribute search. Find is usually intended to get a single record. We might find record 15, for example. Finding can be by *search* or by *browse*. Browse searches record by record, displaying each, until the user finds the one needed. Sorting can sort alphabetically for a field, or numerically for a number. Note that a sort may or may not deal with missing values, and where it places them may be significant.

A *restrict* operation allows the user to retrieve a subset of the total number of records by placing a restriction on the attributes' values. For example, we could restrict a search to all records with a date more recent than 1/1/99, or to cities with a population of more than 100,000 people. A *select* operation allows us to choose what attributes will be taken out from another database to form a new database with fewer "selected" attributes. We usually do this to *join* these records and attributes onto another database in the relational system. As we will see in Chapter 6, a *compute* operation allows us to compute a value for an attribute, to assign a value, or to do mathematical operations between attributes—divide one by another, for example. We can also usually *renumber* an attribute, that is, change the values to our specifications. We might want to find all percentages in an attribute and change them to a zero if they fall below 50% or a one if they are greater, so that we can do a binary combination with another.

For example, in a database of state populations and areas called states, we could use compute to create a new attribute called population_density.

```
compute in states population_density = population / area
<50 records in result>
```

This creates a new attribute, which we can recode into high (3), medium (2), and low (1):

```
restrict in states where population_density > 1000
<20 records selected in result>

recode population_density = 3
<20 values recoded in result>

join result with states replace
<20 records changed in state>

restrict in states where population_density > 100
<12 records in result>

recode population_density = 2
<12 values changed in result>

join result with states replace
<12 records changed>
```

```
compute in states where population_density! = 3 or
  2 population_density = 1
<18 records changed>
```

The recoding is now complete. We can now sort by the new recoded value.

```
list attribute in state population_density

<In database ''state'' attribute values for
  ''population_density''>
<1 18 records>
<2 12 records>
<3 20 records>
<no missing values>

sort result by population_density
<50 records in result>

replace state with result
<50 records changed>
```

In this exchange, note that commands are given one line at a time and often must be used in combination to get the desired effect. Note also that most database systems work by performing their operation on a temporary working set (called `result` in the example) that must be placed back into an existing database when necessary. Many DBMSs use menus, variations on different query languages, and different keywords and commands to accomplish the same results.

Before we leave searching by attribute, consider the problems of using these tools to do even a simple search by distance. Merely to find the distance of each record from one point, we would have to do two subtractions, a multiplication, and a squaring, and then a summation and a square-root operation. Obviously, these database tools, although of immense use, are going to be only humble assistants in our geographical searching needs.

5.3 SEARCHES BY GEOGRAPHY

When we considered searching attributes, we looked at the following search and retrieval commands: *show attributes*, *show records*, *generate a report*, *find*, *browse*, *sort*, *recode*, *restrict*, and *compute*. Moving over to the spatial data within a GIS, some of the operations possible are just spatial equivalents of these, while some are more complex. We discuss the simple retrievals first.

In the map database our records are, instead, features. There are some special attributes specific to the spatial data, and those relate to the coordinates and their measures, plus the characteristics of the lines and polygons. Showing attributes, then, consists of examining the new spatial attributes, such as the actual coordinates themselves, the lengths of arcs, and the areas of polygons. Note that these are already valuable. They could have been the source of the areas used in the population density calculations in the example in Section 5.2. Using the attribute search functions on these attributes now has spatial consequences. For example, we could find all arcs greater than a certain length, or polygons more than 1 hectare.

Show all records in a spatial sense becomes either show all attributes or display all features on a map. Generating a map, which allows us to search for information visually, is a spatial retrieval operation as far as the GIS is concerned. If we wish to generate a report, the spatial equivalent would be to produce a finished map to cartographic standards, including labels, metadata, legend, and so on.

Browsing works on a map by highlighting. We may color-code a specific feature or features. Some GISs allow a displayed single feature to be blinked on and off for visual effect. Finding becomes what many GIS packages call *identify* or *locate*; that is, use a pointing device of some kind such as a mouse to point to a feature. Indicating its selection successfully can then retrieve that feature's attributes from the attribute database (Figure 5.4).

This spatial searching, browsing the map and picking features, like turning over stones at the beach to search for crabs, is a very powerful GIS capability indeed, especially if the GIS is on a portable computer and the feature located is the one you are standing in front of as located by your portable GPS receiver. On the map we can also search by indicating a single feature, all features within a rectangle that we drag out, or all features within an irregular area that we sketch on the screen with a drawing tool.

Sorting has less spatial meaning and is usually given a GIS context by examining the spatial pattern that results from a sort by attribute. For example, we could sort states in the state database used in Section 5.2 by their per capita income, then display on the

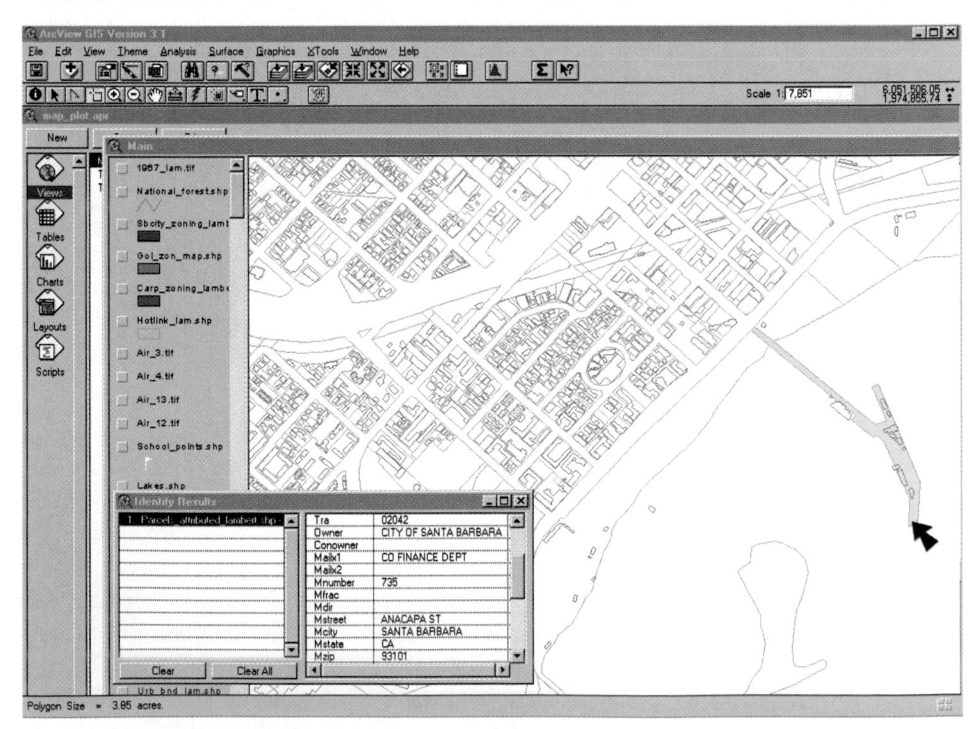

FIGURE 5.4: Spatial search in a GIS using the ArcView identify tool to list the attributes for a polygon chosen by pointing. Parcel and building data are for the City of Santa Barbara. (Courtesy of Ryan Aubry.)

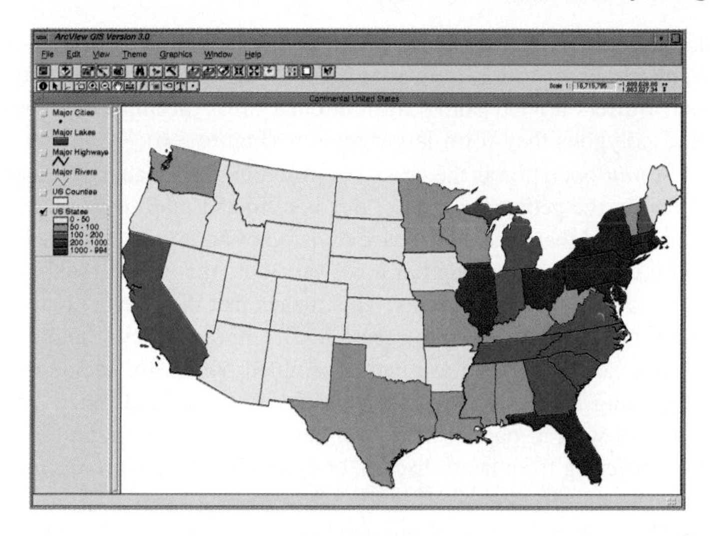

FIGURE 5.5: Recoding (classifying) the attribute values to make a shaded choropleth map showing coterminous United States population density in 1990 using ArcView GIS 3.0.

map the highest and lowest states. Similar to this is recoding by attribute. In the example in Section 5.2, we converted the density values into high, medium, and low values by recoding. We could use this recoded attribute directly to make a choropleth or shade-tone map of the population density, as in Chapter 7 (Figure 5.5).

Operations that perform attribute manipulations, recode, and compute in a spatial sense merely change which features are displayed and how. For example, we could compute and display a new attribute, such as population density in Section 5.2. However, each operation has an equivalent spatial operation. Recoding features spatially, that is, changing the scope of their attributes, is equivalent to a spatial merge. Removing isolated pixels by assigning them to their enclosing or most dominant neighboring region is an example.

Compute can be given a spatial context as distance, length, area, or volume computations and transformations. For example, we could generate a new map that contains the distance to the nearest point feature, the cumulative downstream distance along a stream, or the travel distance along a road network. These new maps could be displayed or used in conjunction with the GIS layers to perform more complex operations, just as we combined the attribute *query commands* in the preceding section to achieve an end result.

Select and join are the remaining operations. *Select* means to extract specific attributes and to reduce the width of the database. By first selecting, we could use only certain themes or layers in a GIS retrieval operation, or we could change the map scale or extent. Picking a subregion, merging quadrangles into a county, aggregating land cover categories from level 2 to level 1 (from seven classes to one class for urban land, for example), picking only major rivers from a full stream network, or generalizing lines on the map for depiction at a broader scale are all examples of a geographic select.

The form of select used most frequently in GIS operations is the buffer operation, that is, to select only those parts of a map or those features that lie within a certain distance of a point, a set of points, a line, or an area. Examples are to restrict a search for malaria victims in a West African nation to those villages more than 10 kilometers from a health clinic, or to limit our search for a summer cottage to one within 200 meters of a lake. Buffers around points form circular areas, around lines they form "worms," and around polygons they form larger regions (Figure 5.6).

A *join* operation is the cross-construction of a database by merging attributes across flat files. In the geographic sense, this is termed a map *overlay*. In a map overlay, a new map is created that shares the space division of both source maps (Figure 5.7). Every new polygon created on the map has a new attribute record in an expanded attribute database associated with the map overlay. This means that the overlay map is searchable by either of the sets of regions used to create it. Examples are city health districts overlain with zip codes, so that we can use data assembled on health, accidents, and so on, with data from mailing lists on population, ethnicity, income, and other variables. By joining the map layers with a map overlay, we can compute, for example, the average income of people suffering from heart disease, or determine by age groups those people at risk from certain age-related diseases (Figure 5.8).

The power of the attribute database manager came from the use of multiple operations in sequence. The same is true of GIS retrieval operations. For example, we could take the overlain health district data in Figure 5.8 and again multiply it with a computed distance map from hospitals represented as points, perhaps geocoded from the

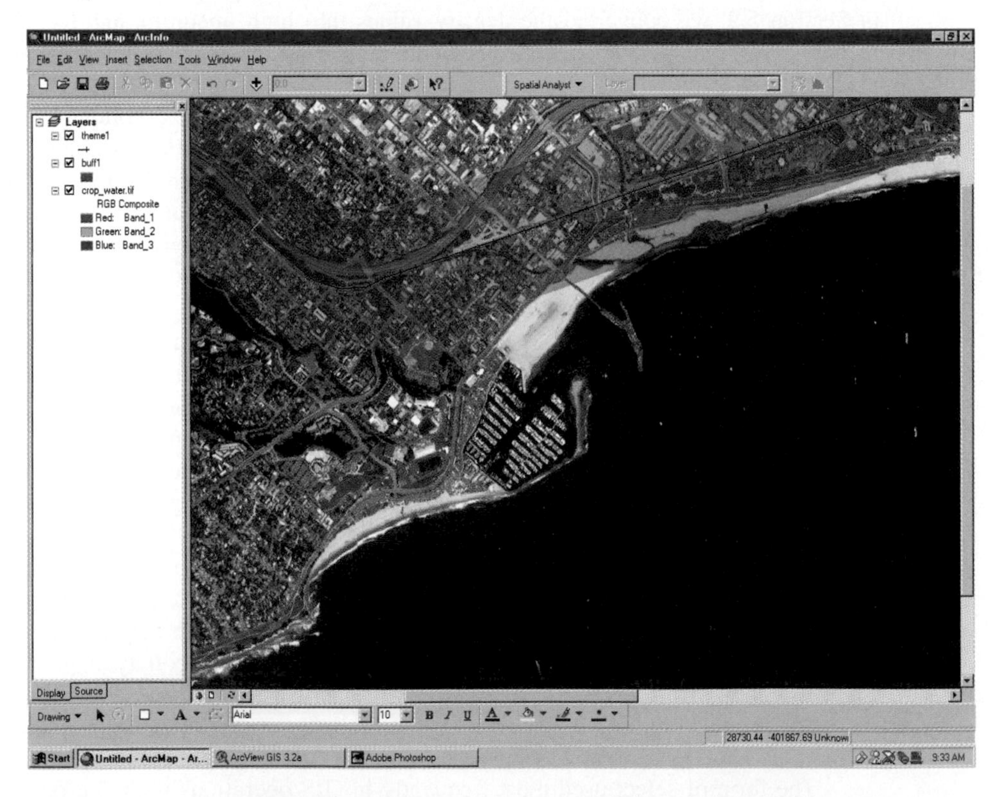

FIGURE 5.6: An example of GIS buffering as a retrieval mechanism. One thousand foot buffer from the railroad tracks in downtown Santa Barbara, shown in transparent red. ArcInfo 8 image by Ryan Aubry.

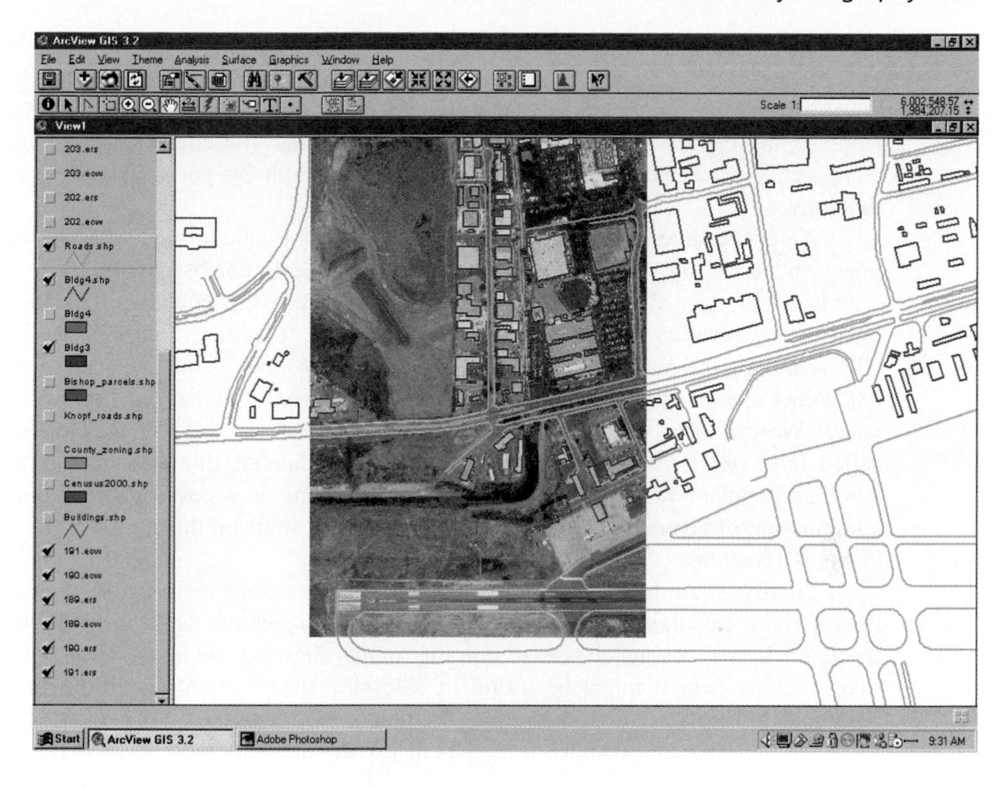

FIGURE 5.7: Map overlay unites different map layers by giving them a common reference base. Here road data and building footprints are overlain onto a high resolution air photo of Goleta, CA.

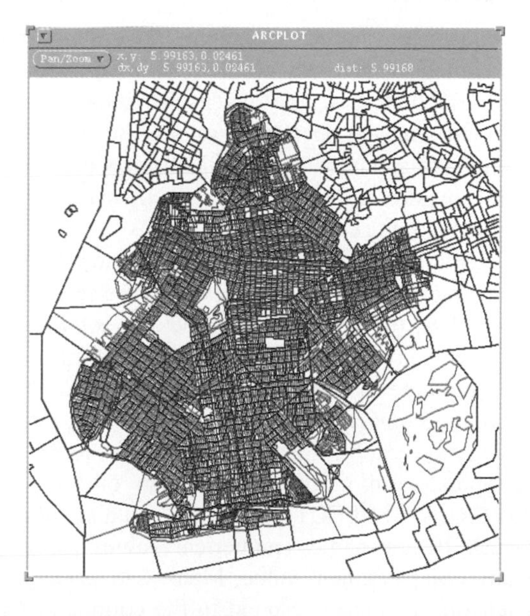

FIGURE 5.8: Map overlay. Example shows health districts and zip codes overlain for parts of New York city. The GIS can combine the two, so that attributes from one database can be cross referenced with the other. (Map by Barbara Tempalski.)

Yellow Pages of the telephone book. This would allow us to display a map showing, perhaps in red, where large numbers of people live long distances from hospitals. Some geographic queries don't fit a clear definition of attribute query operations. A map display can be zoomed into and out of to change the equivalent resolution with comparative ease. Some geographic queries can search by geographic properties and topology.

An entire suite of geographic searches are searches and tests by relations of points, lines, and areas. For example, we can select all points enclosed within one or more regions. Join then allows us to assign attributes from the point features to the area features, say weather statistics from weather stations, to administrative districts. Typical GIS searches are point in polygon, line in polygon, and point distance to a line. If the points are oil storage tanks and the lines are rivers, this can have great analytical value. We can also weight layers in an overlay, perhaps building a composite layer called *land suitability* in the same way that early planners did with overlay mapping. Another popular GIS layer to construct by weighting is a cost surface, built from a combination of layers and distance calculations. Low spots on this map may be good for business locations.

Finally, some highly specific geographic computations are possible. Examples are line-of-sight calculations, computing a viewshed or region that is visible from one place on a map using a base of terrain; maps showing slope and aspect or direction of slope, which might be useful in assessing development suitability or flooding potential; traffic volumes measured on a street network, used to predict traffic jams; or maps showing merged outputs of models or predictions with data, to predict earthquake hazards.

5.4 THE QUERY INTERFACE

Both database management and geographic information management share the fact that the user must somehow interact with the data in an appropriate way. The first generation of DBMS and GIS both used only batch-type interaction with the data, usually closely linked to working with the operating system, the physical management of disk, and so on. This type of interaction dates from the punched card, in that all processes had to be thought out in advance and a file (or stack of cards) produced that could execute the different commands one at a time.

When interactive computing became commonplace, the command line as a query vehicle for data query took over. Commands were typed into the computer one at a time, under the control of the DBMS itself, and the software responded by performing the computations one at a time while the user waited for the command to be completed. Many GISs still use this type of interaction, or permit it to allow the use of macros. *Macros* are files containing commands to be executed one at a time. If an error is detected in a macro, the execution can be stopped and the file modified to correct the mistake.

In the typical form, a command consists of a keyword for the operation such as IMPORT, OVERLAY, SELECT, and so on, and a set of optional or required parameters. Parameters may be file names, numerical values associated with the task, names of options, or any other pertinent value. Many GIS packages will supply defaults for any parameters left out, and most respond to the command without parameters by giving a list of parameters. The largest GIS packages can have hundreds or even thousands of commands, giving more than one way to do a particular task by different routes.

Most GIS packages now are fully integrated with the WIMP (windows, icons, menus, and pointers) interface specified by the operating system, such as Windows or X-Windows. Choices are now most commonly made by menu, with message windows popping up for the user to provide essential parameters when they are required. Values can also sometimes be set by sliders, widgets, and by screen tools such as dials, choice lists, and buttons.

Another fairly recent trend is that most GISs also contain a language or macro tool for automating repetitive tasks. Examples are ArcView's Avenue and Visual Basic, MapInfo's MapBasic, and Arc/Info's AML. In some cases, these languages can interact with the graphical user interface (GUI) tools, presenting a choice as a menu, for example. Thus any GIS user can now become a programmer, establishing his or her own particular task as a query tool for all other users. In large GIS operations when training and employee time are limited, GIS analysts are often employed to automate many other GIS tasks, such as routine queries or database editing.

Finally, there have been efforts to develop a suite of database interaction commands that all users can assume as standard interface to relational databases. The result, the Structured Query Language (SQL), has found a broad acceptance as a much used tool in regular database management, although less use in GIS. Some have argued that almost all GIS operations are possible in SQL, while others have sought to extend its capabilities into the spatial domain. Given the differences between DBMSs, this is a welcome effort.

Most GISs have often lagged somewhat in their attribute database capabilities behind commercial systems for more routine data applications. The more recent interfaces, however, and the broad support for macros and windowing systems have led to a fair amount of convergence in the "look and feel" of GISs, although we are still a long way away from all GISs functioning in much the same way. Given the rich variety of GIS applications, and the rapidly expanding nature of the field, the standardization of GIS operations and query mechanisms is still many years away.

5.5 STUDY GUIDE

5.5.1 Summary

CHAPTER 5: What Is Where?

Basic Database Management (5.1)

- **A GIS can answer the question: What is where?**
- **Retrieval is the ability of the DBMS or GIS to get back on demand data that were previously stored.**
- **Geographic search is the secret to GIS data retrieval.**
- **Many forms of data organization are incapable of geographic search.**
- **GIS systems have embedded DBMSs, or link to a commercial DBMS.**
- **A data model is a logical construct for the storage and retrieval of information.**
- **GIS map data structures are map data models.**
- **Attribute data models are needed for the DBMS.**
- **The origin of DBMS data models is in computer science.**
- **A DBMS contains:**
  - **A data definition language**
  - **A data dictionary**

  - **A data-entry module**
  - **A data update module**
  - **A report generator**
  - **A query language**
- **DBMS queries are sorting, renumbering, subsetting, and searching.**
- **The query language is the user interface for searching.**
- **Historically, databases were structured hierarchically.**
- **Most current DBMSs use the relational model.**
- **Relational databases are based on multiple flat files for records, with dissimilar attribute structures, connected by a common key attribute.**

Searches by Attribute (5.2)

- **Examples of searches by attribute are find and browse.**
- **Examples of data reorganization are select, renumber, and sort.**
- **Compute allows the creation of new attributes based on calculated values.**
- **Attribute queries are not very useful for geographic search.**

Searches by Geography (5.3)

- **In a map database the records are features.**
- **The spatial equivalent of a find is locate; the GIS highlights the result.**
- **Spatial equivalents of the DBMS queries result in locating sets of features or building new GIS layers.**
- **Buffering is a spatial retrieval around points, lines, or areas based on distance.**
- **Overlay is a spatial retrieval operation that is equivalent to an attribute join.**
- **Combinations of spatial and attribute queries can build some complex and powerful GIS operations, such as weighting.**

The Query Interface (5.4)

- **GIS query is usually by command line, batch, or macro.**
- **Most GIS packages use the GUI of the computer's operating system to support both a menu-type query interface and a macro or programming language.**
- **SQL is a standard interface to relational databases and is supported by many GISs.**

5.5.2 Study Questions

Basic Database Management

Make a table listing the component parts of a DBMS. What particular task or tasks does each section of the DBMS do? Add a column to the table giving a brief summary of the role of that component. For example, the data-entry module could have the entry, "Permits user to enter attribute data into the database." Add another column that lists parallels between attribute databases and map databases, such as "report generation" and "map display."

Searches by Attribute

List and define each of the tools that the user of a DBMS has available for searching by attribute. What are the differences between the types of search; for example, *find* versus *browse*?

Searches by Geography

What geographic retrieval tools are available for each of the following: point features, line features, and area features? How can these be used in combination to make complex queries to the GIS?

The Query Interface

What are the major types of user interfaces that the GIS user could face as he or she moves from one GIS software package to another? What are the advantages and disadvantages of each? Which would you prefer as a new GIS user?

5.6 EXERCISES

1. *Using the DBMS within your GIS and a sample data set, work through each of the following attribute-only queries: find, sort, browse, restrict, select, and compute. In each case, use the report generator to print a clean copy of the result, printing only a few records. Write a one-paragraph set of instructions for a novice to do the same thing.*
2. *Use the geographic search commands within your GIS and a sample data set to do the following spatial queries: locate, highlight, and buffer. In every case, generate a simple but well-designed map to show the outcome. How many records does each of the commands isolate, and why?*
3. *Using two different maps that are coregistered within your GIS at the same scale and on the same projection, do a map overlay. Use the report generator of the DBMS to list out the attributes of each separate, and then the composite set of "most common geographic units." How many records are in each data set? Did the overlay require you to deal with sliver polygons? How would your GIS deal with any very small polygons that might have been created? Calculate a per polygon time that the operation took, using a stopwatch as a timer if your computer does not print execution times.*
4. *If you have access to two GIS systems, examine the manuals and the software to see what the user is expected to do to conduct (a) a spatial "locate" and (b) a DBMS-style list of the names of the attributes, and a list of the first few records' attribute values. What sorts of steps are needed? How intuitive are the steps? What sort of user interface does the GIS require you to use (e.g., menus, WIMP, command line, etc.).*

5.7 REFERENCES

Berry, J. K. (1993) *Beyond Mapping: Concepts, Algorithms and Issues in GIS*. Fort Collins, CO: GIS World.

Burrough, P. A. (1986) *Principles of Geographical Information Systems for Land Resources Assessment.* Oxford: Clarendon Press.

ESRI (1995) *Understanding GIS: The Arc/Info Method*. New York: Wiley.

Peuquet, D. J. (1984). "A conceptual framework and comparison of spatial data models." *Cartographica*, vol. 21, no. 4, pp. 66–113.

Huxhold, W. E. (1991) *An Introduction to Urban Geographic Information Systems.* New York: Oxford University Press.

Warboys, M. F. (1995) *GIS A Computing Perspective.* London: Taylor and Francis.

5.8 KEY TERMS AND DEFINITIONS

attribute: A numerical entry that reflects a measurement or value for a feature. Attributes can be labels, categories, or numbers; they can be dates, standardized values, or field or other measurements. An item for which data are collected and organized. A column in a table or data file.

batch: Submission of a set of commands to the computer from a file rather than directly from the user as an interactive exchange.

browse: A method of search involving repeated examination of records until a suitable one is found.

choropleth map: A map that shows numerical data (but not simply "counts") for a group of regions by (1) classifying the data into classes and (2) shading each class on the map.

compute: Data management command that uses the numerical values of one or more attributes to calculate the value of a new attribute created by the command.

data definition language: The part of the DBMS that allows the user to set up a new database, to specify how many attributes there will be, what the types and lengths or numerical ranges of each attribute will be, and how much editing the user is allowed to do.

data dictionary: A catalog of all the attributes for a data set, along with all the constraints placed on the attribute values during the data definition phase. Can include the range and type of values, category lists, legal and missing values, and the legal width of the field.

data entry: The process of entering numbers into a computer, usually attribute data. Although most data are entered by hand or acquired through networks, from CD-ROMs, and so on, field data can come from a GPS receiver, from data loggers, and even by typing at the keyboard.

data model: A logical means of organization of data for use in an information system.

database: Any collection of data accessible by computer.

DBMS (database management system): Part of a GIS, the set of tools that allow the manipulation and use of files containing attribute data.

default: The value of a parameter or a selection provided for the user by the GIS without user modification.

feature: A single entity that composes part of a landscape.

flat file: A simple model for the organization of numbers. The numbers are organized into a table, with values for variables as entries, records as rows, and attributes as columns.

file: Data logically stored together at one location on the storage mechanism of a computer.

find: A database management operation intended to locate a single record or a set of records or features based on the values of their attributes.

geographic search: A find operation in a GIS that uses spatial properties as its basis.

hierarchical data model: An attribute data model based on sets of fully enclosed subsets and many layers.

highlight: A way of indicating to the GIS user a feature or element that is the successful result of a query.

identify: To find a spatial feature by pointing to it interactively on the map with a pointing device such as a mouse.

join: To merge both records and attributes for unrelated but overlapping databases.

key attribute: A unique identifier for related records that can serve as a common thread throughout the files in a relational database.

locate: See **identify.**

macro: A command language interface allowing a "program" to be written, edited, and then submitted to the GIS user interface.

menu: A component of a user interface that allows the user to make selections and choices from a preset list.

overlay: A GIS operation in which layers with a common, registered map base are joined on the basis of their occupation of space.

parameter: A number, value, text string, or other value required as the consequence of submitting a command to the GIS.

query: A question, especially if asked of a database by a user via a database management system or GIS.

query language: The part of a DBMS that allows the user to submit queries to a database.

relate: A DBMS operation that merges databases through their key attributes to restructure them according to a user's query rather than as they are stored physically.

relational model: A data model based on multiple flat files for records, with dissimilar attribute structures, connected by a common key attribute.

renumbering: Use of the DBMS to change the ordering or ranges of attributes.

report generator: The part of a database management system that can produce a listing of all the values of attributes for all records in a database.

restrict: Part of the query language of a DBMS that allows a subset of attributes to be selected out of the flat file.

retrieval: The ability of a database management system or GIS to get back from computer memory records that were stored there previously.

search: Any database query that results in successful retrieval of records.

select: A DBMS command designed to extract a subset of the records in a database.

sort: To place the records within an attribute in sequence according to their value.

SQL (Structured Query Language): A standard language interface to relational database management systems.

subsetting: Extracting a part of a data set.

update: Any replacement of all or part of a data set with new or corrected data.

verification: A procedure for checking the values of attributes for all records in a database against their correct values.

CHAPTER 6

Why Is It There?

6.1 DESCRIBING ATTRIBUTES
6.2 STATISTICAL ANALYSIS
6.3 SPATIAL DESCRIPTION
6.4 SPATIAL ANALYSIS
6.5 STUDY GUIDE
6.6 EXERCISES
6.7 REFERENCES
6.8 KEY TERMS AND DEFINITIONS

What every sceptic could inquire for;
For every why he had a where-fore.
—Samuel Butler. Hubridas, *1663*

Because it's there.
—George Mallory, 1923, when asked by the New York Times *why he wanted to climb Mount Everest.*

Geospatialman. Not always sure where he is, but always ready to tell you where to go.

6.1 DESCRIBING ATTRIBUTES

As we have seen in the chapters leading up to this point, a GIS has at least two parts: the attribute part and the map part. The attribute data, managed by their regular database manager, are little different from any other type of statistical information when it comes to analysis. In this chapter we move away from the construction and management of data in a GIS to actual use of the information. To best understand information in the form of numbers, we must describe geographic data in methodical and quantitative ways, that is, with well-understood statistics. If this was as far as GIS went, however, there would be few advantages to GIS compared to any of the major computer statistical packages available to scientists.

What makes analysis within a GIS different is that the attribute data have established links to maps. Any statistic we can think of to describe the data then automatically has geographic properties and as a result can be placed on a map for visual processing. As we will see later in this chapter, the situation is more fruitful than that, because we can use the geographical properties described in Section 2.4 for statistical inquiry as well. This means that in addition to answering the question "Where?" as far as the features are concerned, we can also ask "Why is it there?" We can come up with some definitive answers to these questions and display the answers to the questions and analyses as maps. As this chapter shows, this can give the user an amazing amount of power when GIS analysis is brought to bear on a problem.

This chapter begins by covering how it is that attribute-type data are described. The visual description of the histogram and the number descriptions of the average and difference from the average are covered. In addition, these simple measures are described in terms of their spatial attributes when the two spatial dimensions or coordinates are used as the numbers under description. As shown, map statistical description leads to an initial ability to place onto a map what the numbers demonstrate. The average (mean) and difference from the average (variance) both have visual and geographical meaning.

6.1.1 Describing One Attribute

To revisit the basic structure of a database, review the beginning of Chapter 2, and the discussion of a database consisting of baseball cards. As we saw, all data can be thought of as structured in a table. The rows of the table are records, and the columns are attributes. Each record for each attribute contains a value, and that value falls into a set of types of data, text, numbers, and so on. For example, the values of an attribute called "date" for the record "357" could be "7/2/2002."

The value here is really three numbers (a month, a day, and a year) but for the purposes of the database is considered as text. An additional condition is that to be a GIS database, at least one of the attributes must be a link to a map. At the most basic level, the eastings and northings of points can fit as two attributes. We start with this simple case, but as we have also already seen, geographic data can be for features that are points, lines, and areas and combinations of them.

Table 6.1 is a listing of a simple spatial data set. It contains a listing of readings taken using two portable GPS receivers, a Garmin GPS40 and a Garmin etrex personal navigator, by University of California, Santa Barbara Geography student Westerly Miller along the coast of California between Goleta and Malibu in May of 2002 (Figure 6.1 and Table 6.1). The structure of Table 6.1 closely resembles a geographic database. The attributes are information about each reading, a description of the place where the reading was taken, which unit was used to take the reading, the latitude and longitude

TABLE 6.1: Sample GPS Data

	LOCATION	Garmin eTrex Personal Navigator					Garmin GPS 48				
		Longitude(DM)	atitude (DM)	Longitude (DD)	Latitude (DD)	Elev (ft.)	Longitude(DM)	atitude(DM)	Longitude (DD)	Latitude (DD)	Elev (ft.)
1	SLIP 1J33, S.B. HARBOR, CA.	-119.41396	34.24263	-119.6899333	34.4043833	19.0	-119.41398	34.24264	-119.6899667	34.4044000	39.0
2	SLIP 4A48, S. B. HARBOR, CA.	-119.41496	34.24389	-119.6916000	34.4064833	5.0	-119.41496	34.24389	-119.6916000	34.4064833	23.0
3	YACHT CLUB, S. B. HARBOR, CA.	-119.41550	34.24209	-119.6925000	34.4034833	-7.0	-119.41551	34.24210	-119.6925167	34.4035000	47.0
4	ROB'S DRY DOCK, S. B. H., CA.	-119.41493	34.24292	-119.6915500	34.4048667	-26.0	-119.41495	34.24291	-119.6915833	34.4048500	28.0
5	SLIP 3B30, S. B. HARBOR, CA.	-119.41555	34.24367	-119.6925833	34.4061167	-18.0	-119.41553	34.24366	-119.6925500	34.4061000	18.0
6	S. B. MARITIME MUSEUM	-119.41622	34.24253	-119.6937000	34.4042167	-4.0	-119.41620	34.24253	-119.6936667	34.4042167	1.0
7	BREAKWATER, S. B. HRBR., CA.	-119.41289	34.24292	-119.6881500	34.4048667	23.0	-119.41286	34.24286	-119.6881000	34.4047667	36.0
8	SEA LANDING, S.B. HARBOR, CA.	-119.41463	34.24478	-119.6910500	34.4079667	-16.0	-119.41461	34.24480	-119.6910167	34.4080000	12.0
9	BEACH HOUSE, S. B., CA.	-119.42896	34.23774	-119.7149333	34.3962333	5.0	-119.42900	34.23773	-119.7150000	34.3962167	119.0
10	LEADBETTER BEACH, S. B., CA.	-119.42149	34.23949	-119.7024833	34.3991500	6.0	-119.42148	34.23951	-119.7024667	34.3991833	45.0
11	STATE ST. FOUNTAIN, S. B., CA.	-119.41338	34.24721	-119.6889667	34.4120167	27.0	-119.41338	34.24724	-119.6889667	34.4120667	50.0
12	END STEARN'S WHARF, S. B., CA.	-119.41102	34.24488	-119.6850333	34.4081333	27.0	-119.41399	34.24491	-119.6899833	34.4081833	45.0
13	CAMPBELL HALL, U. C. S. B., CA.	-119.50727	34.24972	-119.8454500	34.4162000	93.0	-119.50729	34.24972	-119.8454833	34.4162000	100.0
14	01 ELLISON HALL, U. C. S. B., CA.	-119.50703	34.24940	-119.8450500	34.4156667	27.0	-119.50698	34.24940	-119.8449667	34.4156667	74.0
15	06 ELLISON HALL, U. C. S. B., CA.	-119.50717	34.24935	-119.8452833	34.4155833	52.0	-119.50717	34.24938	-119.8452833	34.4156333	100.0
16	BIKESHOP, U. C. S. B., CA.	-119.50962	34.24864	-119.8493667	34.4144000	26.0	-119.50961	34.24869	-119.8493500	34.4144833	79.0
17	BASE STORKE TWR, U. C. S. B.,	-119.50905	34.24747	-119.8484167	34.4124500	82.0	-119.50901	34.24748	-119.8483500	34.4124667	76.0
18	TOP STORKE TWR, U. C. S. B.,	-119.50895	34.24762	-119.8482500	34.4127000	-66.0	-119.50905	34.24771	-119.8484167	34.4128500	66.0
19	RINCON BEACH, CA.	-119.28142	34.22571	-119.4690333	34.3761833	7.0	-119.28142	34.22571	-119.4690333	34.3761833	48.0
20	IN 'N OUT BURGER, VENTURA, CA	-119.16354	34.16099	-119.2725667	34.2683167	15.0	-119.16353	34.16101	-119.2725500	34.2683500	42.0
21	PT. MUGU, CA.	-119.04804	34.05978	-119.0800667	34.0996333	42.0	-119.04805	34.05978	-119.0800833	34.0996333	97.0
22	VENTURA COUNTY LINE, CA.	-118.57759	34.03161	-118.9626500	34.0526833	75.0	-118.57756	34.03165	-118.9626000	34.0527500	75.0
23	TRANCAS MKT., MALIBU, CA.	-118.50567	34.01887	-118.8427833	34.0314500	15.0	-118.50573	34.01890	-118.8428833	34.0315000	54.0
24	ZUMA BEACH, MALIBU, CA.	-118.48108	34.00755	-118.8018000	34.0125833	0.0	-118.49077	34.00757	-118.8179500	34.0126167	-138.0
25	LITTLE DUME BEACH	-118.47648	34.00642	-118.7941333	34.0107000	4.0	-118.47639	34.00647	-118.7939833	34.0107833	89.0
26	PARADISE COVE, MALIBU, CA.	-118.47469	34.00876	-118.7911500	34.0146000	35.0	-118.47471	34.00880	-118.7911833	34.0146667	69.0
27	POINT DUME, MALIBU, CA.	-118.48437	34.00185	-118.8072833	34.0030833	142.0	-118.48453	34.00197	-118.8075500	34.0032833	60.0
28	WESWARD BEACH, MALIBU, CA.	-118.48624	34.00272	-118.8104000	34.0045333	50.0	-118.48623	34.00272	-118.8103833	34.0045333	99.0
29	ZUMA SUSHI REST. MALIBU, CA	-118.48819	34.01139	-118.8136500	34.0189833	118.0	-118.48816	34.01139	-118.8136000	34.0189833	114.0
						758.0					1567.0

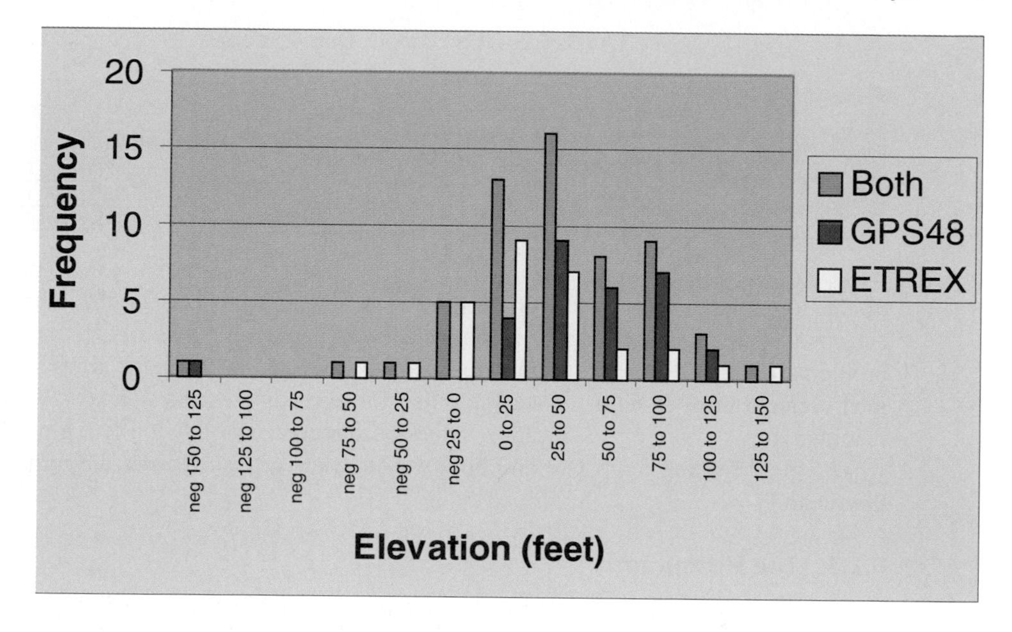

FIGURE 6.1: Histogram of the GPS elevation data from Table 6.1.

of the position as shown by the GPS receiver, and the elevation of the location. Note that the elevations are in feet, the units of data collection, and that the original GPS latitude/longitude readings in degrees-minutes format have been converted (using a spreadsheet) to decimal degrees.

6.2 STATISTICAL ANALYSIS

A first question that might be asked about the database is, What are the extremes of the data? Data extremes are simply the highest and lowest values for all records for one attribute; that is, the high and low within a single column. Table 6.2 lists these extremes and the range, defined as the highest value minus the lowest value, stated in the units of the attribute. The range of latitudes was 0.4013 degrees for the etrex, and 0.4011 for the GPS40. The range of longitudes was 0.8988 degrees for the etrex and 0.8988 for the GPS40, different only at the seventh decimal place. There was more variation in the elevation readings, which are less accurately determined with hand-held GPS units. They varied from 208 feet (63.4 m) for the etrex to 257 feet (78.3 m) for the GPS40. Both units showed elevations below sea level, highly unlikely locations for data collection, and therefore clearly errors.

We concentrate first on the elevation attribute. A first descriptive question about the data beyond the ranges is: What are the elevations of the point that were sampled? Even though most of the readings were taken along the coast, the values for the elevation range considerably. Some of the data are clearly poor readings, outside the range we would normally expect and perhaps due to the positions of the satellites or some other factor. For example, the base of the Storke tower on the UCSB campus was shown as at elevation 82 feet by the etrex and at 76 feet by the GPS48. Yet the elevation

TABLE 6.2: GPS Sample Data Extremes

Data Extremes GPS Information							
	etrex			GPS48			
	Longitude	Latitude	Elev	Longitude	Latitude	Elev	
Minimum	-119.6899333	34.0030833	-66.0	-119.6899667	34.0032833	-138.0	
Maximum	-118.7911500	34.4043833	142.0	-118.7911833	34.4044000	119.0	
Range	0.8987833	0.4013000	208.0	0.8987834	0.4011167	257.0	

of the top of the tower was −66 feet for one unit and 66 feet for the other. These curiously symmetrical readings were both clearly quite wrong! Quite likely, the reception problems of the GPS system had something to do with it, but there is also clearly a measurement error in terms of accuracy that far exceeds the precision of the elevation reading. How can these bad elevation values be screened out? Obviously we need to see what a good reading looks like and how it can be distinguished from the remainder of the readings.

6.2.1 The Histogram

The diagram in Figure 6.1 is called a histogram. It is a plot of the data from Table 6.1 in the elevation columns. On the horizontal, bottom axis, the histogram shows the values of elevation, grouped into categories by increments of 25 feet. The actual data in the table are in feet, given to the nearest foot, which is the precision of both of the GPS receivers. On the left hand, vertical axis, we show how many elevation records fall within each elevation range. The number in each group is called that group's frequency.

FIGURE 6.2: UCSB geography student Westerly Miller collecting the data shown in Table 6.1. Point 9, Shoreline Drive, Santa Barbara, CA. (Used with Permission.)

The sort of histogram shown is very common in GIS and in many other sciences. There is a cluster of values around the middle and a rapid drop-off as we move toward very high and very low values. In terms of elevation, clearly there are two sources of the variation in the readings, the real differences in elevations at the points, and the error in GPS measurements. Since none of the readings were taken underwater (although we may have been below the datum used, NAD83), clearly the error group is a major part of the variation. If we had a histogram showing only the error, then we might expect that the errors were random and would lead to a symmetrical histogram, peaking at the average or mean value. Although we took readings at only 29 points with the GPS receivers, if we had taken millions of them, it is likely that the shape of the error histogram would have become symmetrical about a central axis through an average value. This distinctive shape, called a bell curve because of its shape, is known in statistics as the normal distribution. It is the scatter of values that we get when we take measurements of a number, expecting a single value, but knowing that error is anticipated, that the error is just as likely to give a high measurement as a low one, and that the amount of error is not systematic, such as a misreading of a measurement scale. In other words, the error is random and unpredictable. Many real-world distributions show this form.

Other types of error distributions are shown in Figure 6.3. The histogram could be skewed to one side, which would mean that it is more likely to over- or underestimate a number than the opposite. The values could be equally even and dispersed about the average, implying that the measurement is perfectly accurate or at least perfectly consistently wrong. The error could be the same everywhere along the line, implying that no value is a better reading than any other. We could get a group of errors, perhaps occasional misreadings of a 3 as a 5 and the occasional dropping of a decimal place. This would give us a multi-peaked histogram, as shown in Figure 6.3.

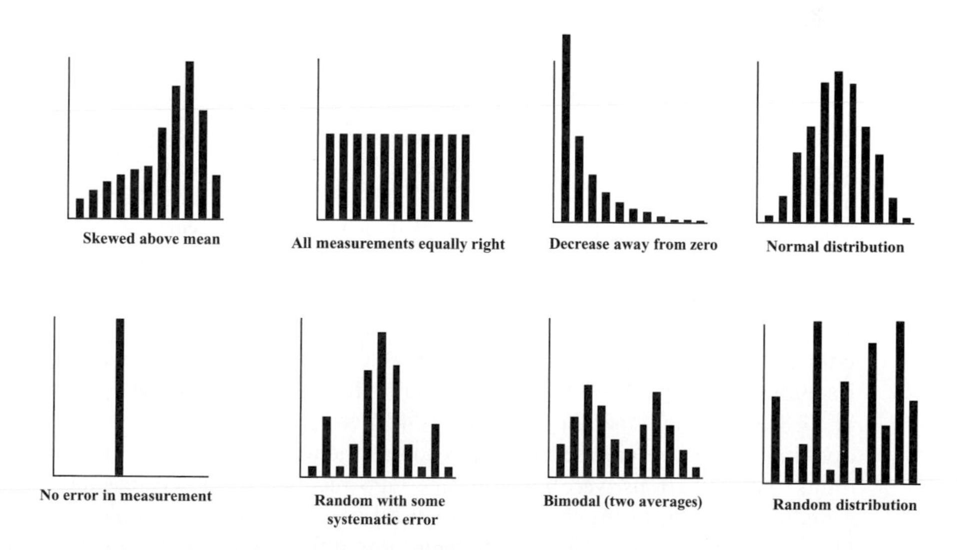

FIGURE 6.3: Some possible alternatives to the normal distribution with a large number of observations.

6.2.2 The Mean

If errors are consistent, we can correct for them. For example, we could reason that all the negative elevations in Table 6.1 were in error. We could safely eliminate these seven numbers and probably correctly believe them to be wrong. We do not have this option if we have only one number or reading—we have no choice but to use it! If we have two readings only, and they disagree, we would probably average them. If we had three readings and they disagreed, we could average them, reject one reading that was obviously wrong (varies by too much), or average the two readings that most closely agree with each other.

The more numbers we have, the more we can see what the typical amount of variation is, that is, how corrupted are the readings by a random amount of error. If this is the case, as in Figure 6.1, then we can go ahead and average the numbers, or at least give expected amounts of error. There are alternatives to averaging. For example, a simple representative value for a group of records can be selected by sorting the elevations by height, which the GIS database manager can do (Table 6.3), and then taking the value of the middle reading. This value is called the median. This works fine for an odd number of elevation readings; the center comes out exactly. An advantage is that this is a "real" typical value because it is an actual part of our data set. If the attribute in the database was, for example, state average salary for GIS professionals in dollars, we could pick the middle or median state and compare our own state to it. This is not so simple if we have an even number of records, in this case 50. We have to take the two center values and average them, losing the attachment to a single data record.

In the case of the GPS data we have a total of 58 elevations, an even number. Table 6.3 sorts these values, and highlights the two at the center. These are 39 and 42 feet, which average to $(39 + 42) \div 2$ or 40.5 feet. The median elevation of the data in Table 6.1, therefore, is 40.5 feet. The median is called a measure of central tendency, because it gives us a value which is descriptive of the group of values as a whole. A nice feature of the median is that it is unaffected by the extreme values at the limits of the distribution. Another is that it is made up of values that actually occur in the data (39 and 42), and is not an abstract number.

Another measure of central tendency is the average, known in statistics as the mean. It is computed rather simply, by adding together the values for all the records and dividing by the number of records. We may have to leave some of the numbers out of the averaging process. For example, we may choose to exclude the negative elevations from our calculations. These values will then be called missing values. It is important not to confuse missing values with zero, which is not a missing value, but a legitimate reading of "0.0."

The sum of the elevations for Table 6.1, shown at the bottom of the elevation column in Table 6.3, is 2325 feet. This gives us an average or mean elevation of 40.086 feet. Excluding the negative elevations gives a higher sum of 2645 feet for a higher average of 51.863 feet (note that the median was 40.5 feet). Because the mean was calculated by dividing, it is more precise than the readings themselves; that is, it has more significant digits. The elevations were given by the GPS receiver to the nearest foot, and we have rounded the mean to the thousandth or third decimal place. We could also say that it is more accurate, because the measurement now reflects many readings, not just one. The more readings we take at each location (we took two), the more accurate the location it should become, unless there is some systematic error of which we are unaware (say, the

TABLE 6.3: Variance in the Elevation Data

All GPS Elevations		Elevation-Mean	(elev - mean)^2
-138.0		-178.086	31714.697
-66.0		-106.086	11254.283
-26.0		-66.086	4367.387
-18.0		-58.086	3374.007
-16.0		-56.086	3145.663
-7.0		-47.086	2217.111
-4.0		-44.086	1943.594
0.0		-40.086	1606.904
1.0		-39.086	1527.732
4.0		-36.086	1302.214
5.0		-35.086	1231.042
5.0		-35.086	1231.042
6.0		-34.086	1161.870
7.0		-33.086	1094.697
12.0		-28.086	788.835
15.0		-25.086	629.318
15.0		-25.086	629.318
18.0		-22.086	487.801
19.0		-21.086	444.628
23.0		-17.086	291.938
23.0		-17.086	291.938
26.0		-14.086	198.421
27.0		-13.086	171.249
27.0		-13.086	171.249
27.0		-13.086	171.249
28.0		-12.086	146.076
35.0		-5.086	25.870
36.0		-4.086	16.697
39.0		-1.086	1.180
42.0		1.914	3.663
42.0		1.914	3.663
45.0		4.914	24.145
45.0		4.914	24.145
47.0		6.914	47.801
48.0		7.914	62.628
50.0		9.914	98.283
50.0		9.914	98.283
52.0		11.914	141.938
54.0		13.914	193.594
60.0		19.914	396.559
66.0		25.914	671.525
69.0		28.914	836.007
74.0		33.914	1150.145
75.0		34.914	1218.973
75.0		34.914	1218.973
76.0		35.914	1289.801
79.0		38.914	1514.283
82.0		41.914	1756.766
89.0		48.914	2392.559
93.0		52.914	2799.870
97.0		56.914	3239.180
99.0		58.914	3470.835
100.0		59.914	3589.663
100.0		59.914	3589.663
114.0		73.914	5463.249
118.0		77.914	6070.559
119.0		78.914	6227.387
142.0		101.914	10386.421
2325.0		0.000	**129618.569**
40.086	Mean		

elevations were on the wrong datum, or the earth's fit to the ellipsoid was pretty poor at this place!).

To be accurate, we must test the value against an independent source of higher fidelity and reliability than these measurements. On a map, such an elevation could be found at a bench mark elevation or by using an extremely accurate GPS system with differential correction to achieve a higher accuracy level. For the sample data, we could plot them on a digital raster graphic version of the United States Geological Survey quadrangle maps at 1 : 24,000, and read off the elevations from the contours. This may, however, be an independent source of a lower authority level because the contour map would require interpolating values between contours.

The error still requires understanding, and this usually comes from a computation of the amount of variance in the data. Normalizing or standardizing this variance allows us to get a number that not only characterizes the amount of error, but also allows us to compare the error between data sets, and to map the error using the GIS.

6.2.3 Variance and the Standard Deviation

Calculation of the standard deviation is a little more complex than the mean but can be seen in action in Table 6.3. In this table, we have extracted the elevation columns from Table 6.1 and added two columns for newly computed information. First, we can calculate the mean elevation again. The line at the end of the records is the total of the elevations in feet (2325). This is divided by 58 to get the mean of 40.086 feet.

In the next column, the mean we just calculated has been subtracted from each elevation. In real terms, some of the GPS elevations were too high and some too low. When we subtract the average, some (too high) give positive remainders, and some give negative (too low). Note that the two values used to compute the median were either side of the zero remainder value, showing that the mean and the median are indeed both measures of central tendency. The remainders, if summed as we did for calculation of the mean, will add up to zero. The effect of rounding may produce a nonzero result. This is not a good measure of difference from the average! We obviously need to get rid of the plus and minus. We need just the difference from the mean in feet. One easy way to do this, and to make numbers a long distance away from the mean stand out, is to square the differences, that is, multiply them by themselves. Any negative value times itself becomes a positive value, which solves our problem.

The squared differences are shown in the third column in Table 6.3. These numbers can now be added up, although as squares they can be quite large numbers. The sum, shown at the bottom of the column (129,618.569), is called the total variance, because it is a single number that shows by how much the values as a whole disagree. If you doubt this, imagine that all of the elevations were the same. The mean would be the same value, and every identical value minus itself would equal zero. We would then square the number zero 58 times, and add up the results, still giving us zero. Thus the variance measures how much the numbers disagree with each other.

The only problem with the total variance is that it becomes higher and higher as more and more readings are taken. To remove this effect, we can divide by the number of records, to average the variance among the records. In reality, we divide not by the number of records but by this number minus one. This is called the degrees of freedom of the value. If we had only one record, the mean would have to be the same as this record's value, and the variance would be zero. We get variation only when the second value is measured, so we say that with two readings we have one degree of freedom. Dividing 129,618.569 by 57 (58 – 1) gives a value of approximately 2274.01.

In addition, we can take the square root of the result so that the units are back as feet rather than as square feet. This number (47.687), standardized in this way, is called the standard deviation. This number is in the same units as the values of the records, in this case feet. Better than just that, it is the average amount by which readings differ from the average and so can be called an expected error of any given reading. Of course, the readings are just as likely to be above the average as below. We could say that the elevation within the point sample is the mean (40.1 feet) plus or minus the expected error of 47.7 feet, and so is most likely to lie between −7.6 feet and 87.8 feet. These limits are called the error band or margin of error.

6.2.4 Statistical Testing

The final step that we can put these elevation values through is to do statistical tests. Important for this stage is the idea that the measurements we took with the GPS receiver are just a few of a large number of possible readings. In statistical jargon, the entire possible set of GPS readings is called a population, and the actual numbers in Table 6.1 are a sample from this population. For most statistical purposes, the sample is considered as only a tiny proportion of the population as a whole. For example, we could consider our sample as representative of all coastal area elevations in Southern California.

The purpose of drawing this division is that we can use the sample's mean and variance to make an estimate of those of the population, and then use the two interchangeably. The reason we divided by 57 and not 58 to get the variance is because we want a population, not a sample estimate. It is useful to us, for example, to know the standard deviation of the elevation values at the location given in Table 6.1, because we can use it to get an estimate of the difference in elevation between the two GPS receivers compared with the overall measurement capability of GPS as a way to record elevation in the sample.

We can use a statistical model of the bell curve, called the standard normal distribution, to estimate how likely any given measurement of the elevation is to be correct. This distribution, provided in most statistical textbooks, allows us to look up the standard deviation and the number of records and to estimate the odds against getting this elevation measurement given the tabulated standard deviation. The numbers in the statistical table are the amount of area beneath the standard normal curve that corresponds to probabilities. This is one way that we can determine whether the largest and smallest measured elevations are reasonable.

For example, if our mean elevation is 40.01 feet and the standard deviation is 47.2 feet, what is the chance of getting a GPS reading of −138 feet? The reading of negative 138 feet is 178.1 feet below the mean, or $178.1 \div 47.2$ of a standard deviation. This number of standard deviations is called a Z-score. In this case it is 3.77. Looking this up in the statistical table shows us that 0.4999 of the curve lies between the mean and this value, and 0.0001 lies beyond it. The normal curve is symmetrical about the mean. These values are shown in Figure 6.3.

The standard normal distribution table tells us that for our 58 readings, the likelihood of a number falling at random between the mean and this value is 49.99%. This is obviously large compared to the 0.01% chance of falling lower. So we could argue that there is a strong probability that this reading is in error. Often, a reading is rejected as highly unlikely if it is more than two standard deviations away from the mean.

Another way that these numbers can help in analysis is to answer the following question: Is there a difference in measured elevation between the two makes of GPS receiver? It would be quite simple to calculate the mean separately for each of the two types of receiver. In the GIS database we could select out those attributes based on their "GPS-type" attribute. Similarly, we can also recalculate the standard deviations for the two data sets separately. When we do this, we get for the Garmin eTrex a mean of 26.14 feet and a standard deviation of 43.71 feet, and for the Trimble GPS, a mean of 54.03 feet and a standard deviation of 48.12 feet.

Now we can return to the question of whether the two values differ from each other. If they were the same, they would have the same means exactly. However, because the two sample means were estimates of the real elevations using 58 points, we have to compare the difference between the two means against the estimated population standard deviation. This is called a test of means. As the number of samples is different, we have to estimate the standard deviation of the overall mean as the square root of the two normalized variances added together. For the eTrex, the sum of squared differences from the mean was 53,497.36, and dividing by the 28 (29 minus 1) samples gives us a normalized variance of 1910.62. Similarly, for the GPS48, the sum of squared differences from the mean was 64,836.8, and dividing by the 28 (29 minus 1) samples gives us a normalized variance of 2315.6. Adding these numbers and taking the square root of the results gives almost exactly 65 feet.

This number is the estimated standard deviation of the mean itself. This implies that we can test the difference between the means against it, computing a Z-score and a probability as before. The two means of 26.14 feet and 54.03 feet differ by 27.89 feet. Dividing by the calculated standard deviation of the mean gives a number of 0.429. This value cannot be looked up in the regular statistical table but can be compared on a scale called Student's T. We also have to look up the T value by its number of degrees of freedom, which here is the total of the two samples minus two, or 50. This is another probability table. At the 5% confidence level, T would have to be about 1.675 or greater to reject the hypothesis that the two means were the same.

At only 0.429, then, we cannot prove statistically that there is any difference between the elevations measured by the two GPS receivers. Thus, even though eTrex has a lower standard deviation, and the GPS48 seems to often give readings that appear high for a coastal area, neither gives elevation values superior to those of the other. The conclusion, then, is that the two GPS receivers are similar in capability, with an expected error in estimating elevations of their "population" average of 65 feet. This is hardly what could be described as cartographic precision. The results may also reflect the influence of a few large errors or blunders. Fortunately, the receivers are better at latitude and longitude.

The standard deviations are also similar, but differ by 4.41 feet. These can be thought of as the accuracy of the elevation for the two GPS receivers. The slightly better (lower) value for the eTrex may be due to several things: the technical characteristics of the receiver, the fact that the receivers have different precision, even perhaps an interaction between the two receivers. Statistics can also allow us to test whether the two standard deviations are significantly different from each other, or whether they could be different because of random causes. This, however, we leave for later, or for a statistics class, because we have yet to look at the geographic, mappable characteristics of these numbers.

6.3 SPATIAL DESCRIPTION

In the preceding section we looked at how to describe a single attribute statistically. The first and most significant factor in dealing with spatial data is that there are at least two spatial measurements, an easting and a northing. We could summarize spatial description, as describing two attributes simultaneously.

In the simplest and most basic way, we can duplicate the attribute descriptions above for the locational data to give spatial descriptions. In this case, we can treat the two separate parts of the coordinates, the eastings and the northings, as if they are each a single attribute, which indeed they are. Just as we began the discussion of describing the values of a single attribute by discussing the concept of a minimum and a maximum value for an attribute and the concept of a range, when the attributes describe coordinates, a first point is described by the minimum easting and the minimum northing, and a second point describes the corresponding maxima. The two points define a rectangle, whose two side lengths are the ranges in easting and northing, respectively, and that encloses all the points.

This is called the bounding rectangle of the points. It can be found by simply sorting the records by easting, and taking the first and last record, and then repeating for the northing. A bounding rectangle for the points in Table 6.1 is included in Figure 6.4.

6.3.1 The Mean Center

In much the same way that we calculated means and standard deviations separately for the two GPS receiver's elevations, so also were they calculated for the latitudes and longitudes. These were first translated into decimal degrees, then summed and divided to find the average latitude and longitude for the eTrex (34.2840575°N, 119.4430961°W)

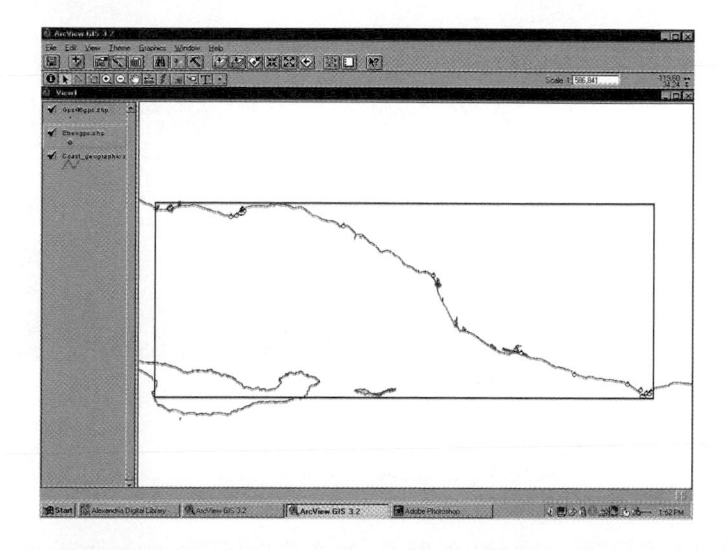

FIGURE 6.4: Bounding rectangle of the GPS data, showing the spatial extremes of the data set.

and for the GPS48 (34.2840879°N, 119.4438299°W). The result of the two means is itself a point, with both a real geographic location and a special geographic name, the mean center. This point is also sometimes called a centroid, a point chosen (in this case statistically) to represent a geographic distribution. Although the GPS data are a set of points, lines and area features can also have a centroid, selected in any one of several ways.

Figure 6.5 is a photograph of a place in Rugby, North Dakota, that claims to be the geographic center of North America. Although this is a fascinating monument and the nearby diner is probably heavily dependent on its visitors for its food business, it should be quite obvious that, unlike for a set of points, an entire continent could have any number of centroids! For example, this could be the point farthest from any coastline, the center of all of the points making up the coastline, the center of the bounding rectangle, or the center of the largest circle that can be drawn inside North America.

The World Almanac lists the geographic center of North America not in Rugby, but in Pierce County, North Dakota, 10 km west of Balta (48°10′ N, 100°10′ W). The mean

FIGURE 6.5: The monument in Rugby, North Dakota. Photographs by Colette Flanagan. (Used with permission.)

center calculation would also change depending on the map projection, datum, and ellipsoid. Judging from the flags seen on a visit to the site, it is not even clear whether Rugby's definition of North America includes Mexico, Alaska, Greenland, or Hawaii. Obviously, one place is as good as any for this type of monument. It would be interesting to know whether the diner predates the monument, or vice versa.

6.3.2 The Standard Distance

Figure 6.6 is a plot of the GPS points in map space, with a different colored symbol for each of the two GPS receivers. Looking at this map, is it possible to see any difference between the two sets of measurements? Since the overall spread of the points exceeds the differences in location between the two receivers, it is very hard to say, even on the zoom of Santa Barbara harbor. Instead, we can compare the distributions statistically, by examining the standard deviations in the easting and northing directions, in this case in latitude and longitude.

Imagine the line between the two GPS locations for each point, with all eTrex points drawn on top of each other. We could look at the bearing of these lines, "rays" stretching out between the two readings for each point. We would expect the bearings and the lengths to be random, but the average length would now give a mean with a real meaning, the expected average distance difference between the two receivers. This can also be calculated from the standard deviations in the easting and northings, calculated as the square root of the sum of the squared distances in the two directions.

Converting to a single number like this is termed normalizing. This parameter, called the standard distance, is a map equivalent of the standard deviation. Again, are there major differences between the two GPS receivers? The standard distance is at last truly a measure of the agreement between the GPS receivers. In degrees, the standard deviation in longitude was 0.003178 and in latitude was zero. Thus the two receivers differed only in longitude, by an amount that converts to a surprisingly high 352.5 meters or 1156 feet. What could have caused this difference? Calculating for each receiver separately, and using standard length tables for the length of a degree at different earth locations, the mean centers are off by 81.4 meters in their easting but only 2.8 meters in

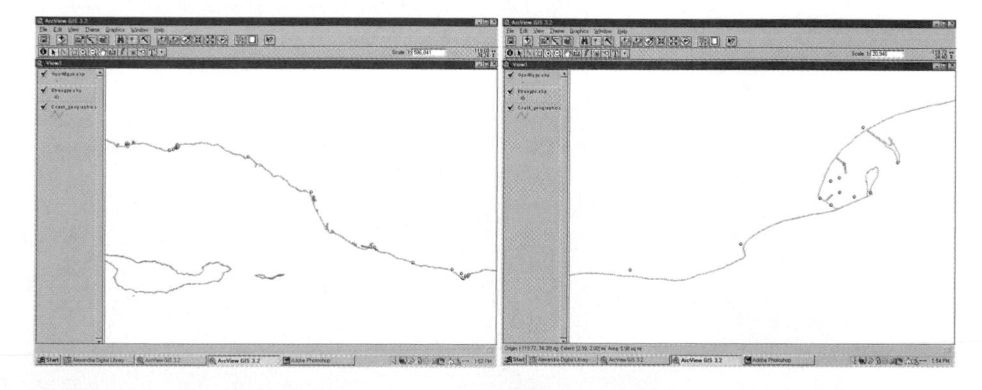

FIGURE 6.6: ArcView display of the data in Table 6.1. Both the GPS48 and the eTrex Garmin GPS receiver data are plotted, but are so close spatially that they are not separable at this scale. Area displayed is Santa Barbara to Malibu, California. Right is a zoom of Santa Barbara harbor area.

their northing. What has caused the difference: chance, systematic error, or some other cause? Just by asking this question and having some data and a method to answer it objectively, we are already doing spatial analysis.

While statistics are useful in demonstrating that an error is present, and that it has an impact on the aggregate statistical descriptors, the GIS can help us to isolate exactly which readings have caused the problem. Figure 6.7 shows a map created in ESRI's ArcView 3.2 made by bringing data from the spreadsheet created to store the GPS data as a table and mapping it. The value plotted for each point is the difference in latitude and longitude for each point, squared to get rid of negatives, divided by the size of the sample, and added for latitude and longitude. The values were converted to meters using the same tables as above, and the square root of the sum taken. This is then a map of the magnitude of the total spatial discrepancy between the two GPS receivers. Two points clearly stand out, point numbers 12 and 24. These two points alone account for almost all of the error in the data. Without them, the two receivers seem to be not only quite accurate, but also in agreement. As implied in the discussion, the largest errors are in longitude. What caused them? It seems unlikely that the errors were caused by the random errors inherent in the GPS receivers, which are of a far lesser magnitude. Could numbers have been inverted (say a 345 became a 543), digits transliterated (a "5" read as an "8"), or perhaps a digit was left out when the values were entered into the spreadsheet?

Overall, it is clear that the statement at the start of the next chapter is correct, that a map is a set of errors that have been agreed upon! On the other hand, the mapping

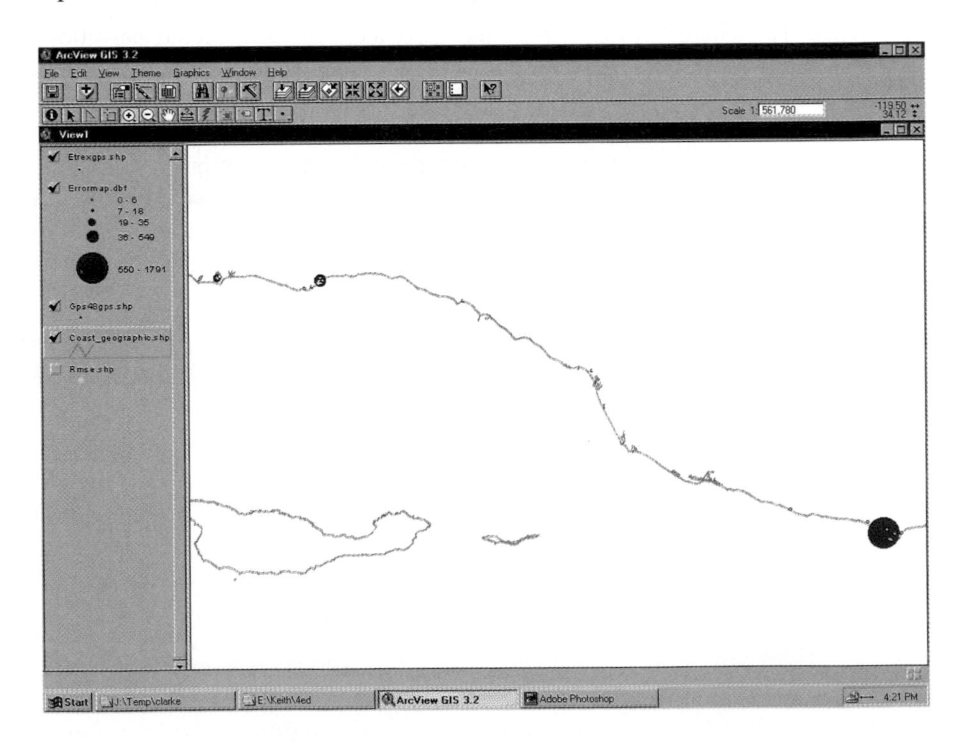

FIGURE 6.7: ArcView map of the GPS data with root mean squared error between the two GPS devices, units rescaled into meters.

capability of the GIS and the power of statistical and spatial analysis are such that we can use our same GIS tools to find errors in our data. However "pure" we believe our data to be, there are always errors. It is often said that eliminating the last 1% of error costs as much as eliminating the other 99%. It is best, therefore, to be both aware of errors and to be able to describe them as far as your own data are concerned, especially spatially.

6.3.3 Geographic Features and Statistics

In Chapter 2 we met the idea that geographic features can be classified into points, lines, and areas by their dimensions on the map. Describing each of these can lead to measuring spatial properties directly from the digital files containing the geocoded representations of the features. We started Chapter 6 with a set of points, the GPS example, because points are the easiest type of feature to describe. Although we have so far used quantitative measures to describe geographical features, many arrangements of features are described verbally.

For example, points are sparse, uneven, random, regular, uniform, scattered, clustered, shotgun, or dispersed. Patterns are regular, patchwork, repetitive, or swirling. Shapes are rounded, oval, oblong, drawn-out, or resemble Swiss cheese. The challenge is to find numbers that say the same thing. The bounding rectangle, the mean center, and measures such as the standard distance can provide excellent descriptors of points, although more complex measures are obviously needed for the higher-dimension features.

Lines have a number of points, a line length, a distance between the start and end points (or nodes), the average length of one of the line segments, and a line direction. A useful description of a line could be the ratio of actual line length divided by the start-to-end node length, what could be called a straightness index. For a straight line, this measure would be 1. For the Mississippi River this would be a far larger number. The direction could be taken as the clockwise angle bearing from north (the overall "trend" of the line), although this value would have a big variance along a curved or wiggly line, too.

Areas are even more difficult to describe. Simplest to measure with a GIS are the area in square meters, the length around the boundary, the number of points in the boundary, the number of holes, and the elongation, taken as the length of the longest line axis divided by the axis at 90 degrees to it. We can also divide the area of the bounding rectangle into the area, a space-filling index with a maximum value of 1. If the area has neighboring areas, we could count them or determine the average length of the area's boundary shared in common with a neighbor.

Not all of these numbers are easy to compute with a GIS. Sometimes a multiple-step process must be used and information created or computed in the attribute database and passed back to the map for display. Almost every GIS has a compute command that allows the performing of math operations like

```
COMPUTE ATTR5 = (ATTR2 + ATTR3)/ATTR4
```

Each new measure, however, can be an intermediate step in the computation of another statistic. For example, we could measure the lengths of streams by district in kilometers, measure the area of the district in square kilometers, and then create a new attribute in the database of the stream density in meters of stream per square kilometer. This might be an interesting data value to then map by district in the GIS. Computations are so common using areas of polygons that many GIS packages compute and save areas when they are created, whether they are wanted or not.

6.4 SPATIAL ANALYSIS

Numbers that describe features are useful, but as we noted in Chapter 2 the purpose of geographic inquiry is to examine the relationships between geographic features collectively and to use the relationships to describe the real-world phenomena that the map features represent. The geographic properties we noted in Figure 2.15 were size, distribution, pattern, contiguity, neighborhood, shape, scale, and orientation.

Each spatial relation begs three fundamental questions: (1) How can two maps be compared with each other?; (2) How can variations in geographic properties over a single area or GIS data set be described and analyzed?; and (3) How can we use what we have learned using the analysis to explain and therefore predict future maps of the geography in question? The third question may be as simple as selecting the best route from A to B on a map, or as complex as modeling the future growth of cities based on their size, shape, and development over time. GIS gives us the capability of doing both, and anything in between. In terms of comparing maps, a simple way is to bring multiple maps into coregistration and then merge their themes to make a composite. This is what is meant by map overlay analysis. An example of map overlay will follow a first discussion of spatial models and how GIS adds to their construction, examination, and use.

6.4.1 U.S. Gender Ratios: An Example

A full set of descriptive statistics for all of the properties listed here is beyond the scope of this book. Instead, two geographic analysis problems will be covered, starting with a simple geographic distribution, and ending with some speculations about prediction.

Consider the data in Table 6.4 and the accompanying maps shown in Figure 6.8. The value shown is the gender ratio, defined as the number of males in the population per 100 females. States with numbers over 100 have more men than women, and those states lower than 100 have more women than men. The numbers are selected from the 2000 census of population for the United States. After selecting the data, an obvious first step is to map the distribution. This map, using a method called *choropleth mapping*, in which states are shaded by gender ratio in groups of values, is shown as Figure 6.8 on the upper left.

Remember that looking at the map, using the full power of the human visual system and just plain thought, is every bit as powerful as any spatial analysis method. This critical step, *plot* and *look*, should never be skipped in GIS analysis, at the risk of using complex methods to prove the geographically obvious. So, first, examine the upper left map in Figure 6.8 carefully and then move on to the next section. How can this distribution be described? The data are ratio values for areas. Each geographic property suggests a question about the data, with a few of these listed below.

SIZE What are the high and low values of the gender ratio? Do large states have characteristic values? What about small states?

DISTRIBUTION Is there a regional difference between numbers? Does the West have all the low values and the East the highs? Are there clusters of similar values?

TABLE 6.4: Ratio of Males to Females Times 100 for the Lower 48 United States. Source 2000 Census

State	Males/100 Females	Longitude	Model	Residual
AL	89.00	-86.2833	95.6930	-2.3930
AR	92.50	-92.2667	96.5535	-1.2535
AZ	93.00	-112.0000	99.3913	0.3087
CA	93.10	-121.5000	100.7575	-1.4575
CO	93.30	-104.9833	98.3822	3.0178
CT	93.40	-72.6667	93.7348	0.1652
DC	93.40	-77.0000	94.3580	-5.3580
DE	93.40	-75.5000	94.1423	0.2577
FL	93.80	-84.2833	95.4054	-0.1054
GA	93.90	-84.3833	95.4198	1.3802
IA	94.30	-93.6167	96.7476	-0.4476
ID	94.40	-116.2000	99.9953	0.5047
IL	94.40	-89.6167	96.1724	-0.2724
IN	94.50	-86.1333	95.6714	0.6286
KS	94.60	-95.6833	97.0448	0.6552
KY	94.60	-84.9167	95.4965	0.1035
LA	94.80	-91.1667	96.3953	-2.5953
MA	94.90	-71.1167	93.5119	-0.5119
MD	95.30	-76.4167	94.2741	-0.8741
ME	95.30	-69.7000	93.3082	1.4918
MI	95.60	-84.5833	95.4485	0.7515
MN	95.90	-93.0833	96.6709	1.4291
MO	96.00	-92.1667	96.5391	-1.9391
MS	96.10	-90.1667	96.2515	-2.8515
MT	96.20	-112.0000	99.3913	-0.0913
NC	96.30	-78.6500	94.5953	1.4047
ND	96.30	-100.7667	97.7759	1.8241
NE	96.30	-96.7167	97.1934	0.0066
NH	96.60	-71.5000	93.5670	3.2330
NJ	96.70	-74.7667	94.0368	0.2632
NM	96.80	-106.0000	98.5284	-1.8285
NV	96.80	-119.7500	100.5058	3.3942
NY	97.20	-73.8333	93.9026	-0.8026
OH	97.60	-83.0000	95.2208	-0.8208
OK	97.70	-97.5333	97.3109	-0.7109
OR	98.10	-123.0500	100.9804	-2.5804
PA	98.40	-76.8333	94.3340	-0.9340
RI	98.50	-71.3833	93.5502	-1.0502
SC	98.60	-81.0000	94.9332	-0.4332
SD	99.10	-100.3333	97.7135	0.7865
TN	99.30	-86.8000	95.7673	-0.8673
TX	99.30	-97.7000	97.3348	1.2652
UT	99.60	-111.8667	99.3721	1.0279
VA	99.70	-77.5000	94.4299	1.8701
VT	100.40	-72.5833	93.7228	2.3772
WA	100.50	-122.8667	100.9540	-1.8540
WI	101.20	-89.3833	96.1388	1.4612
WV	101.40	-81.5833	95.0171	-0.4171
WY	103.90	-104.8167	98.3583	2.8417

FIGURE 6.8: Data for the Gender Ratios example plotted on the USA lower-48 states base map. Upper left: the gender ratio data (2000 Census, Male to Female ratio). Upper right: Longitude of state capitals. Lower Left: The linear model. Lower Right: residuals from the model.

PATTERN Is there a distribution that repeats over the space? Are states with high gender ratios always at some regular distance from those with low gender ratios?

CONTIGUITY Are states with high gender ratios always surrounded by states with low values, or vice versa?

NEIGHBORHOOD Does the gender ratio drop off steeply around several key focal points?

SHAPE Do elongated states have higher gender ratios than those of rectangular or irregular states?

SCALE Would all of the other geographic properties change if we examined the gender ratio using another, higher-resolution, set of districts, such as counties or minor civil divisions?

ORIENTATION Are there connecting lines, perhaps following major highways, between states with high gender ratios going coast to coast? Is there a direction to a general increase in the gender ratio across the map?

Clearly, a whole host of statistics could be used to answer any one of these problems, and many different analyses could be performed. Not all of them will be possible with the GIS that we have, however. We will simply take the last of the questions and learn by example. First, we could divide the map into an eastern and a western half of the country using the traditional dividing line of the Mississippi River. If we do this, the GIS attribute database could be edited to include a new attribute, COAST, with attribute values EAST or WEST. We could then compute the average gender ratio separately for the two parts of the data set. Doing this for the sample data shows an average for the east of 94.65 males per 100 females and for the west of 98.46. This is not really a conclusive answer to the question, however.

6.4.2 Testing a Spatial Model

Perhaps a better way to state the final question in Section 6.4.1 is, Is there a statistical relationship between longitude and the gender ratio? If we had to express this mathematically, we could say that the gender ratio S is a function $f(\ )$ of longitude, or

$$S = f(\lambda)$$

The simplest form that this relationship might take is a linear relationship. You may remember from high school math the formula for a straight line, $y = mx + c$. The y is called the *dependent variable*, because its value depends on what is computed on the right-hand side of the equation and because it is the one we want to predict. The *independent variable*, x, is the one for which we have the data. The value m is the slope of the line, in this case the rate of increase or decrease if the value is negative, of the number of females per 100 males as we move through a degree of longitude. Finally, c is called the *intercept*. It is the value of the gender ratio at the place where longitude is zero. More about this later. In terms of our gender ratio data,

$$S = m\lambda + c \tag{6.1}$$

The first problem is how we select a single longitude to represent each state. We visited this problem earlier in this chapter in the context of Rugby, North Dakota. In this case we simply select a point placed by the highly subjective "eyeball" method—that is, the longitude of a point placed at the visually determined center of the state. Actually, the numbers selected were those used in a mapping program that drew circles to represent state values on a proportional circle map.

Now we can build a geographic database with only two attributes, the two we seek to relate, namely *gender ratio* and *longitude*. These two attributes are mapped in Figure 6.8. With only two variables, we can generate a scatter plot. The space of the scatter plot is "attribute" space, not geographic space. If there were a clear relationship as a straight line, the dots (each one a state) should line up along a sloping line. Take a look at Figure 6.9. Is there such a relationship? Just as we analyzed variation of a single attribute and then two attributes at once in Section 6.2, so also can we analyze variance for the two attributes here. We can use a method called *least squares* to model the scatter with a straight line.

Least squares calculates the variance between the two attributes (multiplied together) as a proportion of the variance in the independent variable. In each case, the total variance is computed as the sum of each attribute's values divided by the number of records (the mean), and this mean is subtracted from each of the attribute values before multiplying. Multiplying the two deviations from the means gives the cross-variance, and this value is divided by the regular squared variance for the independent variable.

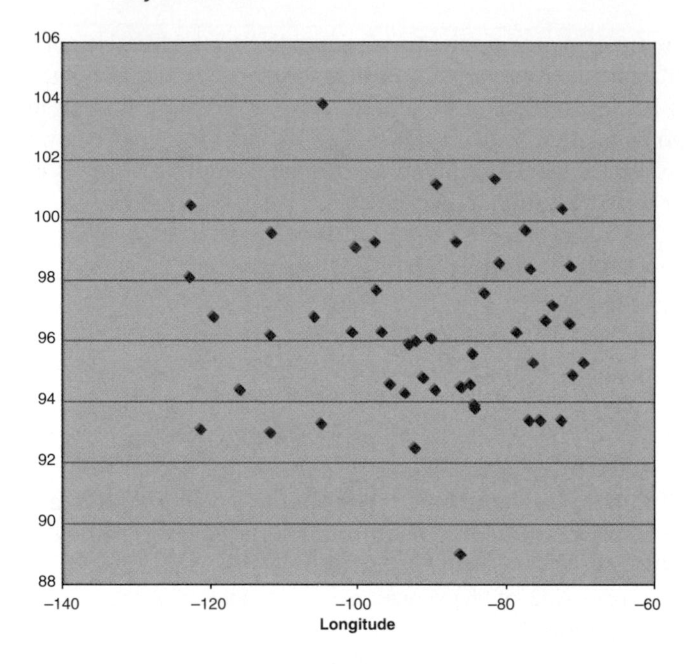

FIGURE 6.9: Scatter plot of Gender ratio data from Table 6.4. Gender ratio (Males per 100 females) is on the y axis, longitude of state capitals on the x axis.

We can use a formula to find values for m and c that minimize the sum of the squared variance. This gives the best-fit straight line through the data. From this we calculate that the formula linking the gender ratio and longitude has the form

$$S = -0.1438\lambda + 83.285 \tag{6.2}$$

The values predicted by this spatial model are mapped in the upper right-hand window in Figure 6.8. The strength of the statistical relationship is given by Pearson's correlation coefficient. The proportion of variance that the relationship "explains," a value called the *coefficient of determination* or r-squared, as a percentage is 61.8%. Obviously, the fit could be better. There must be a part of the geographic relationship that we missed.

6.4.3 Residual Mapping

A common way of seeking a deeper understanding of a spatial relationship is to examine the amount that each record deviates from the current model under analysis. For every record, if we plug the value for the independent variable (x) into the linear equation (6.2), we end up with an amount above or below the line in the up-down (y or dependent variable) direction on the scatter diagram. If we add all these together, they sum to zero, just as when we examined the deviations from the mean in the attribute description in Section 6.1.

These amounts are called *residual*s. Each record has a residual, just as each record has a geographic extent. Again, we can use the compute command or its equivalent in our GIS or database manager to calculate the residual for each state. It is the actual gender ratio for the state with the model value subtracted from it. Then we can use the GIS directly to make a map of the residuals. Just such a map is shown in Figure 6.8 in the lower right window.

This map takes a little interpretation but is a very powerful analytical tool. What the map of residuals shows is the unexplained variation in the simple linear model of gender ratio shown in the same figure on the lower left, and expressed in equation (6.2). Units are males per 100 females. Values in the range -1 to $+1$ mean that the model fits pretty well. High positive residuals are large underestimates, and high negative residuals are overestimates. As an examination of the map shows, by far the highest negative residual from the model is the District of Columbia. Thus the model seriously overestimated the number of males per hundred females in Washington, D.C. and there are far more females in Washington, D.C. than would be expected. The highest positive residuals are in the states of Nevada, New Hampshire, Vermont, Wyoming, and Colorado. In these areas, the model underestimated the gender ratio, i.e., there were more males than expected.

Regionally, the Deep South, the Pacific Northwest, and the "rust-belt" states seem to have negative residuals, meaning that the linear model overestimated their gender ratio, suggesting that they had more females than would be expected. It may be that these states have older populations and that women tend to live longer. States with more positive residuals (underestimates of the gender ratio, implying more males and fewer females in the population than the model predicts) are more scattered, with a dominance in the Mid-West and Mountain states. If we had a more complex linear model, involving both latitude and longitude, perhaps we could explain far more of the variation in the gender ratio. Conversely, some states seem to be exceptional: Nevada, for example. Should these states perhaps be investigated in more detail or at a finer scale? And what would happen if the linear model were applied for other time periods or other countries?

The example we have followed here has a few lessons to offer. First, spatial analysis follows the same path as much of scientific inquiry. We begin by displaying the attributes that we are interested in and looking at their aspatial (e.g., the histogram) and spatial (the map) characteristics. We try to see whether geographic properties have influenced the form of the distribution and if they are able to explain it. Then we formulate a model of the geographic relationship. In the preceding case this was a simple linear model relating the gender ratio and longitude.

We then formulate a way to test the model, usually a measure of goodness of fit between the model and the data that we have. In statistical terms, we suggest a hypothesis about the relationship, propose a counter or null hypothesis, and then devise a test to accept or reject the hypothesis based on the results. This often involves a probability from the normal distribution; in statistics, cutoff probabilities such as 95% or 99% are used to accept or reject a hypothesis.

Spatial analysis then goes further. We seek to explain geographically why the model does or does not fit. If the fit is inadequate, we can choose another model, change the geographic scope of the problem (e.g., the scale or the extent), or expand the model to include more attributes, that is, build a more complex model. Good science dictates that a simple model is preferred over a complex model, but that when a complex model explains a data set successfully, it is acceptable.

6.4.4 Prediction

The final stage of a model's use is to predict rather than to explain. Ideally, the geographic properties themselves have some explanatory basis and can point to a process in action. For example, if we analyzed a disease and found a high concentration in a single district, we could speculate that there was a clustered distribution with a single "source" and an outward diffusion from the source. Proof of this model would be to find a sample of people who had the disease and to show that they all contracted the disease in the single district. Prediction would then follow. An all-out attack on the disease in the one district would be the best strategy to eliminate the disease.

We can return to the gender ratio example to take a closer look at prediction. We can hypothesize that women live longer in the East and account for the linear relationship with longitude. This would obviously be extremely difficult to test by experiment. We could easily, however, use different data to examine the model. We could use county gender ratio data to see if the same model holds at different scales. We could use data for Mexico and the Canadian provinces to see whether the relationship carried over the edges of our geographic extent. We could also take the model to its limits, a very common tool in physics, math, and chemistry.

For example, our model uses longitude as the independent variable. What would the model predict as a gender ratio at the prime meridian, say for England and West Africa? What about for the middle of the Pacific Ocean at the 180th meridian? Clearly, the model has geographic constraints upon its predictive powers, and these can be tested and described. If the model is geographically invariant, this too is of great interest. Virtually every phenomenon on earth, even gravity, varies in some way over geographic space. If this were not the case, geography would not exist as a discipline, and GIS would be no more powerful than a spreadsheet program! There is value in mapping data even if no spatial analysis is intended, but merely as a visual description of a geographic pattern.

As GIS use becomes more commonplace, and as geographic explanations become of more value in managing resources, more of the geographic analyst's time will be spent in searching GIS data for spatial relationships. This geographic detective work can go a long way toward dealing with all the phenomena that have yet to be examined rigorously. This explains why GIS is being so rapidly accepted in disciplines such as archeology, demography, epidemiology, and marketing. In these fields, GIS allows the scientist to look for relationships that could not be seen without the lens of cartography and the integrative nature of GISs turned toward information and data like a high-powered telescope. The eye-opening moment of seeing the data visually for the first time is a common GIS convert's experience. Just as the explorers of the last century mapped out America and the world, so the GIS experts of today are mapping out new geographic worlds, invisible at the surface, but visible and crystal clear to the right tool controlled by the right vision.

In the search for spatial relationships, the GIS analyst is largely alone, however. The tools for searching are only now being integrated into GIS. In cartography, a process called the *design loop* is often used in map design. The map is generated to the specifications, and then the tools of digital cartography are used to make slow incremental improvements until the optimum design is reached. Obviously, not all of the improvements lead to success. Many serve to show what not to do; nevertheless, trial and error are important steps in the map's improvement.

Such a process is often used in GIS. A typical analysis is the result of extensive data collection, geocoding, data structuring, data retrieval, and map display, followed by the sorts of description and analysis highlighted here. The next steps, prediction, explanation, and the use of these in decision making and planning, are the ultimate goal. Without seeing spatial relationships, however, these last prizes will be lost, and the GIS will not have reached its potential as an information management tool.

The GIS spatial relationship search loop consists of the steps outlined above, data assembly, preview, hypothesis design, hypothesis testing, modeling, geographic explanation, prediction, and examining the limitations of the model. It is the automated nature of the GIS and its flexibility that allows effective use of trial and error. In this process the GIS components most used are the database manager, for selecting and reorganizing attributes; the map display module, for display of intermediate and final analyses; and the computational or statistical tools available as part of the GIS. Some of these are sophisticated, although in many GIS packages they are rather simplistic.

One capability now being added to GIS is that of allowing examination of how distributions and relationships vary over time. Time is not particularly simple to integrate into a GIS, because each attribute data set and its maps are best interpreted as a single snap-shot in time. Nevertheless, both the attributes and the map are under constant change, and the geographic phenomena they represent are indeed very dynamic. Even an apparently stable attribute such as terrain is affected by mining, erosion, and volcanoes. Human systems are virtually all in a constant state of change.

We can simply compare two time periods for which we have data. Many human and social data are collected only once a decade, meaning that changes happening more rapidly than this will not be seen. Comparing two time periods allows only a single measurement, by value or map, of change. We can map, for example, on a satellite image all those areas that have changed between the time periods shown in the two images, assuming the same geographic extent at similar map scales. This can give us a direction of change, the addition or loss of wetlands, for example, and the amount of addition or loss, but not a rate of loss or gain. To measure a rate of change needs a minimum of three images or maps.

The most effective tool for time-sensitive geographic distributions is animation (Peterson, 1995). Animation allows the GIS interpreter to see changes as they take place. It is a key part of scientific visualization as far as a GIS is concerned, because usually more can be learned by examining the dynamics of a geographic system than its form. Imagine seeing glimpses of a chess game, for example, at only three stages in a game. The forms remain largely unchanged, although some pieces have moved around and disappeared from the board.

As we get more "frames" in the sequence, we arrive at the stage where every move is visible and the rules of movement of chess pieces become discernible. Finally, a great number of frames allow us to see not only the rules but the players themselves, their strategy, and the drama of the game. Just as there is an appropriate geographic resolution for map data, so is there a suitable time resolution. Like enlargement and reduction in space, time can be made slower or faster than "real" time to reach this appropriate resolution.

Figure 6.10 shows a few individual frames from an animation built with a GIS, in this case the historical growth of the Washington, DC, area. As is obvious, a static textbook cannot bring across the dynamic nature of the sequence. For this, use the MPEG sequence on the World Wide Web at `http://edcwww2.cr.usgs.gov/umap/umap.html`.

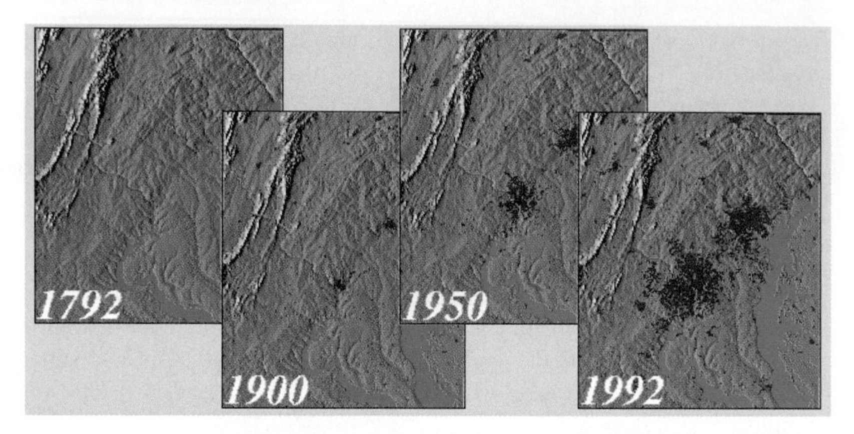

FIGURE 6.10: Frames from an animation of historical urban growth in the Washington, D.C. to Balitimore metropolitan area. Images by William Acevedo, USGS. Used with Permission.

6.4.5 Map Overlay Example

One of the oldest analytical methods used in GIS is map overlay. Map overlay is the set of procedures by which maps with different themes are brought into geometric and scale alignment so that their information can be cross referenced and used to create more complex themes. We have met the method already several times, and should recall that the maps to be overlain must be of the same spatial extent, on the same map projection and datum, be at comparable granularity (that is, the spatial units, whether pixels or polygons, should be about the same average size), and if the layers are to be used with map algebra, at the same raster grid size and resolution.

The power of the GIS is in handling the geometry of the overlay process. Handling and preparing the themes is up to the GIS analyst. Under the simplest possible configuration, GIS layers are all converted to binary maps, and an overlay then sifts the map space to leave open the areas that satisfy the selection criteria in use. This is the case in the simple overlay analysis we met in the early chapters, and duplicates in the GIS methods that were worked out using transparent overlay maps and blacked out areas on the transparencies. Many of these methods date back to the turn of the twentieth century.

One means of map overlay is to intersect all of the layers involved to generate a set of most common geographic units. In map algebra, the raster plays this role. The attributes are then inherited or passed down to subsetted areas, and the attribute table gets longer and longer as more and more units are created. We have already seen the many problems with vector map overlay, including sliver polygons. Blind map overlay will happily assign attributes to very small sliver polygons, and use them in further analysis. A solution to this problem is to first process each layer to reduce the number of solution classes that will find their way into the final map. A selective query from each layer is one simple way to do this. A third overlay method is to find some common unit into which all values can be transformed. One GIS project that the author worked on solved the apparent incompatibility of the GIS layers for an ocean GIS by converting all of the themes into dollars, and adding them together to yield a composite. This is

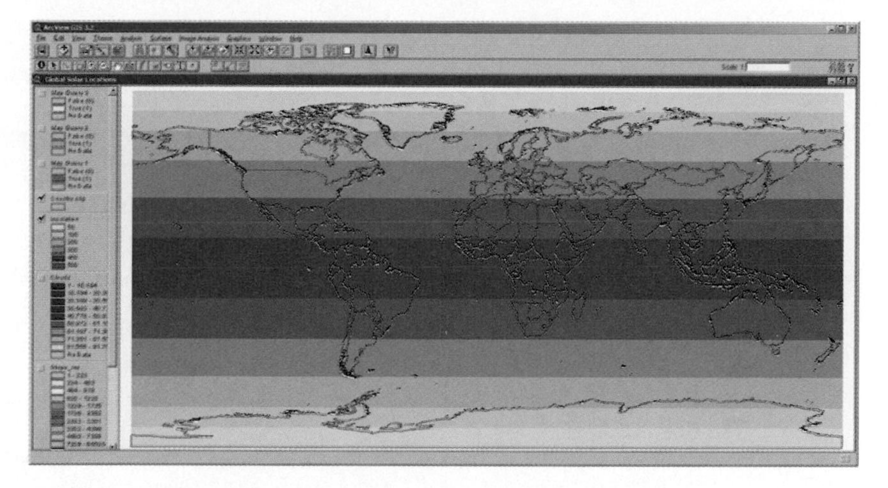

FIGURE 6.11: ArcView GIS layer of global insolation. Derived from a figure in *Geosystems*, by Robert Christopherson.

not always possible, of course, and far more common is to weight the overlays involved by some prechosen set of values reflecting the relative, not the absolute, importance of the layers.

For example, Figure 6.11 shows a world map of the factors considered for large scale siting of global solar power production. Information was searched for on the World Wide Web, downloaded and then processed into ESRI's ArcView so that map layers could be registered and converted to a common map projection using ArcView's projection extension. The themes used were those thought to relate to the possible supply and demand for solar-generated power. On the supply side, layers were the maps of incoming solar radiation, the expected average cloud cover, and global topography. On the demand side, a global population layer was used to buffer the solution set and limit it to those areas located close to major population agglomerations.

The overlay exercise consisted of creating a query to the GIS that essentially converted each of the layers into a binary map, a thresholding operation. The query requested insolation levels greater than 200 Watts per square meter, three levels of cloud coverage on average less than 65, 70, and 75%, and elevations lower than 5000 feet (1524 m), since although incoming solar power increases as the atmosphere becomes less dense with height, large scale construction favors lower and flat topography (Figures 6.12 and 6.13). These were combined by selecting those areas with population densities greater than 50 people per square kilometer (Figure 6.14). The resultant map shows that largest contiguous areas suitable for large scale solar power generation to include the Southwestern United States, Northern Chile, South Africa, the edges of the African Sahara, the Arabian peninsula, and Pakistan (Figure 6.15).

Having done the analysis with three different thresholds as far as cloud cover is concerned, two issues related to overlay analysis are shown. First, the different criteria are subjective, and need to be weighted to reflect their relative importance in the GIS solution. For example, it may be that total insolation is far more, say 10 times, more important than elevation in deciding where solar power generation is located. This can

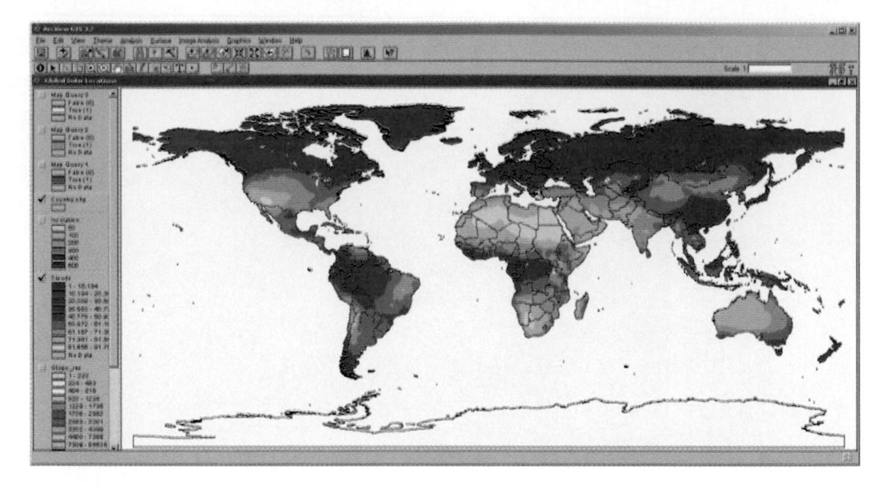

FIGURE 6.12: GIS layer of global average cloud cover. Source: UNEP Grid.

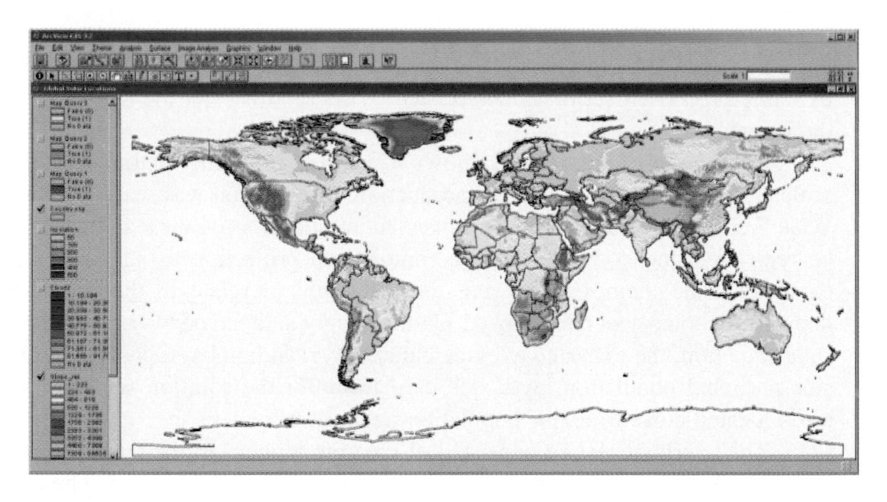

FIGURE 6.13: GIS layer of global topography. Source: Data set GTOPO30, United States Geological Survey, EROS Data Center.

be accommodated by multiplying the weights by the binary layers, each factored so that the sum of the weights is one, and then summing them across the layers. The final map will reflect the critical factors and their importance. However, selecting the weights can be a very complex process.

Second, in this case the solution area is highly influenced by only one of the layers, the cloud cover. This is why three levels of cloud cover were used in the final map. This "most sensitive" or critical layer in the map overlay process often marginalizes the contribution of other layers, perhaps even eliminating them from the final solution space altogether. This layer sensitivity is especially important to understand. Often a small amount of testing can reveal the critical layer. Some research suggests treating the layers as fuzzy, combining the factors together as a smooth field and creating margins of error on the solution map. Another important factor that affects analysis

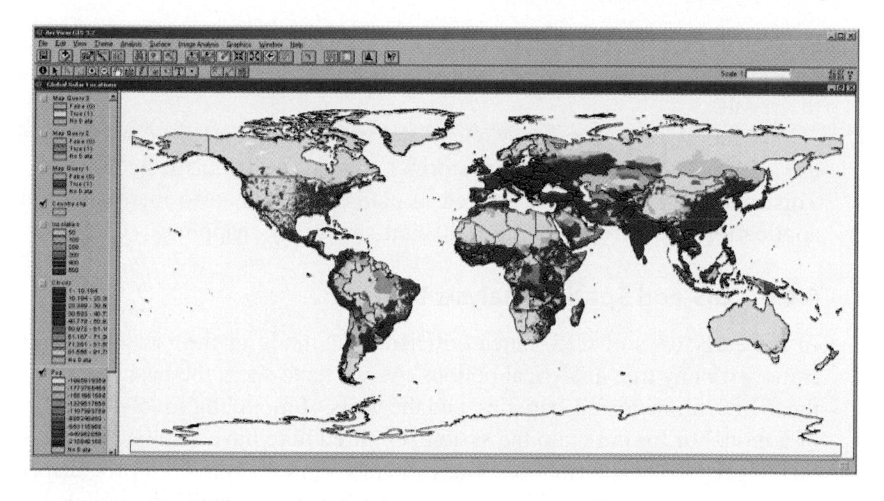

FIGURE 6.14: GIS layer of global population density. Source: NCGIA.

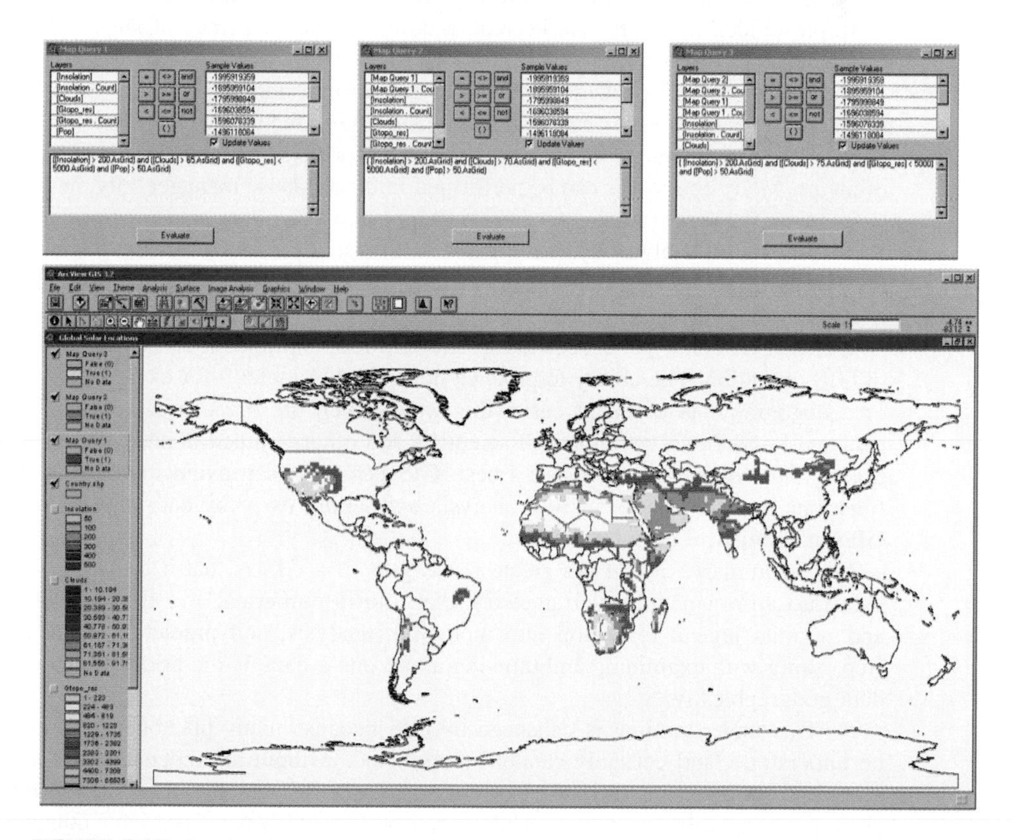

FIGURE 6.15: Solution map for the map overlay example. Global suitability for solar power production. Above the map are the three queries that generated the increasingly narrow solution sets. Figures 6.11 to 6.15 by Jeff Hemphill and Westerly Miller.

is how the error, for example caused by different generalization on each input layer, impacts the final decision by propagating through the overlay analysis and influencing the result.

Map overlay remains as one of the most common forms of GIS analysis. With the use of buffers and distance transforms, some very sophisticated analysis can be done. This method is extensively applied in planning, but is also increasingly used in all GIS applications, for fire modeling to habitat suitability mapping.

6.4.6 GIS and Spatial Analysis Tools

In the early days of GIS, much criticism was made of the fact that GIS software rarely came with any true analytical options. As we have seen, the basic tools of description are those of arithmetic and statistics, and the tools of modeling involve allowing the encoding of a model or formula into the system. Omitted here have also been models based entirely on the geographic distribution. Many models work on network flows, dispersion in two- or three dimensional space, hierarchical diffusion, or probabilistic models based on weights determined by buffers, and so on. This sort of model is manageable in a GIS using the tools of retrieval: overlay, buffering, and the application of spatial operators. Even a simple model, however, can become a quite lengthy sequence of steps for the GIS's user interface.

Almost all GIS packages allow operations to be bundled together as macros or as sequences of operations as part of a model, such as in the GISMO options. Although this goes a long way toward routine analysis, exploratory GIS data analysis is still something of an art. Many operations can be performed in the database manager only, and often GIS users move the data from the database manager into a standard statistical package such as SAS (Statistical Analysis System) or SPSS (Statistical Package for the Social Sciences) for analysis. One GIS (Arc/Info) offers a direct link to another statistical package (S-Plus) as an option.

Most GIS analysts use statistical and GIS tools in tandem during the analysis stage of GIS operation. The ability to produce nonspatial graphics—for example, a scatter plot or a histogram—is often far easier this way. Given the broad acceptance of statistical packages, and the large number of scientists and others trained in and familiar with their use, a compromise solution seems best. GIS packages can avoid duplicating the many functions necessary for statistical analysis by making two-way data movement between GIS and statistical software easy.

In summary, one of the greatest strengths of a GIS is that it can place real-world data into an organizational framework that allows numerical and statistical description and permits logical extension into modeling, analysis, and prediction. This important step, along with examining and thinking about one's data, is the bridge to understanding data geographically.

This understanding is enhanced by GIS because many phenomena simply cannot be understood, and certainly cannot be predicted, without an understanding of the geographic forces at work and their expression among the map's features as imprints of the principal geographic properties. Unfortunately, most GIS packages have contained only rudimentary tools for spatial analysis. However, GIS practitioners have filled the gaps with standard statistical software, and great strides are now being made as GIS contributes to the new models that are resulting in a host of different applications beyond the traditional scope of geography.

6.5 STUDY GUIDE

6.5.1 Summary

CHAPTER 6: Why Is It There?

Describing Attributes (6.1)

- **GIS data description answers the question: Where? GIS data analysis answers the question: Why is it there?**
- **GIS data description is different from statistics because the results can be placed onto a map for visual analysis.**

Statistical Analysis (6.2)

- **The extremes of an attribute are the highest and lowest values, and the range is the difference between them in the units of the attribute.**
- **A histogram is a two-dimensional plot of attribute values grouped by magnitude and the frequency of records in that group, shown as a variable-length bar.**
- **For a large number of records distributed with random errors in their measurement, the histogram resembles a bell curve and is symmetrical about the mean.**
- **The mean is the sum of attribute values across all records, divided by the number of records. It is a representative value, and for measurements with normally distributed error, converges on the true reading.**
- **A value lacking sufficient data for computation is called a missing value.**
- **Accuracy is determined by testing measurements against an independent source of higher fidelity and reliability.**
- **The standard deviation is the average amount by which record values differ from the mean.**
- **The total variance is the sum of each record with its mean subtracted and then multiplied by itself.**
- **The standard deviation is the square root of the variance divided by the number of records less one.**
- **A sample is a set of measurements taken from a larger group or population. Sample means and variances can serve as estimates for their populations.**
- **The standard deviation in a spatial sense is a good descriptor of the accuracy of measurements.**
- **A mathematical version of the normal distribution can be used to compute probabilities associated with measurements with known means and standard deviations.**
- **A test of means can establish whether two samples from a population are different from each other, or whether the different measures they have are the result of random variation.**

Spatial Description (6.3)

- **For coordinates, data extremes define the two corners of a bounding rectangle.**
- **For coordinates, the means and standard deviations correspond to the mean center and the standard distance, both of which are good descriptors of spatial properties.**

- A centroid is any point chosen to represent a higher-dimension geographic feature, of which the mean center is only one choice.
- The standard distance for a set of point spatial measurements is the expected spatial error.
- Descriptions of geographic properties such as shape, pattern, and distribution are often verbal, but quantitative measures can be devised, although few are computed by GIS.
- GIS statistical computations are most often done using retrieval options such as buffer and spread, or by manipulating attributes with arithmetic commands.

Spatial Analysis (6.4)

- Geographic inquiry examines the relationships between geographic features collectively to help describe and understand the real-world phenomena that the map represents.
- Spatial analysis compares maps, investigates variation over space, and predicts future or unknown maps.
- Many GIS systems have to be coaxed to generate a full set of spatial statistics.
- A linear relationship is a predictable straight-line link between the values of a dependent and an independent variable. It is a simple model of the relationship.
- A linear relation can be tested for goodness-of-fit with least-squares methods. The coefficient of determination r-squared is a measure of the degree of fit, and the amount of variance explained.
- Differences between observed values of the dependent variable and those predicted by a model are called residuals.
- A GIS allows residuals to be mapped and examined for spatial patterns.
- A model helps explanation and prediction after the GIS analysis.
- A model should be simple, should explain what it represents, and should be examined at the limits before use.
- Overlay analysis is a very common form of GIS analysis.
- Overlay analysis requires maps on a common geometry with compatible themes.
- Themes can be weighted to reflect their relative importance.
- Often one theme dominates in controlling the final solution set to an overlay problem.

Searching for Spatial Relationships (6.5)

- Tools for searching out spatial relationships and for modeling are only lately being integrated into GIS.
- Statistical and spatial analytical tools are also only now being integrated into GIS, and many people use separate software systems outside the GIS.
- Real geographic phenomena are dynamic, but GISs have been mostly static. Time-slice and animation methods can help in visualizing and analyzing spatial trends.
- GIS places real-world data into an organizational framework that allows numerical description and lets the analyst model, analyze, and predict with both the map and the attribute data.

6.5.2 Study Questions

Describing Attributes

Separate all of the readings in Table 6.1 taken east of Santa Barbara. Does this group represent a statistically different location and elevation from the remainder of the readings? Why or why not?

Write out step-by-step instructions for a child to calculate a mean and a median from a list of 10 numbers. Modify the instructions for 11 numbers. Write a one-paragraph explanation of what the numbers mean.

Spatial Analysis

Why is spatial analysis different, though related to, statistical analysis? Give examples.

Copy from a map such as a 1 : 24,000 series USGS map a group of objects, including a set of points with known elevations, some rivers, and forested areas. List as many measurements as you can devise to characterize the basic geographical properties of each feature, classified by dimension. Which measures are the easiest to calculate and why?

Draw a flow diagram of the stages of scientific inquiry surrounding analysis in a GIS. What is necessary before analysis can be conducted? What does a successful analysis lead to? What might prevent the ease of movement through the flow in the diagram?

Why have statistical and spatial analytical methods not been incorporated into all GIS packages?

6.6 EXERCISES

1. *For a GIS data set of your choosing, use only the tools available within the system to compute the extremes, the mean, and the standard deviation for all the attributes in the data set. What problems do you encounter along the way?*
2. *How might the length of a line and the area of a polygon be calculated (a) in a vector GIS and (b) in a raster GIS? Why might they be expected to give different results?*
3. *Design a model that might account for the risk of wildfire for a GIS data set consisting of layers for vegetation type and condition, soils, streams, topography, and wind direction. How might the model be tested?*
4. *Use a GIS to overlay a map of topography with polygonal districts such as counties. Using any method available, compute and then map the variance or the standard deviation of the elevation values within each district. Explain the distribution on the map.*
5. *Repeat the gender ratio example in the chapter for data on mean annual precipitation by state. Does the spatial analysis result in a stronger relationship? Why might this be?*
6. *Trace a feature from a map, or select a single polygon feature within a GIS. Choose as complex a feature as you can find. Using the GIS's capabilities, locate as many centroids by as many methods as you can devise. Do any of these points fall outside the polygon? Which methods give similar results?*

6.7 REFERENCES

Campbell, J. (1993) *Map Use and Analysis*, 2nd ed. Dubuque, IA: Wm. C. Brown.

Earickson, R. and Harlin, J. (1994) *Geographic Measurement and Quantitative Analysis*. New York: Macmillan.

Peterson, M. P. (1995) *Interactive and Animated Cartography*. Upper Saddle River, NJ: Prentice Hall.

The World Almanac and Book of Facts. New York: Pharos Books. Published annually.
Unwin, D. (1981) *Introductory Spatial Analysis*. London: Methuen.

6.8 KEY TERMS AND DEFINITIONS

analysis: The stage of scientific inquiry when data are examined and tested for structure in support of hypotheses.

attribute: An item for which data are collected and organized. A column in a table or data file.

bearing: An angular direction given in degrees from zero as north, clockwise to 360.

bell curve: A common term for the normal distribution.

bounding rectangle: The rectangle defined by a single feature or a collection of geographical features in coordinate space, and determined by the minimum and maximum coordinates in each of the two directions.

centroid: A point location at the center of a feature used to represent that feature.

compute command: In a database manager, a command allowing basic arithmetic on attributes or combinations of attributes, such as summation, multiplication, and subtraction.

converge: The eventual agreement of measurements on a single value.

data extremes: The highest and lowest values of an attribute, found by selecting the first and last records after sorting.

dependent variable: The variable on the left of the equals sign in a formula model, whose values are determined by the values of the other variables and constants.

difference of means: A statistical test to determine whether or not two samples differ from each other statistically.

error band: The width of a margin plus and minus one standard error of estimation, as measured about the mean.

expected error: One standard deviation in the units of measure.

goodness of fit: The statistical resemblance of real data to a model, expressed as strength or degree of fit of the model.

gradient: The constant of multiplication in a linear relationship, that is, the rate of increase of a straight line up or down. See also **slope**.

histogram: A graphic depiction of a sample of values for an attribute, shown as bars raised to the height of the frequency of records for each class or group of value within the attribute.

hypothesis: A supposition about data expressed in a manner to make it subject to statistical test.

independent variable: A variable on the right-hand side of the equation in a model, whose value can range independently of the other constants and variables.

intercept: The value of the dependent variable when the independent variable is zero.

least squares: A statistical method of fitting a model, based on minimizing the sum of the squared deviations between the data and the model estimates.

linear relationship: A straight-line relationship between two variables such that the value of the dependent variable is a gradient times the independent variable plus a constant.

mean: A representative value for an attribute, computed as the sum of the attribute values for all records divided by the number of records.

mean center: For a set of points, that point whose coordinates are the means of those for the set.

median: The attribute value for the middle record in a data set sorted by that attribute.

missing value: A value that is excluded from arithmetic calculations for an attribute because it is missing, not applicable, or is corrupted, and has been signified as such.

model: A theoretical distribution for a relationship between attributes. A spatial model is an expected geographic distribution determined by a given form such as an equation.

normal distribution: A distribution of values symmetrically about a mean with a given variance.

normalize: To remove an effect biasing a statistic; for example, the influence of the size of the sample.

null hypothesis: The state opposite to that suggested in a hypothesis, postulated in the hope of rejecting its form and therefore proving the hypothesis.

Pearson's product moment correlation coefficient: A measurement of goodness of fit computed as the square of the sum of the cross-variance of two variables divided by the sum of the variance in the independent variable. When squared (the coefficient of determination) the value is the proportion or percentage of the variable "explained" by the model under test.

population: The total body of objects from which a sample is taken for measurement.

prediction: The ability of a model to provide information beyond that for which measurements are available.

***r*-squared:** A common term for the coefficient of determination.

random: Having no discernible structure or repetition.

range: The highest value of an attribute less the lowest, in the units of the attribute.

record: A set of values for all attributes in a database. Equivalent to the row of a data table.

residual: The amount left when the observed value of the dependent variable has subtracted from it that predicted by a model, in units of the dependent variable.

sample: A subset of a population selected for measurement.

slope: The constant of multiplication in a linear relationship; that is, the rate of increase of a straight line up or down. See also **gradient**.

sort: To place the records within an attribute in sequence according to their value.

standard deviation: A normalized measure of the amount of deviation from the mean within a set of values. The mean deviation from the mean.

standard distance: A two-dimensional equivalent of the standard deviation, a normalized distance built from the standard deviations of the easting and northings for a set of points.

table: An arrangement of attributes and records into rows and columns to assist display and analysis.

test of means: Hypothesis test to establish whether two samples with their own means and standard deviations are drawn from the same overall population.

units: The standardized measurement increments for values within an attribute.

variance: The total amount of disagreement between numbers. Variance is the sum of all values with their means subtracted and then squared, divided by the number of values less one.

PEOPLE IN GIS

Brenda G. Faber Owner, Fore Site Consulting

Brenda Faber owns and manages Fore Site Consulting, Inc., a private consulting firm located in Loveland, CO. Fore Site Consulting specializes in land planning decision tools for sustainable development and resource management. Brenda has over 15 years' experience in the development and application of GIS decision software. She holds advanced degrees in Electrical Engineering and Mathematics from North Carolina State University. Brenda is widely recognized for her pioneering research in collaborative spatial negotiation systems in cooperation with the Terrestrial Ecosystem Regional Research and Analysis (TERRA) Laboratory. She was also responsible for concept design and implementation of "Smart Places," a resource modeling application for interactive negotiation of land use proposals. She is currently leading the development of CommunityViz, a Planning Support System that includes 3D exploration, impact analysis, and predictive forecast modeling.

KC: What got you started in GIS?

BF: My graduate degree was in Electrical Engineering and Image Processing. I went to work for IBM after graduate school to develop Robotic Vision techniques. Later, a job transfer within IBM led me to a group doing volumetric GIS research. I found that my background in image processing was a natural fit for scientific and raster GIS work.

KC: What is your current role in the GIS industry?

BF: I currently own a small consulting company that develops Planning Support Systems for municipalities and federal land management agencies. Planning Support Systems are a relatively new class of planning tools that extend the capabilities of traditional GIS to include impact, simulation, and visualization options. These systems are primarily concerned with exploring the implications of land use alternatives.

KC: What in your educational background prepared you for a career in GIS?

BF: I have degrees in mathematics and image processing. As I mentioned before, image processing was a natural fit for my on-the-job training in GIS while working for IBM. I believe my more technical background has enabled me to introduce unique simulation and modeling concepts into the geographic context of GIS.

KC: What is CommunityViz?

BF: CommunityViz is a Planning Support System. It is an integrated suite of GIS extensions that provide a customizable framework for evaluation of land use proposals, including 3D exploration, impact analysis, and predictive forecast modeling. CommunityViz is unique among Planning Support Systems, because it integrates three unique planning perspectives into one multi-dimensional environment. The development of CommunityViz has been sponsored by the Orton Family Foundation to enhance collaboration and participation among planners, officials, and citizens. I have worked as a consultant to the Foundation for several years now as a primary developer of CommunityViz.

KC: How is visualization integrated into ArcView in the CommunityViz software?

BF: CommunityViz 3D Visualization allows you to "experience" proposed land scenarios by taking a virtual stroll through a 3D GIS scene. Realistic scenes can be quickly created directly from standard GIS data sets including terrain data (such as a DEM or TIN), orthophotographs, and 3D representations of common GIS features such as buildings, tree canopies, roads, rivers, and fences. You can move about in this virtual scene to explore not only existing landscapes, but the visual impacts of proposed land changes as well.

KC: What needs to be in place before CommunityViz can work?

BF: CommunityViz works as an extension to ESRI's ArcView GIS. Impact modeling in CommunityViz requires basic GIS data sets such as landuse, infrastructure, jurisdictional boundaries, and locally-based impact formulas such as water usage per square foot of commercial floor space. CommunityViz can then be used to monitor potential impacts on resources, citizens, or wildlife due to proposed development or land management changes. 3D visualization in CommunityViz requires a terrain model, orthophotographs, and GIS data layers indicating location of man-made and natural features. Forecasting in CommunityViz requires data sets for local zoning, parcels, land cover, infrastructure, census demographics, local taxes, attitudes, and preferences.

KC: What is your experience working with GIS with planning groups around the country?

BF: I love the natural balance that my work allows between technical development and direct involvement with planning groups. I think one of my greatest strengths is the ability to act as a liaison between the very technical world of computer modeling and real folks who want to make informed decisions about the future of their communities. So many important decisions today are made by "gut feel." By making GIS and analysis techniques more accessible to citizens and policy makers, we will have fewer unanticipated consequences down the road.

KC: What developments do you think will impact GIS in the future?

BF: Clearly, the accelerating rate of faster, smaller, cheaper processors will have a positive influence on GIS, removing the hurdles of processing and analyzing large data sets. Also, increasing bandwidth making it possible to provide interactive GIS analysis (not just static maps) over the web will have tremendous significance for a broader acceptance and use of GIS technologies.

KC: What would you advise someone to do who would like to follow in your career footsteps?

BF: Exposure to a wide array of technical disciplines and experience with real-world project implementation are both very valuable. I believe that innovative concepts are rarely "new", but most often evolve from unique combinations of existing concepts from diverse fields of study or experience. A GIS background combined with additional talents in programming, public speaking, engineering, biology, psychology, etc. can make for a dynamite career in GIS research and development.

KC: Brenda, thanks for the information. (Used with permission.)

CHAPTER 7

Making Maps with GIS

7.1 THE PARTS OF A MAP
7.2 CHOOSING A MAP TYPE
7.3 DESIGNING THE MAP
7.4 STUDY GUIDE
7.5 EXERCISES
7.6 BIBLIOGRAPHY
7.7 KEY TERMS AND DEFINITIONS

"What is the use of a book," thought Alice, "without pictures or conversations?"
—Lewis Carroll, Alice's Adventures in Wonderland *(1865)*

Oh, the vision thing...
Some reporters said I don't have any vision. I don't see that.
—George Bush, quoted in Time, *January 26, 1987*

A map is a set of errors that have been agreed upon.
—Anonymous

Y3K. Cartographers team with geologists, and finally eliminate the problem of continental drift.

7.1 THE PARTS OF A MAP

A *map* can be defined as a graphic depiction of all or part of a geographic realm in which the real-world features have been replaced by symbols in their correct spatial location at a reduced scale. Maps, as we have already seen, are the paper storehouses of spatial information that we use as sources of data for GIS. They are also the final stage in GIS work, the means by which the information being extracted, analyzed, and reconstructed using the powers of the GIS is at last communicated to the GIS user or the decision maker who relies on the GIS for knowledge. Maps within a GIS can be temporary, designed merely for a quick informative glance, or permanent, for presentation of ideas as a substitute for a picture or a report. Whatever the map's context, as in Chapter 2, we must return again to the cartographic roots of GIS for a discussion of the critical information that the GIS practitioner needs to use the map display part of a GIS correctly.

In either case, the map has a structure. Just as a sentence in the English language needs to follow grammar and syntax to be understood, so a map has to follow its own visual grammar. As a starting point, we cover in this chapter what the various parts of a map are called and which of them are essential. This is followed by a description of the methods used by GIS systems for displaying maps and how the choice of a type of map is sometimes made for us by the sorts of attributes and the geographic character of the data in the GIS. We then cover map design, summarizing some of the rules that cartographers have developed for selecting map symbols, such as colors, and then using them to assemble effective maps.

Just as a map has a structure, so that structure can vary according to which media we use for map display. GISs usually use the computer monitor to display a map, rather than the traditional paper. Only now, after many years of computer mapping, are cartographers beginning to understand how map design depends on the display medium. The GIS has been a major reason why this has become an important consideration.

First, however, we should define the terms used in describing maps. Figure 7.1 shows a set of cartographic *elements*. A cartographic element is one of the building blocks that make up a map and from which all maps are assembled. Each element should be present, although in some cases exceptions are possible. The two basic parts of the map are the *figure* and the *ground*. The figure is the body of the map data itself and is the part of the map referenced in ground coordinates. Almost all other parts of a map are located on the map using *page coordinates*, defining locations on the map layout itself rather than the piece of the world it shows. The *graticule* or *grid*, or often a *north arrow*, is the reference link between the two coordinate systems on the final map. Also part of the figure, the ground, and the legend are the *symbols*. The *legend* translates the symbols into words by locating text and the symbols close to each other in the page coordinate space.

The *border* is the part of the display medium (paper, window, computer screen, or other medium) that shows beyond the *neat line* of the map (Figure 7.1). In special circumstances, additional information can be provided in this space, such as the map copyright, the name of the cartographer, or the date. The *neat line* is the visual frame for the map and is usually a bold single or double line around the map that acts as a rectangular frame. From a design standpoint, the neat line provides the basis for the page (i.e., cartographic device) coordinate system, in display units such as inches or centimeters on the page.

Text information is an integral part of a map, and no map is complete without it. Text is contained in the *title* (whose wording sets the theme and the "feeling" for the map), in *place-names*, in the *legend*, and in the *credits* and *scale*. The scale is a visual expression of the relationship between the ground coordinate space and that of the map page space. Because representative fractions change as the map is projected or rendered onto different display devices or windows, a graphic scale is preferred. Place names follow a strict set of placement rules, both on the figure and related to features and within the map space. Point, line, and area features have different placement rules, and there are also rules for dealing with the overlap of names and feature symbols. Some of the rules are shown in Figure 7.2.

Finally, an *inset* is either an enlarged or a reduced map designed to place the map into geographic context or to enlarge an area of interest whose level of detail is too specific for the main map scale. An inset should have its own set of cartographic

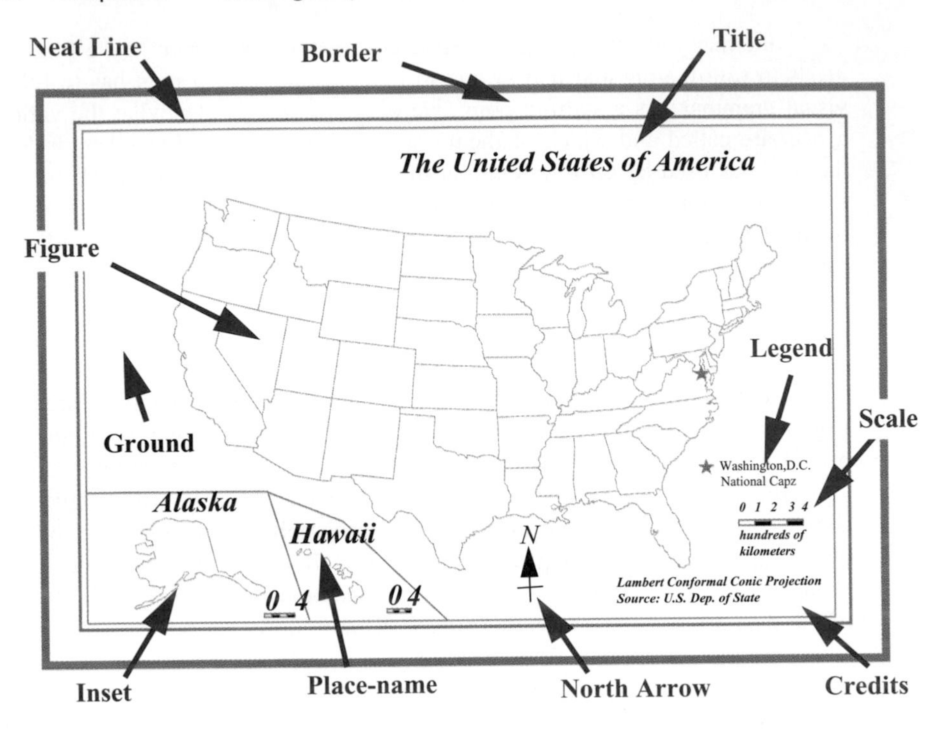

FIGURE 7.1: Cartographic elements.

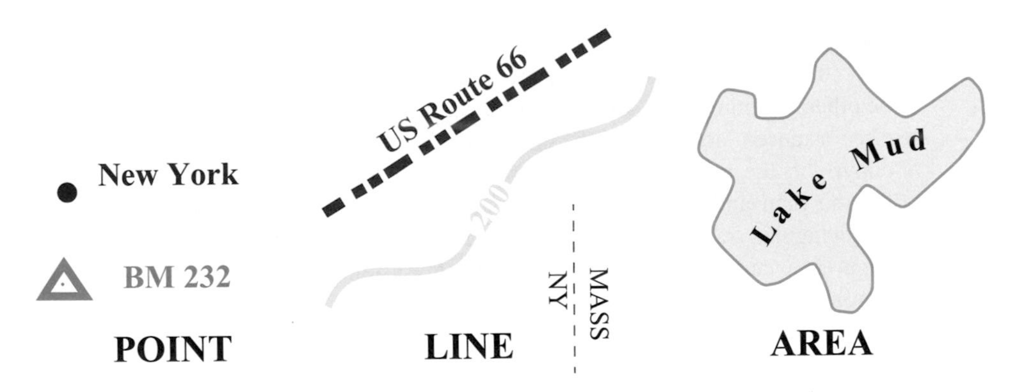

FIGURE 7.2: Some cartographic label placement conventions. Points: right and above preferred with no overlap. Lines: Following the direction of the line, curved if a river. Text should read up on the left of the map and down on the right. Areas: On a gently curved line following the shape of the figure and upright.

elements, although it is usually highly generalized and many elements may be omitted. To avoid confusing the main map with the inset, the inset should be clearly distinguishable from the figure and ground. Many Americans believe that Alaska and Hawaii are small islands off San Diego!

7.2 CHOOSING A MAP TYPE

Over 3000 years of cartographic history, cartographers have designed numerous ways of showing data on a map. One way to divide up the methods is to look at those that show attributes by their geometric dimension, so that we can have point maps, line maps, and area maps, plus maps that show a three-dimensional view. Many maps show some or all of the types of features at the same time. These are often called *general purpose maps*. *Thematic maps* show just one or two themes or layers of information, often coded, colored, or grouped for convenience. In this section we take a look at the breadth of map types available.

A basic outline or *reference map* (Figure 7.3) shows the simplest properties of the map data. An example is a world outline map, with named continents and oceans. A general reference map, usually showing a suite of features including terrain, streams, boundaries, roads, and towns, is called a *topographic map* (Figure 7.4). Topographic maps are often used as reference information behind GIS map layers.

A *dot map* (Figure 7.5) uses dots to depict the location of features and may show a distribution such as population against a base map. A *picture symbol map* (Figure 7.6) uses a symbol, such as the silhouette of a skier, to locate point features such as ski resorts. The *graduated symbol map* (Figure 7.7) is the same, except that the symbol size is varied with the value of the feature. Typically, geometric symbols such as circles, squares, triangles, or shaded "spheres" are used.

A *network map* shows a set of connected lines with similar attributes. A subway map, an airline route map, and a map of streams and rivers are examples. The *flow map*

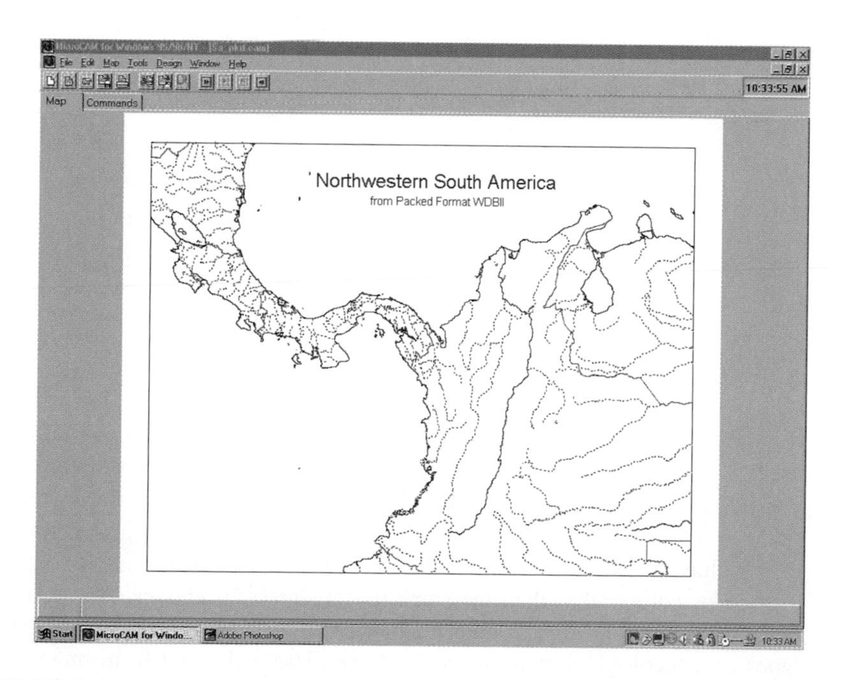

FIGURE 7.3: Reference Map. Central America from the Digital Chart of the World. Plotted from Microcam (`http://www3.ftss.ilstu.edu/microcam`)

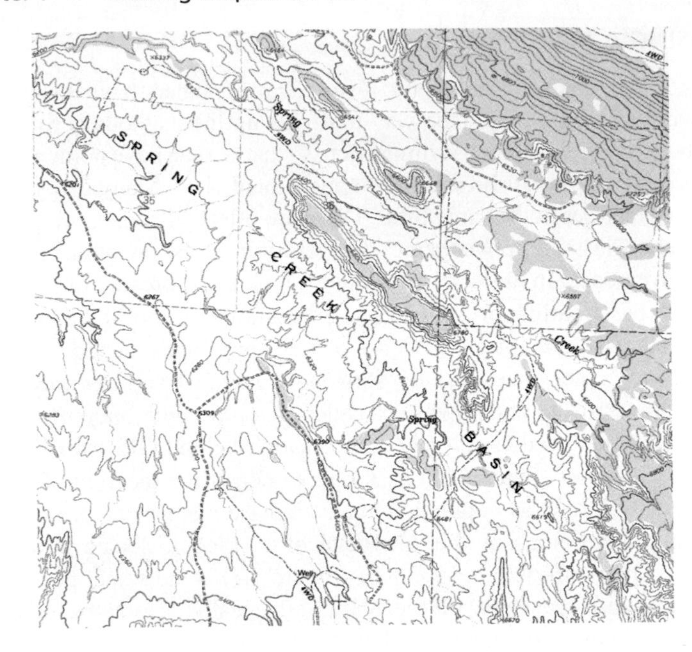

FIGURE 7.4: Topographic map. Section of USGS Digital Raster Graphic for the Dawson Draw, CO 1 : 24,000 quadrangle.

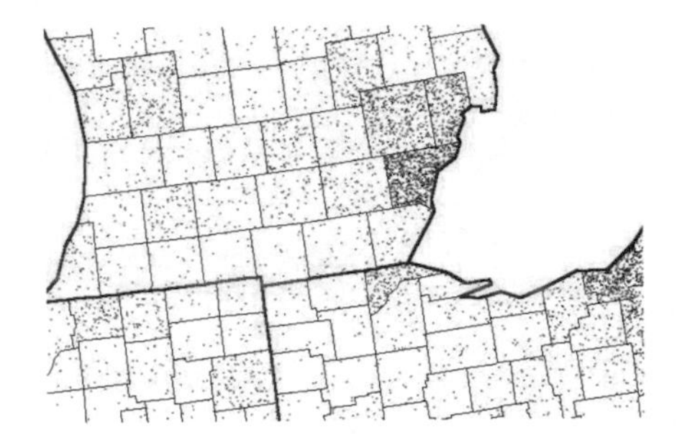

FIGURE 7.5: Dot map of population in the upper mid-west. Map by students of Sandra Arlinghaus, University of Michigan. (Used with Permission.)

(Figure 7.8) is the same, but it uses the width of the line to show value—for example, to show the air traffic volume or the amount of water flow in a stream system.

A *choropleth map* is the familiar shaded map where data are classed and areas such as states or countries are shaded or colored more or less densely according to their value. A variation on this, the unclassed choropleth, uses a continuous variation in tone or color rather than the steps that result from classes. An *area qualitative map* (Figure 7.9) simply gives a color or pattern to an area—for example, the colors of rock types on a geological map, or the land-use classes derived from image classification in remote sensing.

Volumetric data can be shown in several ways. Discontinuous data are often shown as *stepped statistical surface* (Figure 7.10), a block-type diagram viewed in perspective. The standard *isoline map* (Figure 7.11) is a map with lines joining points of equal value.

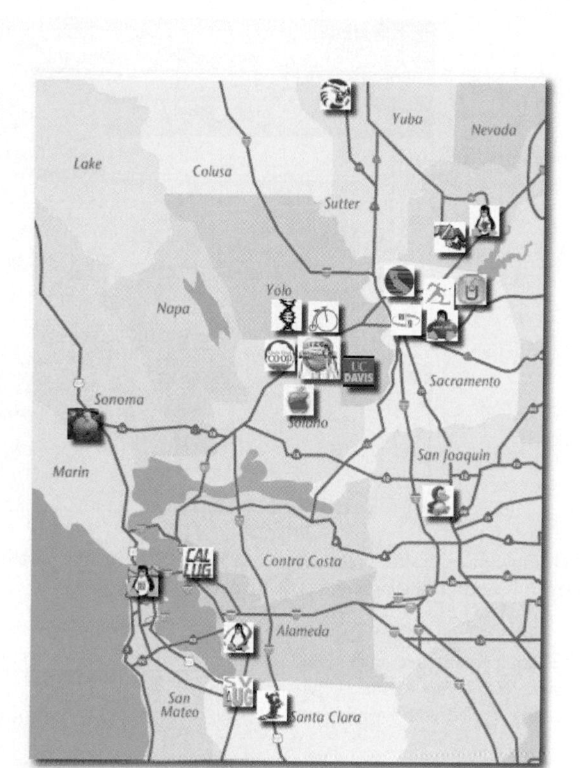

FIGURE 7.6: Picture symbol map. Image taken from Linux Users Group of Davis, CA web site http://www.lugod.org. (Used with permission.)

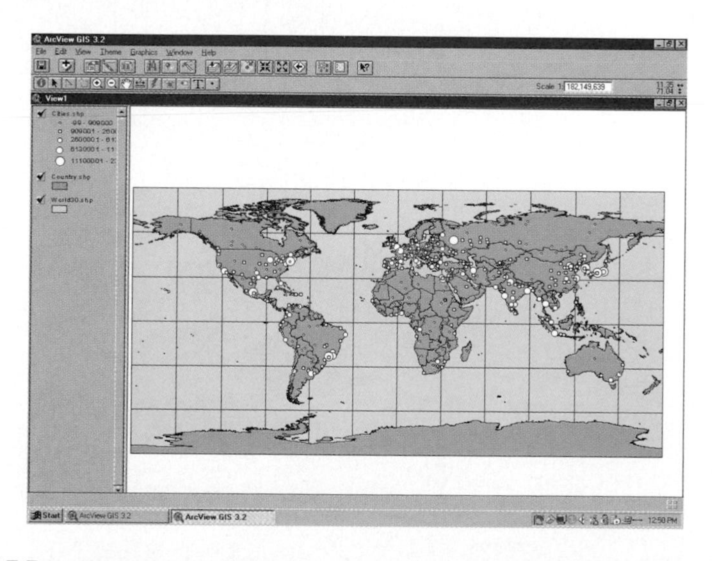

FIGURE 7.7: Graduated symbol map. Proportional circle map of world cities by population, 1995.

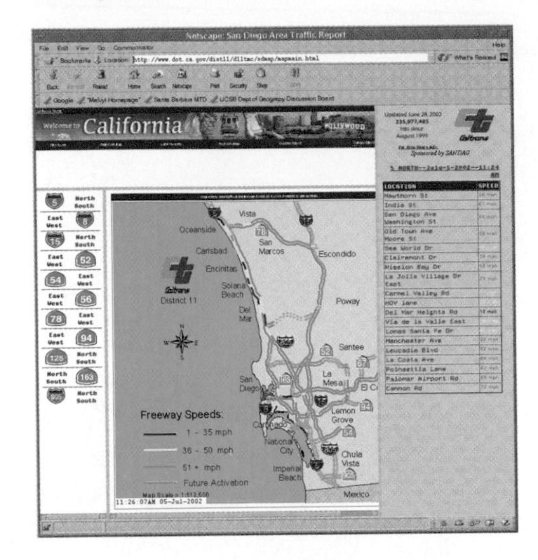

FIGURE 7.8: Flow map. Traffic speed in the San Diego, CA area as posted on the World Wide Web by the California Department of Transportation. (Used with Permission.)

FIGURE 7.9 Area qualitative map. Land use in Spokane, WA. Map by the USGS EROS Data Center. (Used with Permission.)

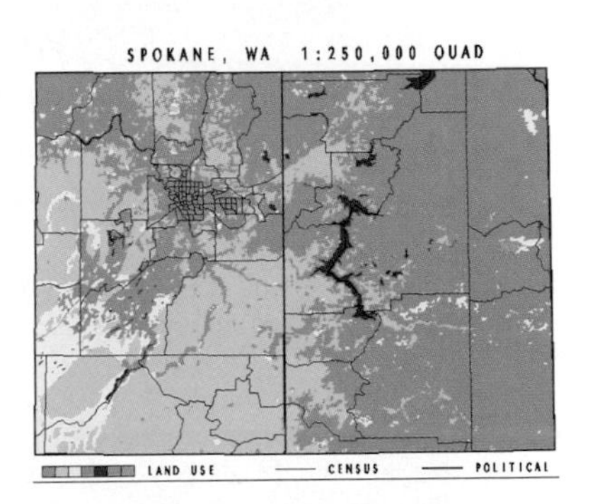

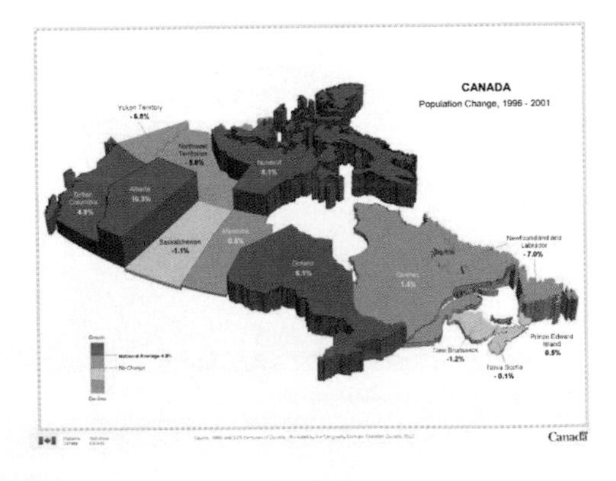

FIGURE 7.10: Stepped statistical surface. Map of population change in Canada, 1995–2001. Source: Statistics Canada. (Used with permission.)

FIGURE 7.11 Isoline map. Contour map of part of the McCall, ID 1 : 24,000 quadrangle.

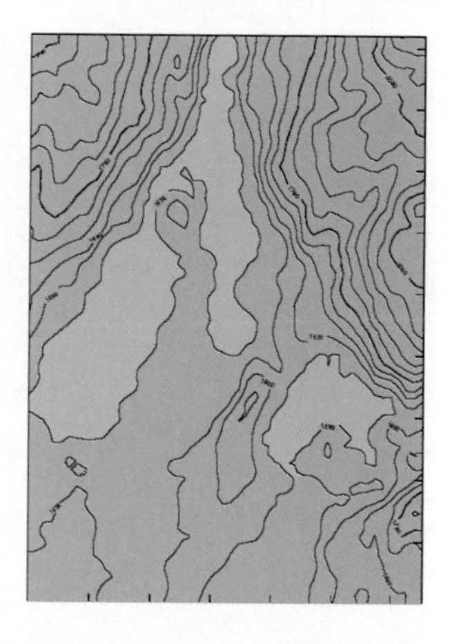

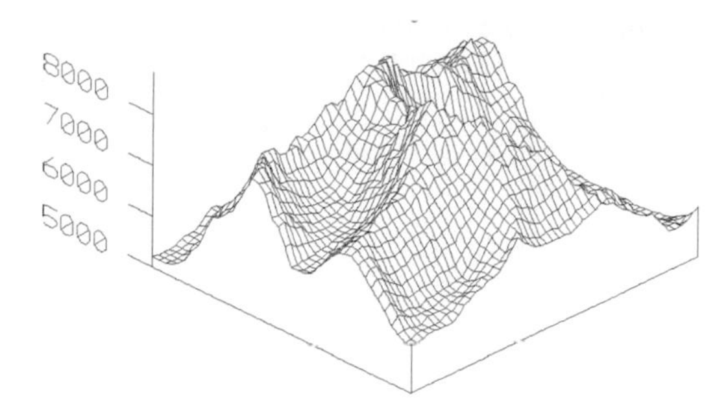

FIGURE 7.12: Fishnet or gridded perspective view. Area shown is the summit area of Mt. Everest. Source: Analytical and Computer Cartography, 2 ed. by the author.

Surface continuity is assumed, meaning that sharp breaks are usually smoothed. The terrain equivalent is the contour map, with its characteristic datum and *contour interval*. A variant is the *hypsometric map* in which the space between contour lines is filled with color using a sequence designed to illustrate variation. Image maps and schoolroom topographic maps use this type.

Three-dimensional views of surfaces rendered in perspective can be either a *gridded fishnet* (Figure 7.12), where a grid is distorted to give the impression of three dimensions, or a *realistic perspective* (Figure 7.13), when an image or shaded map is draped over the surface rather than a grid. The latter technique is often used in animations. Map views of terrain are often represented using *simulated hill shading* (Figure 7.14), where illumination of shadowing is simulated by the computer, and a gray scale or a colored map is used to show the surface. A variant is *illuminated contours*, in which the shading algorithm is applied only to the contours themselves. The final map type considered here is the *image map* (Figure 7.15), in which a value is depicted as variation in tone on a color or monochrome grid. Most raw and false-colored satellite image maps fall into this category, as does the orthophoto map.

So far we have covered the various map types. The GIS user should think of these as a set of possible methods, to be used when the GIS data to be shown have a given

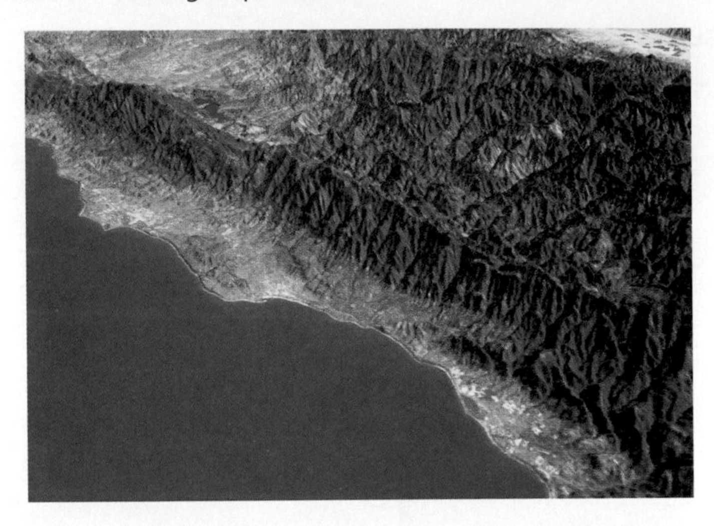

FIGURE 7.13: Realistic perspective view. Image is Landsat Thematic mapper data draped over the USGS DEM data. Image by Martin Herold and Jeff Hemphill. (Used with permission.)

FIGURE 7.14: Hill-shaded relief map. Western United States. Source: the author.

set of characteristics. Earlier in the book we classified features on a map into those that are points, lines, areas, and volumes. Obviously, the nature of the map data in the GIS is different for each of these. A three-dimensional location, for example, usually needs latitude, longitude, and elevation. In addition, the type of attribute information determines what mapping methods can be used.

Figure 7.16 places the mapping methods covered in this section into a framework of a division of the number of dimensions of the GIS features being shown. Similarly, the types of maps make certain assumptions about the nature of the attributes themselves, not just their graphic representation. For example, a reference map that shows cities has point information and text attributes—the names of the cities. The proportional circle map requires that for every point the attribute must be an integer or a floating-point number. A choropleth map requires a floating-point number that has been grouped into shade categories. These data requirements are also given in Figure 7.16.

FIGURE 7.15: Image map. Digital orthophoto quad. (Courtesy of USGS.)

POINT	LINE	AREA	VOLUME
Dot Map [1] Picture Symbol Map [1] Graduated Symbol Map [2]	Network Map [1] Flow Map [2]	Choropleth Map [2][3] Area Qualitative [3] Stepped Statistical Surface [2] Image map [1]	Isoline Map [2] Hypsometric Map [2][3] Gridded Fishnet [2] Realistic Perspective [2] Hill Shaded Map
← Reference Map [1][4] Topographic Map [1][2][3][4] →			

FIGURE 7.16: Types of maps sorted by dimension of features and type of attribute. [1] Feature present, [2] number attribute, [3] categorical attribute, [4] text present.

Particularly common mistakes in choosing mapping methods are choosing choropleth maps for data that are simply "counts," such as population totals, rather than values, rates, or percentages (instead, proportional symbol mapping should be used at the area centroids) because larger areas look like "more" simply because of their geographic size; choosing methods that classify attributes when it is unnecessary; or using too many classes for symbolization. Most other problems are problems of design and are covered in Section 7.3.

7.3 DESIGNING THE MAP

The last stage in the mapping process is the conversion of the GIS data into a map design. Note that for any map type we can have an almost infinite number of choices of symbols, fonts, colors, line thicknesses, and so on. Selecting the "best" design can make an enormous difference in the effectiveness of the map. If a map has taken a large amount of work to generate, it is well worth the GIS user's effort to make doubly sure that the design is sound.

7.3.1 Basics of Design

Some characteristics of the design are predetermined by the choice of the type of map. Primarily, the design stage consists of devising a balanced and effective set of cartographic elements to make the map. A trial-and-error interaction between a map design and a set of symbols or colors comes into play, called the *design loop*. The GIS makes this process possible in the first place by supplying the tools to create, modify, and re-create the map.

It is important to place the map elements correctly. Placement of the elements is usually in one of two ways: first, by having the GIS draw a map, then passing it to a graphic design program and interacting with the map in a design loop; and second, especially in a GIS, by editing a set of macro-like commands that move elements to specific places in the map space. This technique is less efficient and involves many traverses through the design loop.

Most cartography texts state that the cartographer should aim for harmony and clarity in the composition—visual balance and simplicity. This comes from experience and an aesthetic sense that can take years to perfect. For the beginner in GIS, MacEachren (1994) and Dent (1996) give fine summaries of the design experience of professional cartographers.

Text is an important design element. Map text should be clear, correctly and tersely worded, and the words should be positioned as the graphic elements they are. It is easy to make a map title or legend labels either too small or too big, unnecessarily grasping the map reader's attention. Map text should be edited carefully. Many a map in final form has retained a typographical error that should have been eliminated at first glance, or has misspelled a foreign name that should have been checked.

Facts to bear in mind to balance the map elements are (1) that the "weight" of the elements can change when a symbol set (line widths, colors, text fonts, etc.) is chosen; (2) that the elements act in concert with each other in a visual hierarchy, that is, some of the elements naturally stand out from or "above" others, and that using deliberately exaggerated contrast to enhance this hierarchy is usually most effective; and (3) that the combined effect of all the elements is to draw the eye to the center of gravity of the elements. Theory implies that the "visual center" of the map be placed 5% of the map height above the geometric center.

7.3.2 Pattern and Color

The symbolization aspect of design has been studied by cartographers in detail, and more than a few rules of thumb exist. Some symbolization methods are simply not suitable for certain types of maps and certain map data configurations. For example, a frequent misuse of color is on choropleth maps, especially when the computer gives access to thousands of possible colors. Choropleth maps usually establish value by shading, pattern, or color intensity, but rarely by color as such. Thus a sequence from light yellow to orange with a slight color change looks right, but a sequence from red to blue across the rainbow makes the map look like a decorated Easter egg! Color changes are appropriate to distinguish between opposites on the same map, such as a surplus/deficit, above/below a statistical average, or two-party election results.

When only monochrome is to be used, the equivalent applies. Shade sequences should be even, flowing from dark to light, with dark usually being high, and light being

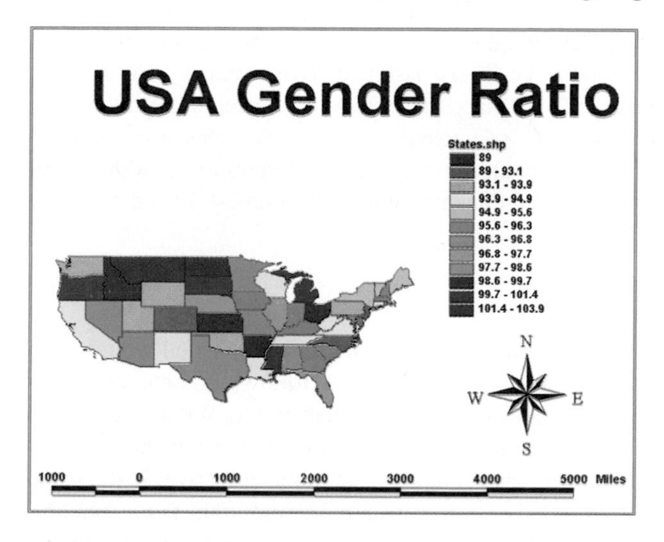

FIGURE 7.17: Some typical map symbolization errors with a GIS. These are: map title too large; map elements unbalanced within neat line; figure too small; too many classes (5 to 7 is best); class shades should not vary in hue; class shades should follow light to dark in sequence; class breaks should not overlap; compass rose adds nothing; and the scale bar is too long. Worst of all, it is impossible to see the spatial pattern in the data!

low. Don't forget that white or blank can be a shade tone, leaving the map looking less cluttered as a result. Another issue is pattern. Combinations of crosshatching, dot patterns, and so on can be extremely confusing to the map reader (Figure 7.17). Combining unmatched patterns can create undesirable optical illusions.

Even on general-purpose maps, color balance is essential. Computer displays use pure color, to which the eye is not usually subjected. Less saturated colors, if available, are more suitable for mapping. In addition, *cartographic convention* should be followed. Ground colors are usually white, gray, or cyan, not black or bright blue. Contours are frequently brown, water features cyan, roads red, vegetation and forest green, and so on. Failure to follow these conventions is particularly confusing to the map reader. Imagine, for example, a globe with green water and cyan land! Map colors can also look completely different on a white rather than a black background, and even on different monitors and plotters.

Color is a complex visual variable. Colors are often expressed as red, green, blue triplets (RGB) or sometimes as *hue*, *saturation*, and *intensity* (HSI). These values are either determined by the hardware device (e.g., 8-bit color allows a total of 256 colors from any of $256 \cdot 256 \cdot 256$ combinations of individual values of RGB) or are decimal values of HSI between zero and one. For example, in RGB, a mid-gray would correspond to [128,128,128]. It is possible to translate directly between the RGB and HSI representations of color. Whereas RGB values are simply the degree to which the respective colored phosphors of the monitor emit light, HSI is closer to the way in which people perceive color.

Hue corresponds to the wavelength of light, going from red at the long-wave end of the visible light spectrum to blue at the other end. *Saturation* is the amount of color per unit display area, and *intensity* is the illumination effect or brightness of the color. Cartographic convention dictates that hue is assigned to categories and that saturation or intensity is assigned to numerical value. When several hues appear in juxtaposition on a map, the colors are perceptually altered by the eye, a phenomenon known as *simultaneous contrast*. Thus maps that use several hues, even as background and line color, should be designed with caution. In addition, the eye's ability to resolve contrast varies significantly with hue, highest in red and green and lowest in yellow and blue.

7.3.3 Summary

The design of a map is a complex process. Good design requires planning, achieving visual balance among map elements, following conventions, employing the design loop, and correctly using symbols and map types. Without consideration of design, and certainly without having all the required map elements, however impressive it may look on a computer screen, the product is just not as effective. If the map is the result of a complex GIS process, good design is even more important to the person who will have to interpret the map. As we have seen, the relationship between cartography and GIS is a close one. While making a map is often given little thought in the GIS process, it is nevertheless an important stage because it is using maps that particularly distinguishes GIS as being a different scientific approach, and it is the map that has the primary visual impact on the GIS user or decision maker using GIS. Just a little extra care and attention to detail at this final stage can lead to immense improvements in the finished GIS product and to the perception that the entire information flow used in the GIS process is professional and complete.

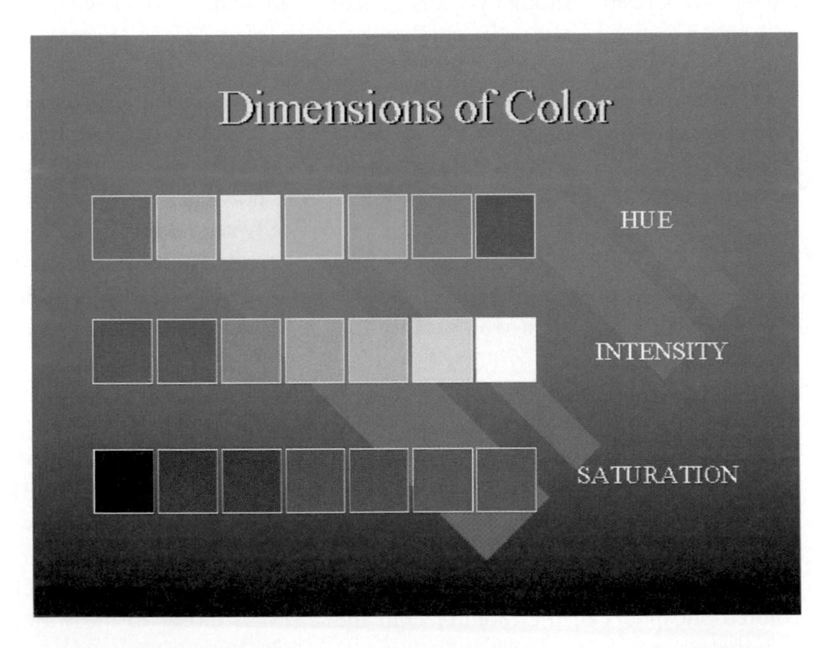

FIGURE 7.18: The dimensions of color. Hue or chroma defines the wavelength of the light reflected or transmitted by the map. Intensity is the level of brightness associated with the color. Saturation is the density of the color on the map surface, expressible as the proportion of the surface area covered in the color. Hue communicates different classes and things, while saturation and intensity communicate numerical level or value.

7.4 STUDY GUIDE

7.4.1 Summary

Chapter 7: Making Maps with GIS

The Parts of a Map (7.1)

- **A map is defined as a graphic depiction of all or part of a geographic realm in which real-world features have been replaced with symbols in their correct spatial location at a reduced scale.**
- **In a GIS, maps can be produced as temporary output; to check a result, answer a query, and so on; or as permanent output as a fully featured cartographic product.**
- **To appear professional and avoid errors, GIS maps should reflect cartographic knowledge about map design.**
- **A map has a visual grammar or structure that must be understood and used if the best map design is desired.**
- **The selection of a map type is often determined by the geographic properties of the data and the attributes.**
- **A map is composed of a set of basic cartographic elements, including the neat line, scale, border, figure, ground, labels, insets, credits, legend, and title.**
- **The figure is that part of the map shown in ground coordinates rather than laid out in page coordinates.**
- **Map text, especially labels on features, follows a set of placement rules that are specific to the dimension of the feature and properties of the map.**

Choosing a Map Type (7.2)

- **There are many different types of maps, divided by purpose into thematic and general purpose, and by dimension.**
- **Types of point maps found in GIS software are dot maps, picture symbol maps, and graduated symbol maps.**
- **Types of line maps common in GIS systems include network maps and flow maps.**
- **Some types of volume maps that GIS can produce are isolines, hypsometric maps, gridded fishnet, realistic perspectives, and hill-shaded maps.**

Designing the Map (7.3)

- **A GIS map is designed in a process called the design loop.**
- **Good map design requires that map elements be placed in a balanced arrangement within the neat line.**
- **Visual balance is affected by the "weight" of the symbols, the visual hierarchy of the symbols and elements, and the location of the elements with respect to each other and the visual center of the map.**
- **Symbols, especially colors, are subject to the constraints of cartographic convention (e.g., forests should be green).**
- **Color is a complex visual variable, and in a GIS is specified by RGB or HSI values.**

- **Design errors are common with GIS use, and these include incorrect selections of map type and symbolization errors.**
- **When a GIS map is the result of a complex analytical or modeling process, good design is essential for understanding.**
- **The map is what distinguishes GIS as a different approach to the management of information, so extra care should be taken to improve the final maps that a GIS generates in a GIS task.**

7.4.2 Study Questions

The Parts of a Map

Using a map that you have found in a newspaper or magazine, identify the map elements listed in Figure 7.1 and label them on the map. Are any of the elements missing? Could the map have been improved by adding any of the elements listed in the figure?

Using a USGS topographic quadrangle or any other general reference map such as a wall map, a road map, or an atlas map, copy onto a diagram examples of label placement for point, line, and area features. Are there any examples where the "rules" of text placement have been violated? How has the cartographer dealt with the problem that in dense label areas, features names would overlap?

Name six items that could legitimately be found within a map's border and outside the neat line (not coffee stains!).

Choosing a Map Type

Make a list of the different types of maps listed in Section 7.2. Using Figure 7.16, verify that the classification of feature dimension is correct. Which map types cross categories? Can you find examples of cartographic methods that cross the boundaries of these types?

Make a set of conditions for data to be suitable for display on a choropleth map.

Designing the Map

Give three simple rules for a complete GIS novice to keep in mind when using a GIS to produce a map.

What design issues should be kept in mind when making a choropleth map?

Annotate Figure 7.17 with each of the mistakes labeled in the caption. Can you see any other mistakes? Can you find any mistakes in other figures in the book?

7.5 EXERCISES

1. *Carefully read the documentation for a GIS to which you have access, and compile a list of what map types the software is capable of producing, compared to the map types listed in Section 7.2. Is the subset of map types suited to a particular dimension of data attributes, such as areas?*
2. *Use your GIS package to draw a simple choropleth map. What tools are available within the GIS to assist in the choice of classes for the choropleth data? Does your GIS allow you to make choropleth maps using values other than ratios or percentages? Is there any guidance for choice of colors, shades, or a map layout? How might the documentation for the system be improved to guide the new cartographer better?*

3. *Make two different maps with your GIS of the same data, one in which you choose a design to enhance differences in the data and one where you try to hide them. Show the maps to some friends or colleagues and ask them about the distributions. Can their opinions about the data be shaped by the choice of symbols for a single map type? Repeat the task using the same data and two sets of symbols, say gray tones versus shading, or red hues versus green hues.*
4. *Using a topographic map or any map you choose, perhaps from the documentation for your GIS, analyze the placement of labels on the maps. Check a cartography textbook for the conventional cartographic rules of label placement. Can the GIS change placement of the labels?*

7.6 BIBLIOGRAPHY

Dent, B. D. (1996) *Cartography: Thematic Map Design*, 4th ed. Dubuque, IA: Wm. C. Brown.

Imhof, E. (1975) "Positioning names on maps," *The American Cartographer*, vol 2, pp. 128–144.

MacEachren, A. M. (1994) *SOME Truth with Maps: A Primer on Symbolization and Design.* Washington, DC: Association of American Geographers Resource Publications in Geography.

Robinson, A. H., Sale, R. D., Morrison, J. L., and Muehrcke, P. C. (1984) *Elements of Cartography*, 5th ed. New York: Wiley.

7.7 KEY TERMS AND DEFINITIONS

area qualitative map: A type of map that shows the existence of a geographic class within areas on the map. Colors, patterns and shades are generally used. Examples are geology, soils, and land-use maps.

border: The area between the neat line and the edge of the medium or display area on which a map is being displayed. Occasionally, information can be placed within the border, but this area is usually left blank.

cartographic convention: The accepted cartographic practice. For example, water is usually cyan or light blue on a world map.

cartographic elements: The primitive component part out of which a map is assembled, such as the neat line, legend, scale, titles, figure, and so on.

choropleth map: A map that shows numerical data (but not simply "counts") for a group of regions by (1) grouping the data into classes, and (2) shading each class on the map.

clarity: The property of visual representation using the absolute minimum amount of symbolism necessary for the map user to understand map content without error.

color balance: The achievement of visual harmony between colors on a map, primarily by avoiding colors that show simultaneous contrast when adjacent to each other.

contour interval: The vertical difference in measurement units such as meters or feet between successive contour lines on a contour map.

contour map: An isoline map of topographic elevations.

credits: A cartographic element in which the sources, authorship, and ownership of the map and the map attributes are cited, often including a date or reference.

design loop: The iterative process in which a GIS map is created, examined for design, improved, and then replotted from the modified map definition until the user is satisfied that a good design has been reached.

dot map: A map type that uses a dot symbol to show the presence of a feature, relying on a visual scatter to show spatial pattern. Most often used where point features are the GIS data, but dots can be scattered at random throughout areas.

figure: The part of a map that is both referenced in the map coordinate system rather than the page layout coordinates and that is the center of the map reader's attention. The figure is contrasted against the ground, or background. For example, on a map of New York State, the state is the figure, and surrounding states, though shown and labeled, are part of the ground and may be toned down.

flow map: A linear network map that shows, usually by proportionally varying the width of the lines in the network, the amount of traffic or flow within the network.

fonts: A consistent design for the display of the full set of English or other language characters, including special characters such as punctuation and numbers.

graduated symbol map: A map type that varies the size of a common geometric symbol to show the amount of an attribute at points or at centroids of areas. For example, cities could be shown with circles of area proportional to population, or census tracts could have a proportional circle divided as a pie chart at a representative point inside the tract.

gridded fishnet map: A map of a three dimensional surface showing a set of profiles, often parallel to the x, the y, or the viewer's axis so that the surface appears three dimensional, as a raised fishnet viewed in perspective.

ground: The part of the body of the map that is not featured in the figure. This area can include neighboring areas, oceans, and so on. The ground should fall lower than the figure in the visual hierarchy.

harmony: The property by which the elements of a map work together to create a balanced aesthetic whole.

HSI: A system for color, specified as values for hue, saturation, and intensity, respectively.

hue: A color as defined by the wavelength of the light reflected or emitted from the map surface.

hypsometric map: A map of topography involving a color sequence filling the spaces between successive contours, usually varying from green through yellow to brown.

image map: A map that in two dimensions shares many of the characteristics of a map, that is, cartographic geometry, some symbols, a scale and projection, and so on, but is a continuous image taken from an air photo, a satellite image, or a scanner. A scanned paper map used as a backdrop in a GIS becomes an image map.

inset: A map within a map, either at a smaller scale to show relative location, or a larger scale to show detail. An inset may have its own set of cartographic elements, such as a scale and graticule.

intensity: The amount of light emitted or reflected per unit area. A map that has high intensity appears bright.

isoline map: A map containing continuous lines joining all points of identical value.

label: Any text cartographic element that adds information to the symbol for a feature, such as the height number label on a contour line.

label placement rules: The set of rules that cartographers use when adding map text, place-names, and labels to features. Some rules are generic to the map as a whole, while others relate to point, line, and area features specifically. Well-designed maps follow the label placement rules and use them to resolve conflicts between the labels, as labels should never be plotted over each other.

legend: The map element that allows the map user to translate graphic map symbols into ideas, usually by the use of text.

line thickness: The thickness, in millimeters, inches, or other units, of a line as it appears on a map.

map: A graphic depiction of all or part of a geographic realm where the real-world features have been replaced with symbols in their correct spatial location at a reduced scale.

map design: The set of choices relating to how a map's elements are laid out, how symbols such as colors are selected, and how the map is produced as a finished tangible product. The process of applying cartographic knowledge and experience to improve the effectiveness of a map.

map title: Text that identifies the coverage and content of a map. This is usually a major map element and can be worded to show the map theme or the map's content.

map type: One of the set of cartographic methods or representation techniques used by cartographers to make maps of particular types of data. Data, by their attributes and dimensions, usually determine which map types are suitable in a map context.

neat line: A solid bounding line forming the frame for the visually active part of a map.

network map: A map that shows as its theme primarily connections within a network, such as roads, subway lines, pipelines, or airport connections.

orthophoto map: An image map that is an air photo, corrected for topographic and other effects. A specific type of mapping program, at 1 : 12,000, by the USGS.

page coordinates: The set of coordinate reference values used to place the map elements on the map and within the map's own geometry rather than the geometry of the ground that the map represents. Often, page coordinates are in inches or millimeters from the lower left corner of a standard-size sheet of paper, such as A4 or $8\frac{1}{2}$ by 11 inches.

permanent map: A map designed for use as a permanent end product in the GIS process.

picture symbol map: A map type that uses a simplified picture or geometric diagram at a point to show a feature type. For example, on a reference map, airports could be shown with a small airplane stick diagram, or picnic areas by a picnic table diagram.

place-name: A text cartographic element that links text for a geographic place to a feature by placing it close to the symbol to which it corresponds, such as a city name as text next to a filled circle.

realistic perspective map: A map of a three-dimensional surface showing a colored or shaded image draped over a topographic surface and viewed in perspective.

reference map: A highly generalized map type designed to show general spatial properties of features. Examples are world maps, road maps, atlas maps, and sketch maps. Sometimes used in navigation, often with a limited set of symbols and few data. A cartographic base reference map is often the base layer or framework in a GIS.

RGB: The system of specifying colors by their red, green, and blue saturations.

saturation: The amount of color applied per unit area. Perceptually, saturated colors appear rich or solid, whereas low-saturation colors look washed out or pastel-like.

scale: The part of the map display that shows the scale of the map figure as either an expression of values (the representative fraction as a number) or as a graphic, usually a line on the map labeled with an equivalent and whole-number length on the ground, such as 1 kilometer or 1 mile.

simulated hill-shaded map: A map in which an apparent shading effect of raised topography is produced by computer (or manually) so that the land surface appears differentially illuminated, as it would in low sun angles naturally.

simultaneous contrast: The tendency for colors at the opposite ends of the primary scale to perceptually "jump" when placed together; for example, red and green.

stepped statistical surface: A map type in which the outlines of areas are "raised" to a height proportional to a numerical value and viewed in apparent perspective. The areas then appear as columns, with a column height proportional to value.

symbol: An abstract graphic representation of a geographic feature for representation on a map. For example, the feature could be a canal, the symbol a blue line of a given thickness.

symbolization: The full set of methods used to convert cartographic information into a visual representation.

temporary map: A map designed for use as an intermediate product in the GIS process and not usually subjected to the normal map design sequence.

topographic map: A map type showing a limited set of features but including at the minimum information about elevations or landforms. Examples: contour maps. Topographic maps are common for navigation and for use as reference maps.

visual center: A location on a rectangular map, about 5% of the height above the geometric center, to which the eye is drawn perceptually.

visual hierarchy: The perceptual organization of cartographic elements such that they appear visually to lie in a set of layers of increasing importance as they approach the viewer.

CHAPTER 8

How to Pick a GIS

8.1 THE EVOLUTION OF GIS SOFTWARE
8.2 GIS AND OPERATING SYSTEMS
8.3 GIS FUNCTIONAL CAPABILITIES
8.4 GIS SOFTWARE AND DATA STRUCTURES
8.5 CHOOSING THE "BEST" GIS
8.6 STUDY GUIDE
8.7 EXERCISES
8.8 REFERENCES
8.9 KEY TERMS AND DEFINITIONS

Was there ever in anyone's life span a point free in time, devoid of memory, a night when choice was any more than the sum of all the choices gone before?
—Joan Didion, Run River, *1963*

You pays your money and you takes your choice.
—Punch, 1846

"Can't act. Can't sing. Balding. Can dance a little."
—MGM executive report on Fred Astaire's screen test

"That's all well and good, bud, but does it come with a remote?"

8.1 THE EVOLUTION OF GIS SOFTWARE

One of the first tasks a GIS user faces is deciding which GIS software to use. Even if a GIS has already been purchased, installed, and placed right in front of your nose, it is very natural to wonder whether some other GIS system might be better, faster, easier to use, have clearer documentation, or be better suited to the actual task you are working on. This chapter gives some of the background necessary to make an intelligent GIS selection. There is quite a history to learn from, including some excellent accounts of spectacular failures, but also many examples of clear statements of how things went right. Examples from the early days of GIS are the papers by Tomlinson and Boyle (1981) and Day (1981). The philosophy here is that the educated consumer is the best GIS user, and an effective user soon becomes an advocate and sometimes a GIS evangelist. This chapter is not intended to tell you which GIS to buy or use. Rather, it is hoped that, it will help you to decide this for yourself.

As is often the case, a good education begins with a little history. Chapter 1 introduced overall GIS development in terms of the distant origins of geographic information science as a whole. This was difficult to do without mentioning specific GIS software packages. Now it is appropriate to discuss the development of software in more detail.

8.1.1 The Ancestors of GIS Software

GIS software did not suddenly appear as if by magic. There was a lengthy period leading up to the first real GISs during which the breed evolved rather rapidly. As we saw in Chapter 1, the intellectual ancestry included the creation of a spatial analysis tradition in geography, the quantitative revolution, and dramatic technological and conceptual improvements in the discipline of cartography.

An early GIS landmark was an international survey of software conducted by the International Geographical Congress in 1979 (Marble, 1980). This survey had three volumes, one of which was entitled *Complete Geographic Information Systems*, although in fact few true GIS packages were represented. This volume was influential in deciding on the name "GIS" because many alternatives were in use at that time. Just as important were the two volumes *Cartography and Graphics* and *Data Manipulation Programs*. Together, these three volumes encapsulated the state of geographic data processing in the 1970s (Brassel, 1977). Most cartographic programs were single-purpose FORTRAN programs to do individual GIS operations such as digitizing, data format conversion, plotting on a specific hardware device such as a pen plotter, map projection transformations, or statistical analysis of data. None of these packages were integrated; a typical use would be to apply a series of one-at-a-time geographic operations to arrive at a final result or map.

Some of the early computer mapping systems had already devised many GIS functions by this time, however. Among these were SURFACE II by the Kansas Geological Survey, which could do point-to-grid conversions, interpolation, surface subtraction, and surface and contour mapping; CALFORM, a package that could produce thematic maps; SYMAP, a sophisticated analytical package from the Harvard Laboratory for Computer Graphics and Spatial Analysis that nevertheless ran only on mainframe computers and gave line-printer plots; and the Central Intelligence Agency's CAM, which made plots from the World Data Bank outline maps with different map projections and features.

By 1980 the first computer spreadsheet programs had arrived, led by the VisiCalc program, a very early microcomputer software "killer app." VisiCalc contained only a few of the capabilities of today's equivalent packages, yet for the first time gave the ability to store, manage, and manipulate numbers in a simple manner. Above all, data could be seen as active in a spreadsheet rather than as a static "report" that consisted of a pile of computer printout. The links to statistical graphics, now common in packages such as SASGRAPH and Harvard Graphics, were a natural extension of this capability.

The ancestry of GIS is completed by the first advances in database management systems. Early systems for database management were based on the less sophisticated data models of the hierarchical and related data models. A landmark was the beginning of the relational database managers in the early 1970s. Relational database managers quickly became the industry standard, first in the commercial world of records management and later in the microcomputer world.

8.1.2 The Early GISs

By the late 1970s all of the necessary parts of a GIS existed as isolated software programs. The largest gap to be filled was between the relational database manager and the programs that dealt with plotting maps. The specific demands of hardware devices from particular manufacturers kept this as a constantly evolving field, with frequent rewrites and updates as systems and hardware changed. Later, the device independence attributable to common operating systems such as Unix and computer graphics programming standards such as GKS, Core, and PHIGS led to a narrowing of this chasm, to the point where today it remains as barely a discernible dip in the GIS ground. The scene was set for the arrival of the first true GISs.

As we saw in Chapter 1, one of the earliest civilian systems to evolve all the capabilities of a true GIS was the CGIS (Canadian Geographical Information System), mostly because this system was the first to evolve from an inventory system toward doing analyses and then management. Essential to the emergence were the georeferencing and geocoding of the data, database management capability, a single integrated software package without separate, stand-alone elements, and a single user interface.

At first, GIS packages had unsophisticated user interfaces, and many actually made the user write short computer program–like scripts or to type highly structured formatted commands one at a time into the computer in response to prompts. As the GIS software evolved, the need for upward compatibility—that is, the need for existing users to be satisfied with a new version because things still work in much the same way as before—meant that many systems preserved elements of these older user interfaces long after they had been replaced by better tools.

The second generation of GIS software included graphical user interfaces, usually involving the use of windows, icons, menus, and pointers. In the typical configuration today, the windows are standardized by the operating system and function in the same way that it does, "inheriting" its characteristics. A first generation of GIS software used windows custom-built by the vendor. Later, after the broad distribution of windowing systems such as X-Windows and Microsoft Windows, the graphical user interface (GUI) tools that are part of the operating system became accessible to software designers and programmers.

The typical system has pop-up, pull-down, and pull-right menus for selecting choices. Choices and locations are indicated with a mouse, although some systems use track balls or light pens. Similarly, the typical GIS can support multiple windows—for example, one for the database and one to display a map—and the tasks can be opened and closed as needed. While closed, they function in the background while they are graphically represented on the screen as an icon or small picture.

8.2 GIS AND OPERATING SYSTEMS

Early GIS was heavily influenced by the types of operating systems in use. Early operating systems were quite unsophisticated but were used with GIS nevertheless. Among these were IBM's mainframe operating systems, MSDOS by Microsoft, and DEC's VMS. These were rapidly replaced as the various GUI-based operating systems came into operation and as the microcomputer and workstation took over from the minicomputer and mainframe.

In the microcomputer environment, the GUI-based operating systems include Windows, Windows-NT, and Windows 95. The unified user interface, revolutionized by the Apple Macintosh's GUI and desktop metaphor, quickly took over as the dominant microcomputer operating environment, although others, such as IBM's OS/2, have remained popular also. These operating systems added two critical elements to the microcomputer's capabilities: *multitasking* (allowing many simultaneous work sessions) and *device independence*, meaning that plotters and printers could be taken out and assigned to the operating system instead of the GIS package, in somewhat the way that printing and screen fonts are handled centrally, rather than duplicated in every Windows package.

One system that had encompassed these capabilities since its inception, and that swept the workstation environment, was Unix. Unix is a very small and efficient central operating system that is highly portable across computer systems. It has been the dominant workstation environment for two reasons: first, because it has complete integrated network support, and second, because several full GUIs exist for Unix in the public domain, the most important being the X-Windows system. X-Windows implementations of most leading GUIs exist, including OpenLook and the Open Software Foundation's MOTIF interface. In many Unix systems, the user can switch the GUI to suit particular needs or applications. A full GUI programming tool kit, including such tools as Xt, Xview, and the X-Windows libraries Xlib, is part of the X-Windows release.

As a final benefit, several versions of Unix and all of the GUI systems run extremely efficiently on microcomputers, including shareware Unix releases such as Linux, not only outperforming the Windows-type GUIs, but being available free or as shareware on the Internet or from inexpensive suppliers on CD-ROM. A key element here has been the Free Software Foundation's releases, including GNU (GNU is NOT UNIX) versions of virtually every key element of Unix.

Thus, two main avenues for GISs have evolved as far as operating systems are concerned. On the microcomputer platform a lingering set of DOS applications is rapidly being rewritten for the updated versions of Microsoft's Windows. In this GIS environment, the number of systems installed, the mobility of laptop and subnotebook computing, and the low cost of software have been major strengths. On the workstation platform, Unix and X-Windows, often with MOTIF as the GUI, reign supreme. This work environment has led to high-end applications, large data sets, networking, depth of software, and high-quality graphics. Both are healthy and prospering workplaces for GIS.

8.3 GIS FUNCTIONAL CAPABILITIES

A GIS is often defined not for what it is but for what it can do. This *functional definition* of GIS is very revealing about GIS use, because it shows us the set of capabilities that a GIS is expected to have. A minimal set of capabilities can be outlined and each GIS package held up to see whether it qualifies. A thorough examination of GIS capabilities is the critical step in how to pick a GIS, because if the GIS does not match the requirements for a problem, no GIS solution will be forthcoming. In contrast, if the GIS has a large number of functions, the system may be too sophisticated or elaborate for the problem at hand—a sledgehammer to crack a nut.

The functional capabilities can be grouped by the categories we have used in this book, closely following the Dueker definition of GIS. These are capabilities for data capture, data storage, data management, data retrieval, data analysis, and data display. These "critical six" functions must always be present for the software to qualify as a GIS. We examine each of these in turn.

8.3.1 Data Capture

As we saw in Chapter 4, getting the map into the computer is a critical first step in GIS. Geocoding must include at least the *input* of scanned or digitized maps in some appropriate format. The system should be able to absorb data in a variety of formats, not just in the native format of the particular GIS. For example, an outline map may be available as an AutoCAD DXF format file. The GIS should at a minimum be capable of absorbing the DXF file without further modification. Similarly, attributes may already be stored in standard database format (DBF) and should be absorbable either directly or through the generic ASCII format.

Before a map can be digitized, however, it needs to be prepared. Different GIS packages handle the amount of preparation required in quite different ways. If the package supports scanning, the map needs to be clean, fold-free, free of handwritten annotation and marks, and on a stable base such as Mylar. If the map is digitized by hand it may need to be cut and spliced if the package does not support *mosaicing* (Figure 8.1), and control points with known locations and coordinates need to be marked for registering the map onto the digitizing tablet. Some GIS packages have extensive support for *digitizing* and sophisticated *editing* systems for detecting and eliminating digitizing errors. Others have few or none.

We also saw in Chapter 3 how essential it was to edit the maps after they have been captured. This requires the software to have an editing package or module of some kind. For a vector data set, at the minimum we should be able to delete and reenter a point or line. For a raster, we should be able to modify the grid by selecting subsets, changing the grid spacing, or changing a specific erroneous grid value.

Other functions typical of an editor are *node snapping*, in which points that are close to each other and that should indeed be the same point, such as the endpoints

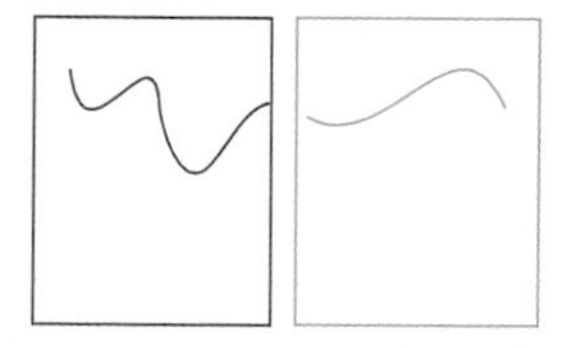
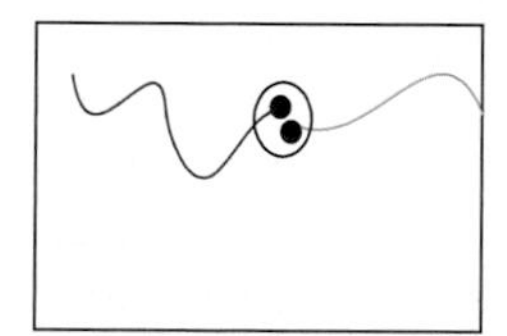
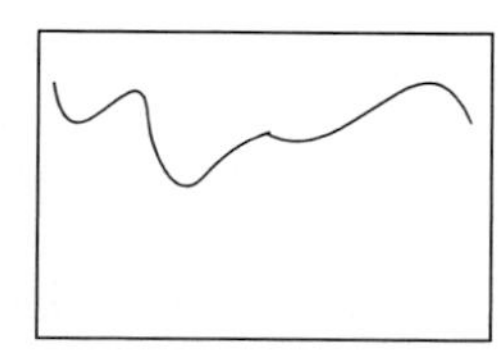

FIGURE 8.1: Steps in mosaicing. Left: Two maps show one feature, but there is a gap. Center: Map edge is merged; nodes are snapped to "zip" feature. Right: Mosaiced map with continuous feature and dissolved map edge.

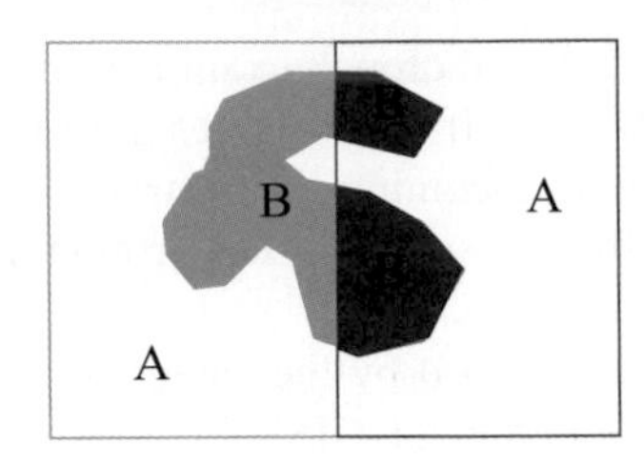

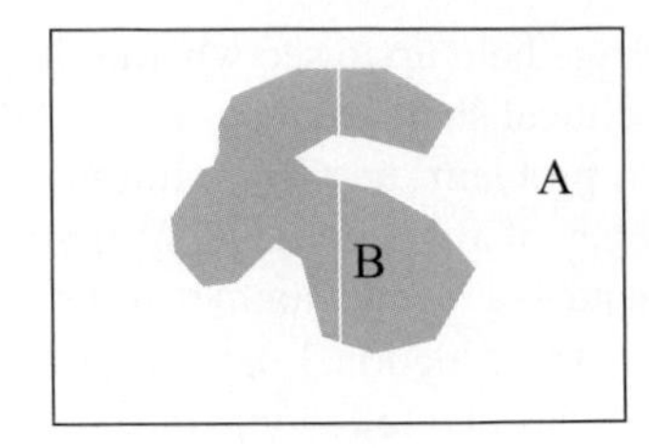

FIGURE 8.2: Steps in the dissolve operation. Left: Two maps show one feature, split across an edge. Attribute and graphic database have three records for type "B". Right: After dissolve, edge lines are removed and the three type "B" records are amalgamated.

FIGURE 8.3: Map generalization includes eliminating all points along a line falling within a band or buffer around a simplification of the line, with a width sometimes called the "fuzzy tolerance." This way, minor errors such as sliver polygons can be eliminated when data have been captured from maps at different scales.

of a line segment, are automatically placed into the graphic database with the identical coordinates; *dissolve*, when duplicate boundaries or unnecessary lines (e.g., the digitized edges of adjacent category-type maps) are eliminated automatically or manually; and *mosaicing* or "zipping," in which adjacent map sheets scanned or digitized separately are merged into a seamless database without the unnecessary discontinuities caused by the lack of edge matching of the paper maps (Figure 8.2). For example, a major road that crosses two map sheets does not need to be represented as two separated features in the final GIS database.

Another important editing function is the ability to deal with map *generalization*. Many digitizing modules of GIS systems, and certainly scanning, generate far more points than are necessary for the use of the GIS. This extra detail can complicate data reformatting and display, slow the analysis process, and lead to memory problems on the computer. Many GIS packages allow the user to select how much detail to retain in a feature. Most will retain points that have a minimum separation and snap together all points within a fuzzy tolerance (Figure 8.3).

For point data sets, most GIS packages will eliminate or average duplicate points with the same coordinates. Some will allow *line generalization*, using any one of many algorithms that reduce the number of points in a line. Common methods include extracting every nth point along the line (where n can be 2, 3, etc.) according to the amount of generalization required, and Douglas–Peucker point elimination, which uses a displacement orthogonal to the line to decide whether a point should be retained (Figure 8.4).

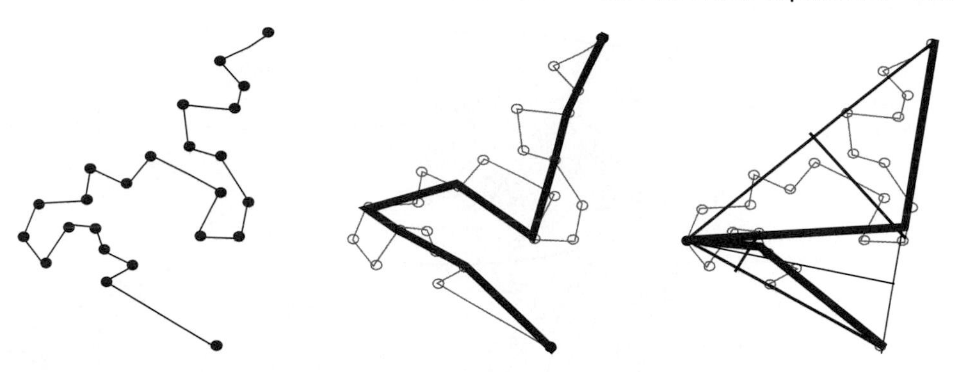

FIGURE 8.4: Line generalization alternatives. The line (left) can be resampled by retaining every nth point (center), or by repeatedly selecting the most distant point from a line between end nodes (right) and redividing the line until a minimum distance is reached, the Douglas–Peucker method.

Area features can be eliminated if they become too small, or can be grouped together, a process many GIS packages call *clumping*. It is also possible to generalize in the attributes, joining classes together, for example.

To be useful, a GIS must provide tools above and beyond the editor to check the characteristics of the database. Checking the attributes is the responsibility of the database manager. The database system should enforce the restrictions on the GIS that are specified during the data definition phase of database construction and stored in the data dictionary. Most of this checking is done at data-entry time. It checks to determine that values fall within the correct type and range (a percentage numerical attribute, for example, should not contain a text string and should have a record of less than or equal to 100).

More intricate and demanding are checks on the map data. Some GIS packages, which do not support topological structuring, do not enforce any restrictions on the map. Some simply check ranges; for example, every grid cell should have a data value between 0 and 255 in an image map. These systems run the risk of lacking a match between the attributes and the space they represent. No part of the map, for example, should fall into two separate areas—that is, the areas on a polygon map should not overlap or leave gaps. This happens when maps are captured at different scales or from inaccurate sources.

Topological GIS systems can check automatically to ensure that the lines meet at nodes and that the entire map area is covered by polygons without gaps or overlaps. Beyond simply checking, many GIS packages allow automatic cleaning of topology, snapping nodes, eliminating duplicate lines, closing polygons, and eliminating slivers. Some systems simply point out the errors and ask the user to eliminate them with the editor. Some go ahead and make the corrections without user intervention. The GIS user should be careful when using automatic cleaning, for the tolerances may eliminate important small features or move the features around in geographic space without accountability. A specific GIS package may or may not be able to deal directly with GPS data conversion, with survey-type data from COGO (coordinate geometry) systems, or with remotely sensed imagery. Some GIS packages have both functions—that is, they serve as GIS and image processing systems. Among these are Idrisi, GRASS, and ERDAS.

Essential to geocoding capabilities, because GIS allows maps from many sources to be brought into a common reference frame and to be overlaid, is the geocoding software's ability to move between coordinate systems and map projections. Most GIS packages

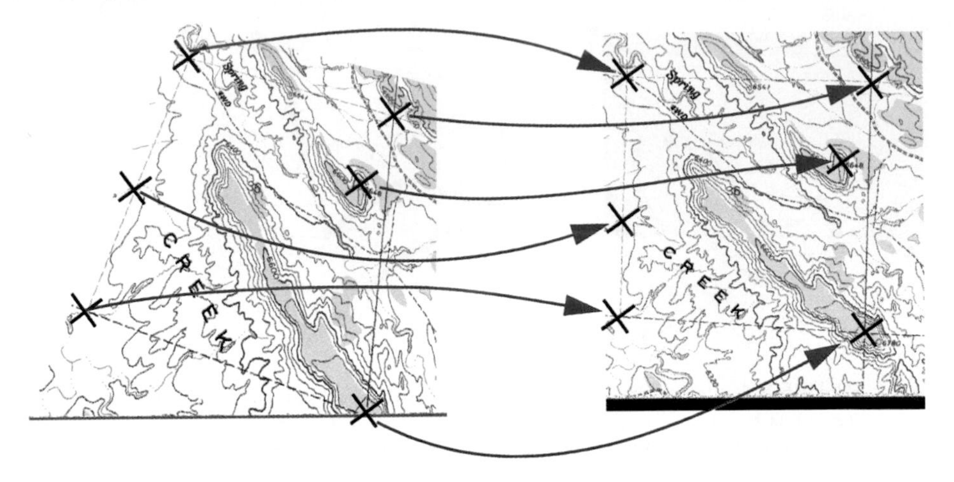

FIGURE 8.5: The rubber sheeting method. A map with unknown geometry (say an air photo or scanned map) can be distorted so that its geometry matches that of another map. Pairs of points must be available both on the image and on the map showing the same place or feature location, called control points. Within the GIS, rubber sheeting warps the geometry statistically into that of the map, so that the two geometries match.

accomplish this using *affine transformations*. Affine operations are plane geometry; they manipulate the coordinates themselves by scaling the axes, rotating the map, and moving the coordinate system's origin.

In some cases, when no good control is available, maps must be statistically registered together, especially when one layer is a map and one an image or photograph. The statistical method known as *rubber sheeting* or *warping* is used for this and is a function inside many GIS packages (Figure 8.5).

8.3.2 Data Storage

Data storage within a GIS has historically been an issue of both space—usually how much disk space the system requires—and access, or how flexible a GIS is in terms of making the data available for use. The massive reductions in the cost of disk storage, new high-density storage media such as the CD-ROM, and the integration of compression methods into common operating systems have made the former less critical and the latter more so.

Current emphasis, therefore, is upon factors that improve data access. This has been a consequence also of the rise of distributed processing, the Internet, and the World Wide Web. As a result, many GIS packages are now capable of using *metadata*, or data about data, in an integrated manner. Metadata support might include a system for managing a single project as a separate entity, to managing many projects with multiple versions, to full support for exchangeable metadata stored in common formats and searchable through online "clearinghouses." The USGS's Global Land Information System, NASA's master directory, and the Federal Geographic Data Committee's Spatial Data Clearinghouse are all examples. Participation in the common library entails both standardizing the metadata to make it searchable and agreeing to make the data available either on or offline.

Other larger issues around GIS use, most essential to the degree of user friendliness of the system, concern the mechanism for user interaction with the software's functionality. Virtually all GIS software allows user interaction via command lines and/or windows within a GUI. The GUI interface is tedious, however, without some way of "*batching*" commands so that they can be executed either at another time, as a background task while the user gets on with another job, or for design-loop editing to change minor aspects of the process. Most systems, therefore, also contain a "language" for the user to communicate with the system. This allows users to add their own custom functions, automate repetitive tasks, and add features to existing modules. These languages are usually command-line programs or *macros*, but they can also be enhancements of existing programming languages such as Basic and Smalltalk.

Although disk storage is less critical than in the past, it can still be a constraint. GIS software on a microcomputer can occupy tens of megabytes even without data, and on a workstation perhaps hundreds of megabytes. As data become higher resolution, as more raster layers are used, and as finer and finer detail becomes available, many GIS data sets can easily move into the gigabyte range in size.

This implies that not only is supporting multiple resolutions important—for example, using coarse browse images as samples of the real thing—but also that data compression should be supported. This can vary all the way from partitioning data sets to meet constraints (such as a maximum number of polygons) to supporting compressed data formats and structures such as JPEG, run-length encoding, or quadtrees.

Also of great importance from a user perspective is the degree to which the system itself provides help to users, either via the operating system or as part of the software. Integration with online manuals, such as in Unix versions, support for context-sensitive hypertext help systems, such as the Windows help feature, and, ideally, an online interactive hypertext help system can be critical for the new user. These help systems can be used only when needed rather than encumbering the advanced user with unnecessary basic information.

Support for data formats is important to a GIS when data are to be brought in from outside (e.g., public-domain data from the Internet). Ideally, the GIS software should be able to read common data formats for both raster (DEM, GIF, TIFF, JPEG, Encapsulated PostScript) and vector (TIGER, HPGL, DXF, PostScript, DLG). Some GIS packages have import functions only into a single data structure, usually either an entity-by-entity structure or a topological structure.

For three-dimensional data, these systems usually support only the triangular irregular network. Others support only raster structures based on the grid, including the quadtree, and either convert all data into this structure or just ignore it. Some GIS packages continue to support only data in a proprietary format, available only at cost from the software vendor. A rather critical GIS function is the ability to convert between raster and vector data, an absolutely essential feature for the integration of multiple data sources such as GPS data and satellite images.

Of increasing interest in recent years has been the development of GIS functions that support data in standard exchange formats. At the national and international levels, several data transfer standards have now been developed, such as the Spatial Data Transfer standard and DIGEST. As these standards become mandated, and as the role of data exchange increases, led by the Internet, most GIS systems will develop support for inputting and outputting data in these standard formats. The 2000 census, with its support for the federal information processing standard for data exchange (FIPS 173), will probably drive GIS vendors to support this necessary next step for GIS.

8.3.3 Data Management

Much of the power of GIS software comes from the ability to manage not just map data but also attribute data. Every GIS is built around the software capabilities of a database management system (DBMS), a suite of software capable of storing, retrieving selectively, and reorganizing attribute information. The database manager allows us to think that all the data are available, that the data are structured in a simple flat-file format, and that they constitute a single entity. In fact, the database manager may have partitioned the data between files and memory locations and may have structured it in any one of several formats and physical data models.

A database manager is capable of many functions. Typically, a DBMS allows data entry, and data editing, and it supports tabular and other list types of output, sometimes independent of the GIS. Retrieval functions always include the ability to select certain attributes and records based on their values. For example, we can start with a U.S. database, and select out all records for states containing cities with over 1 million inhabitants, forming a new database that is wholly enclosed by the original and that duplicates part of it. We can also perform functions such as sorting data by value, and retrieving a selected record by its identification, such as a name or a number.

Address matching involves taking a listed street address, such as "123 Main Street," and using the GIS's existing data to match the address with a geographic region in the GIS. The key to this capability is usually the TIGER files from the U.S. Census Bureau, which contain a topologically connected street and block network, referenced to house numbers. The address match finds the street and then moves along the street's individual blocks until the house number lies within the block and on the correct side of the street.

Many operations on data are very important from a mapping perspective. For example, very often maps captured from different sheets must be merged together, or sometimes a *mask* must be placed over the data to exclude features entirely from the GIS. Examples of masks are private lands within national parks, water bodies, or military bases. Similarly, sometimes data must be assembled in one way, by topographic quadrangle, and then *cookie cut* into another region such as a state or a city boundary. Even more complex, sometimes line features such as the latitude/longitude grid, a river, or a political boundary must be sectioned up or have points added as new features or layers are introduced. This feature, called *dynamic segmentation*, can be done automatically by the GIS (Figure 8.6).

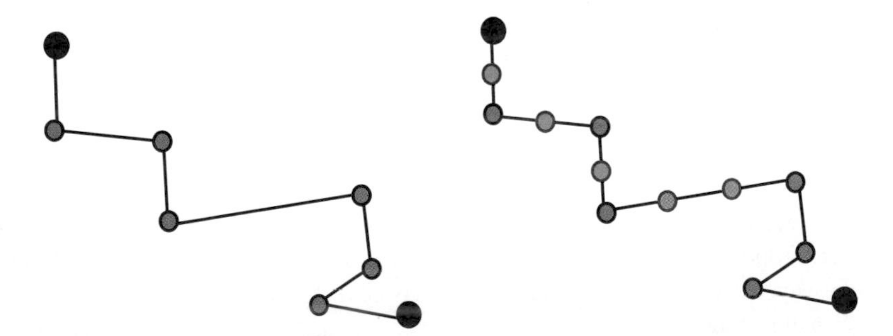

FIGURE 8.6: Using dynamic segmentation, the GIS can create as many segments along a feature as are necessary for analysis or display by adding new nodes (shown in magenta). Each new segment can have its own attributes. For example, it may be necessary to establish a new point to mark every mile measured along a river, and to attach river flow, or pollution data, to the points.

8.3.4 Data Retrieval

Another major area of GIS functionality is that of data retrieval. As we saw in Chapter 5, a GIS supports the retrieval of features by both their attributes and their spatial characteristics. All GIS systems allow users to retrieve data—they wouldn't qualify as a GIS if they did not! Nevertheless, among systems some major differences exist between the type and sophistication of GIS functionality for data retrieval.

The most basic act of data retrieval for a GIS is to show the position of a single feature. This can be by retrieving coordinates as though they were attributes, or more commonly by displaying a feature in its spatial context on a map with respect to a grid or other features. For line features, the same goes, with the exception that line features have the attribute of length, and polygon features have the attribute of area. The GIS should be able to calculate and store these important basic properties as new attributes in the database. For example, for a set of counties we may want to take a polygon attribute such as an area of forest and divide it by the county area to make a percentage density of forest cover. Another common measurement we may want is to count features. For example, with the same database we could count the number of fire stations within the same counties by doing a point-in-polygon count from a separate database of municipal utilities and then relate the forest cover to the fire-prevention capabilities.

We have seen that a GIS has the critical capability of allowing the retrieval of features from the database using the map as the query vehicle. One way, indeed the most basic way, of doing this is to support the ability to point at a feature, using a device such as a mouse or a digitizer cursor, to see a list of attributes for that feature. Again, the ability to select by pointing to a location virtually defines a GIS. If it cannot do this, the system is probably a computer mapping system, not a GIS. Just as critical is the database manager *select-by-attribute* capability. This is normally a command to the database query language that generates a subset of the original data set. For example, we could find all houses in a real estate GIS that had been listed on the market in the last year. Similarly, we could find all houses built after 1990. All GIS systems and all database managers support this capability.

As we saw in Chapter 5, GISs allow a set of retrieval operations based on using one or more map features as handles to select attributes of those features. Although some of them are very simple, these operations are also a real litmus test for establishing whether or not a software package is a GIS. A GIS should allow the user to select a feature by its proximity to a point, a line, or an area. For a point, this means selecting all features within a certain radius. For a line or a polygon, we have used the term *buffering*. Buffering allows the GIS users to retrieve features that lie within perhaps 1 mile of an address, within 1 kilometer of a river, or within 500 meters of a lake (Figure 8.7). Similarly, weighted buffering allows us to choose a nonuniform weighting of features within the buffer, favoring close-by instead of distant points, for example.

The next form of spatial retrieval is map *overlay*, when sets of irregular, nonoverlapping regions are merged to form a new set of geographic regions that the two initial sets share. In the new attribute database it is possible to search by either set of units. A GIS should be able to perform overlay as a retrieval operation since to support the many spatial analyses based on map combination and weighted layer solutions, as discussed in Chapter 6. Vector systems usually compute a new set of polygons by adding points to and breaking up the existing sets, and in raster systems we allow *map algebra*, direct addition or multiplication of attributes stored in cells. Map overlay is an important part

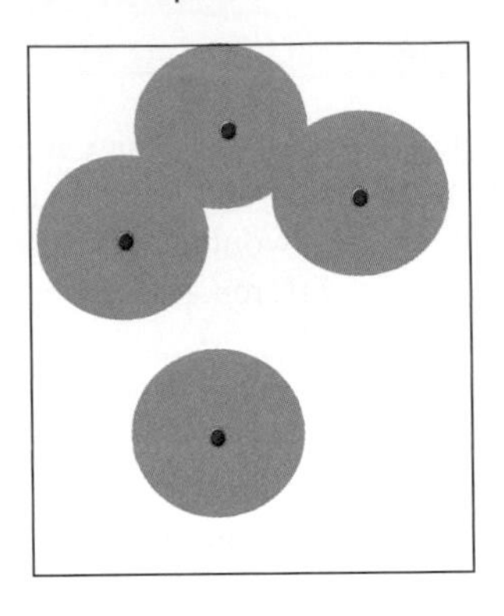
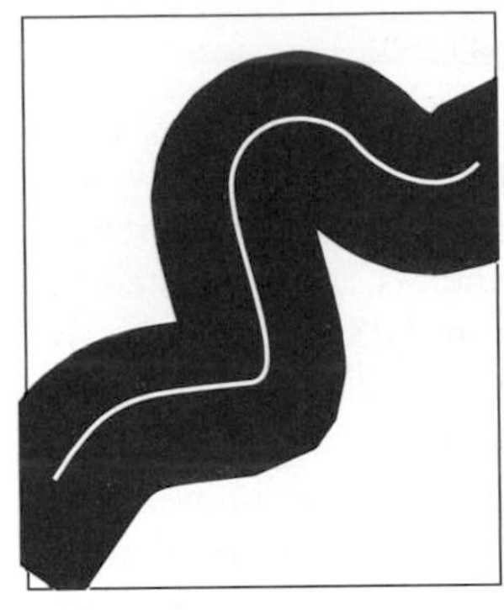
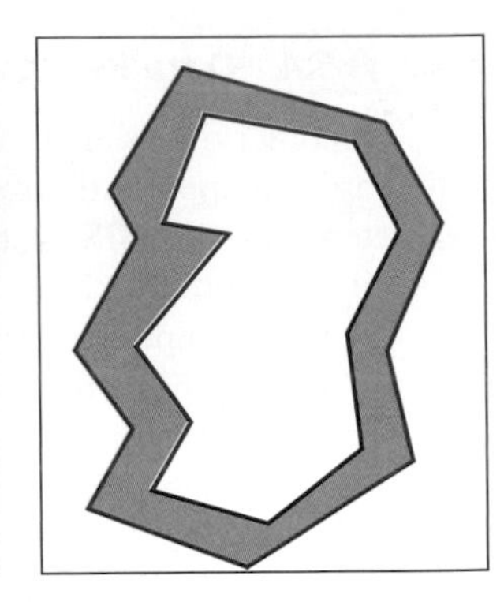

FIGURE 8.7: Buffers can be created around points (left), lines (center), or areas (right). Buffers can be set to a specific distance, such as 1 kilometer, or made a continuous layer of distances from each point on the map to the feature or features in question.

of a major GIS function, that of redistricting, in which new districts can be drawn and the data restructured into the regions so that tests and analyses can be performed by trial and error—for example, to see whether the new districts conform to the federal Voting Rights Act.

Another important set of retrieval options, especially in facilities mapping and hydrological systems, are those that allow networks to be constructed and queried. Typical networks are subway systems, pipes, power lines, and river systems. Retrieval operations involve searching for segments or nodes, adding or deleting nodes, redirecting flows, and routing. Not all GIS systems need these functions, but if the purpose is to manage a system usually abstracted as a network, such as a highway or rail system, a power supply system, or a service delivery system, obviously the GIS should then have this feature.

Dana Tomlin (1990) has elegantly classified the operations that a raster GIS can perform into a structure called *map algebra*. In map algebra, the retrieval operations used are Boolean, multiply, recode, and algebra. *Boolean* operations are binary combinations. For example, we can take two maps, each divided into two attribute codes "good" and "bad" and find a binary AND solutions layer where both layers are "good" (Figure 8.8). *Multiply* allows two layers to be multiplied together—for example, two sets of weights to be combined. In *recode* operations a range of computed attribute codes can be reorganized. An example is taking percentages and converting them to a binary layer by making all values greater than 70% a "1" and all else a "0." Map *algebra* allows compute operations, such as map-to-map multiplication for a binary AND over the space of a grid.

Two truly spatial retrieval operations are the ability to *clump* or aggregate areas, and to *sift*. For example, all areas of saturated soils surrounding swamps could be added to the swamps and recoded as wetlands, making a new, broader category of attribute. Sifting simply eliminates all areas that are too small, individual cells falling between two larger areas, or a tiny sliver polygon. Finally, some complex retrieval operations require the GIS to be able to compute numbers that describe shape. Common shape values are the length of the perimeter of a polygon squared, divided by its area, or the length of a line divided by the straight-line distance between the two endpoints.

1	1	0	0	0
1	1	1	0	0
0	1	0	1	0
0	0	0	0	0
0	0	0	0	0

AND

0	0	0	0	0
0	1	0	1	1
0	1	1	1	1
0	0	0	1	1
0	0	0	0	0

=

0	0	0	0	0
0	1	0	0	0
0	1	0	1	0
0	0	0	0	0
0	0	0	0	0

FIGURE 8.8: Map algebra in its simplest form: Two binary images are ANDed together to give a common area of overlap. Many other operations are possible, such as add, multiply, divide, select maximum, eliminate isolated values, etc.

8.3.5 Data Analysis

The analysis capabilities of GIS systems vary remarkably. Among the multitude of features that GIS systems offer are the computation of the slope and direction of slope (aspect) on a surface such as terrain; interpolation of missing or intermediate values; line-of-sight calculations on a surface; the incorporation of special break or skeleton lines into a surface; finding the optimal path through a network or a landscape; and the computations necessary to calculate the amount of material that must be moved during cut-and-fill operations such as road construction.

Almost unique to GIS, and entirely absent in other types of information systems, are geometric tests. These can be absolutely fundamental to building a GIS in the first case. These are described by their dimensions, point-in-polygon, line-in-polygon, and point-to-line distance. The first, point-in-polygon, is how a point database such as a geocoded set of point samples is referenced into regions. Thus a set of locations for soil samples, generated at random, could be point-in-polygon merged with a digitized set of district boundaries so that a sample list can be sent to each soil district manager. Other more complex analytical operations include partitioning a surface into regions, perhaps using the locations of known points to form proximal regions or Voronoi polygons, or by dividing a surface into automatically delineated drainage basins.

Some of the most critical analytical operations are often the simplest. A GIS should be able to do spreadsheet and database tasks, compute a new attribute, generate a printed report or summarize a statistical description, and do at least simple statistical operations such as computing means and variance, performing significant testing, and plotting residuals.

8.3.6 Data Display

Most of the display capabilities of GISs have been covered in Chapter 7. GIS systems need to be able to perform what has become called *desktop mapping*, generating geographical and thematic maps so that they can be integrated with other functions. GISs typically can create several types of thematic mapping, including choropleth and proportional symbol maps; and they can draw isoline and cross-sectional diagrams when the data are three dimensional.

Almost all GIS packages now either allow interactive modification of map elements—moving and resizing titles and legends—or allow their output to be exported into a package that has these capabilities, such as Adobe Illustrator or CorelDraw. A very limited few GIS packages include cartographic design help in their editing of graphics, defaulting to suitable color schemes, or notifying the user if an inappropriate map type is being used for the data. This would be a desirable feature for many of the GISs on today's market and could avoid many tasteless or erroneous maps before they were created.

8.4 GIS SOFTWARE AND DATA STRUCTURES

In the preceding discussion, the focus was on what functional capabilities the typical GIS offers. It should not be forgotten that many GIS features are predetermined by the GIS's particular data structure. As we saw in Chapter 4, at the very least the underlying data structure that the GIS uses, typically raster or vector but potentially also TIN, quadtree, or another model, such as object-based, determines what the GIS can and cannot do, how operations take place, and what level of error is involved.

In general, the driving force for the choice of structure should be not only what type of system can be afforded, but more critically, what model is most suitable to a particular application, what retrieval and analysis functions will be used most, and what is the acceptable level of resolution and error.

Some examples where particular structures are favored include extensive land characterization applications such as forestry, where detailed data are not required (favors raster); applications involving irregular polygons and boundary lines, such as political units or census tracts (favors vector); applications that require the ability to register all features accurately to ground locations (favors vector); applications making extensive use of satellite or terrain data (favors raster); or applications where image processing functions and analyses such as slope and drainage analysis are to be conducted (favors raster). In many cases, the raster to vector conversion is done outside of the GIS in specialist conversion software, so that care can be taken to avoid the most common types of error, and so that the user can be brought in to resolve cases where the software is unable to solve a rasterization problem.

Increasingly, of course, many GIS systems allow the user to input and keep data in both raster and vector form. The GIS user should realize, however, that virtually all cross-structure retrieval and analysis requires one (or both) of the layers to change structure, and that this transformation often stamps itself irretrievably on the data's form, accuracy, and suitability for further use.

8.5 CHOOSING THE "BEST" GIS

The term "best" is extremely subjective where GIS is concerned. Some systems have extremely loyal followings who advocate their system over others. A "best" system implies that one solution is best for all problems, which is of course largely meaningless. The following subset of GIS systems, most available commercially, is intended to illustrate the breadth and depth of systems on the market today and some of the major and minor differences among these systems.

No endorsement is intended, and the list is provided to further the GIS "consumer's" education. Research has shown that these "big eight" packages account for the majority of those used in educational, and many professional, settings. In some cases, different GIS software packages are used in combination or along with other software for statistical analysis, graphical editing, or database management.

8.5.1 The Big Eight

ArcGIS. ArcGIS, the latest version of Arc/Info (Figure 8.9), is a long-lived, full-function GIS package that has been ported to the microcomputer, the workstation, and the mainframe. Arc/Info and ArcGIS are used to automate, manipulate, analyze, and display geographic data, and the software incorporates hundreds of sophisticated tools for map automation, data conversion, database management, map overlay and spatial analysis, interactive display and query, graphic editing, and address geocoding. The ArcInfo software includes a relational database interface for integration with commercial database management systems and a macro language called AML (ARC Macro Language) for developing customized applications. Since release 8, ArcGIS has instead used Visual Basic as its macro and programming language. ArcGIS uses a generic approach to geographic information systems that is not application specific, allowing the software to address virtually any geographic application. The software runs both on higher-end microcomputers and is available on several Unix workstations and for Windows NT. ArcGIS runs only on Windows NT.

ESRI is broadly accepted as a market leader in GIS, with many thousands of users in a variety of organizations worldwide. The software is used by federal, state, and local government organizations; and by businesses, utilities, and universities to address applications in planning, cartography, transportation, research, telecommunications, oil and gas, forestry, and many other disciplines. Release 8 of the program, in 1999, was a substantial modification of the program's user interface and functionality. Object-modeling capability and links to the Spatial Data Base Engine and other relational database management systems such as Oracle are included. With the latest versions of the software, the compatibility between ArcGIS and ArcView has been increased. The software uses the Windows COM component based software architecture, and is compatible with many other Window-based software tools.

ArcView. ArcView (Figure 8.10) is available for Windows, Macintosh, and a variety of Unix platforms. It is a desktop system for storing, modifying, querying, analyzing,

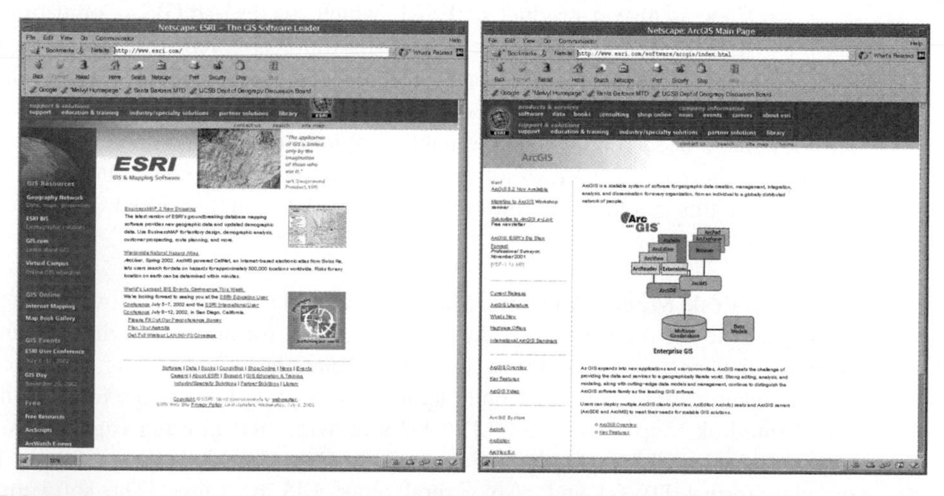

FIGURE 8.9: Arc/Info is a full-function GIS package. Left: The WWW home page for Arc/Info version 8 from the vendor, the Environmental Systems Research Institute at `http://www.esri.com`. (Used with permission, ESRI, Redlands, CA.)

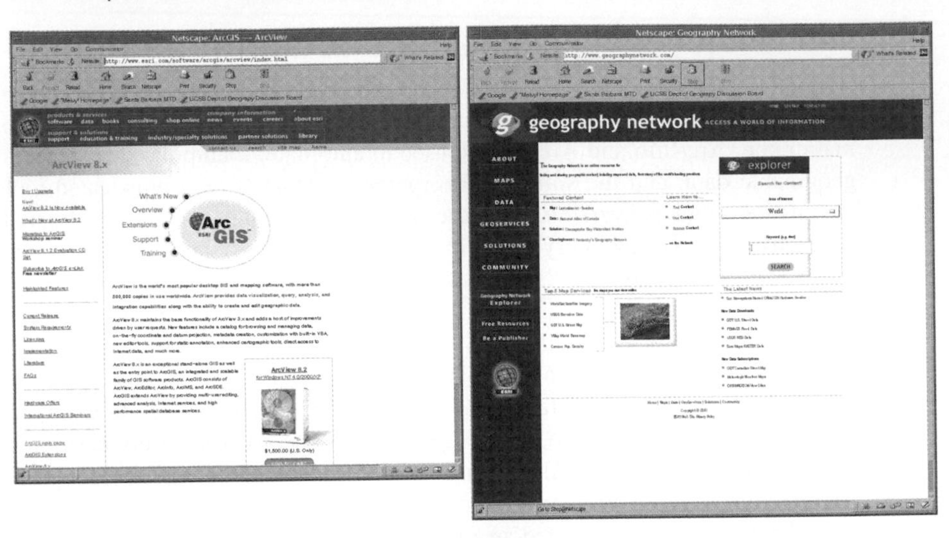

FIGURE 8.10: The ArcView GIS from ESRI Left: ArcView 8.1 product Web page at `http://www.esri.com/base/products/arcview.html`. Right: The ESRI geography network, source of many contributed shared data sets for use with ESRI software at `http://www.geographynetwork.com`.

and displaying information about geographic space. An intuitive graphical user interface includes data display and a viewing tool. Support for spatial and tabular queries, "hot links" to other desktop applications and data types, business graphics functions such as charting, bar and pie charts, and map symbolization, design, and layout capabilities are supported. Geo-coding and address matching are also possible. The Spatial Analyst tool kit makes working with raster data such as terrain and DEMs possible. Other extensions permit network analysis, allow Web activation of ArcView maps, and support advanced display features such as three-dimensional data visualization. ArcView GIS since version 8 has been more compatible with ArcGIS.

ArcView is also a product of ESRI, which makes ArcGIS. Compatibility exists between the two systems, with ArcView being more oriented toward map display than database management. Maps and data files are easily exchangeable between the formats used in the two systems, shape files, grid, images, and coverages. Outdated versions of ArcView have been placed into the public domain and are available over the Internet. A copy of the basic ArcView software is contained on the CD-ROM in the sleeve in the back of this book. More information is available at the ESRI Web site, cited in Figure 8.10.

Autodesk Map. Autodesk Map (Figure 8.11) is a GIS software suite built on the capabilities of the substantial AutoCAD software for automated drafting and design. Because this package is extensively used in planning, engineering, and architectural offices, many people can easily build upon their existing knowledge to enter the field of GIS. Autodesk Map uses AutoCAD 2002's drawing and plotting capabilities. Multiple data formats can be input, including those of AutoCAD (exchange format DXF and drawing format DWG) and also several other GIS packages. The software supports topology, query using Oracle and SQL, data management, and thematic mapping. The Autodesk Raster Design module supports grids and images and the Autodesk Onsite

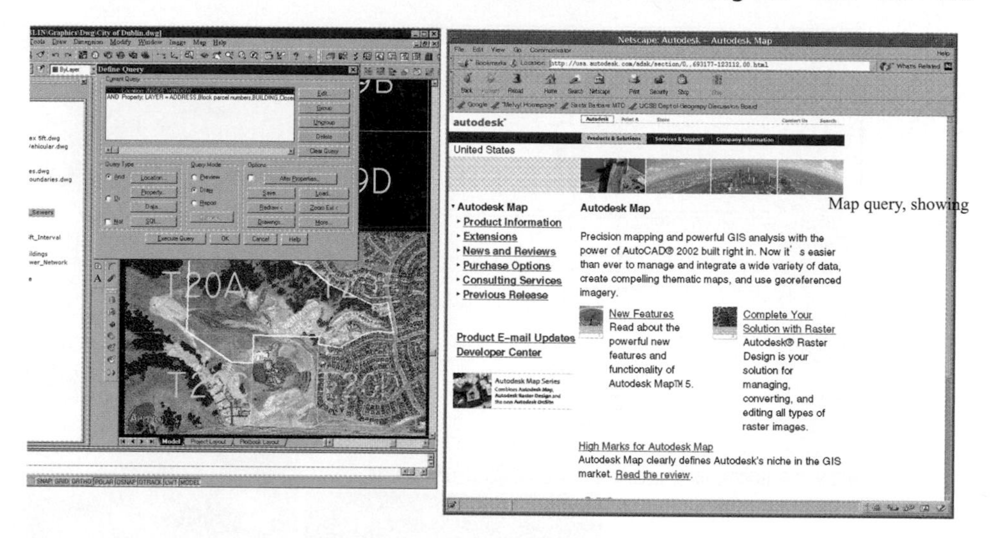

FIGURE 8.11: Autodesk Map from Autodesk. Home page at `http://www.autodesk.com` Left: Sample Autodesk Map query, showing use of Autodesk Map GIS functions within AutoCad 2002 CAD software.

module handles all of the standard GIS data operations. There are extensive tools for coordinate conversion and specification, rubber-sheeting, and map editing and digitizing. The software uses the C++ programming language as a development tool. Output control and plotting support are strong, relying on AutoCAD's capability.

GRASS. The U.S. Army Construction Engineering Research Laboratories (CERL) developed a public-domain software called the Geographic Resources Analysis Support System (GRASS). GRASS (Figure 8.12) is raster based, was the first Unix GIS software, and has been considerably enhanced by the addition of user contributions—for example, in hydrologic modeling. The Web site states that GRASS is an open source, free software GIS with raster, topological vector, image processing, and graphics production functionality that operates on various platforms through a graphical user interface and shell in X-Windows. The source code for the program is available under the GNU General Public License. The latest version, 5.0.0, the development version 5.1, and most prior versions are available free over the Internet. Many users run GRASS on PCs under the Linux version of Unix, although a Windows port is now complete. Since 1985, CERL has released upgrades and enhancements to GRASS and provided technical user support. However, CERL terminated GRASS-related work in the spring of 1996. Public domain user support has been very strong, and highly international.

Since 1996, the headquarters for GRASS support, research, and development has been at Baylor University, within the Department of Geology. The development currently under way is extensive, including releases of new manuals and documentation, and fosters continued research on GIS and visualization using GRASS at International conferences and through user support groups. The GRASS GIS uses a standardized command line input designed to resemble the Unix shell command language, but also uses a GUI under X-Windows. Unix compatibility allows users and programmers to create new applications and link GRASS to other software packages. Connections to the Unix shell and the C programing language allow simple extension and control.

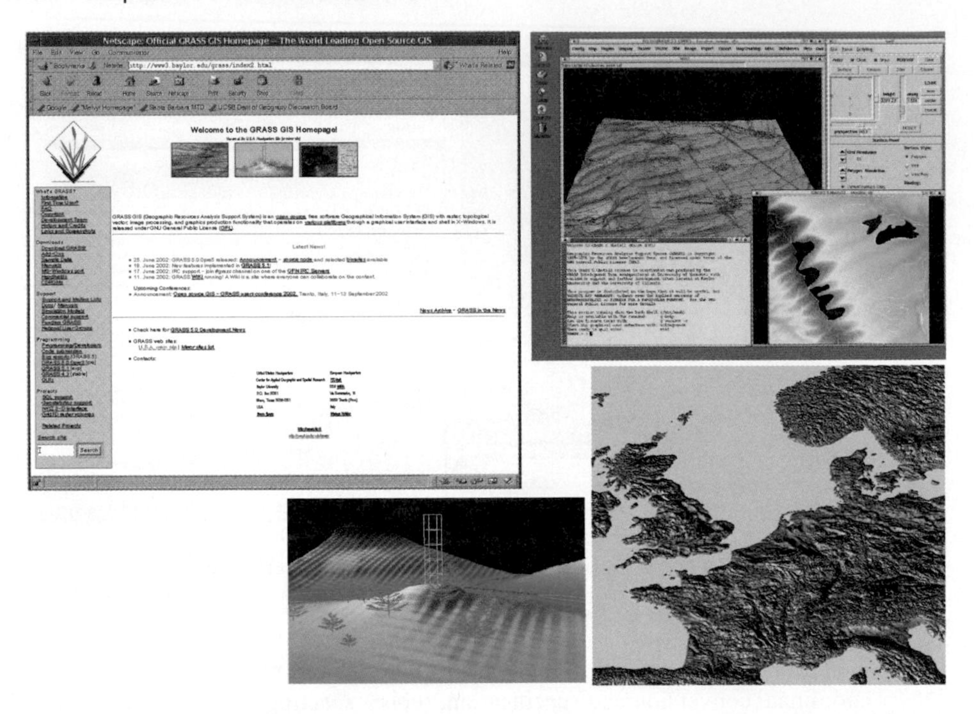

FIGURE 8.12: Left: The GRASS GIS Web page. (http://www3.baylor.edu/grass/index2.html) Right: The GRASS user interface: Bottom: Sample applications from the image galley.

IDRISI. The Idrisi (Figure 8.13) GIS software has been developed, distributed, and supported on a not-for-profit basis by the Idrisi Project, Clark University Graduate School of Geography. To date, there are many thousands of registered users of Idrisi software in almost every country in the world, making it the most broadly used raster GIS in the world. Idrisi is designed to be easy to use, yet provide professional-level GIS, image processing and spatial statistics analytical capability on both DOS- and Windows-based personal computers. It is intended to be affordable to all levels of users and to run on the most basic of common computer platforms. Expensive graphics cards or peripheral devices are not required to make use of the analytical power of the system, which is designed with an open architecture so that researchers can integrate their own modules.

Idrisi for Windows, first released in 1995, added a graphical user interface, flexible cartographic composition facilities, and an integrated database management system to the analytical tool kit. The more recent Idrisi32 is fully Windows and COM compliant and exploits object-oriented methods. Special routines for change and time-series analysis, spatial decision support, and uncertainty analysis and incorporation are included. A stand alone cartographic product, CartaLinx, allows topological editing and database development. Idrisi32 comes with a set of tutorial exercises and data that guide the new user through the concepts of GIS and image processing while also introducing the features of Idrisi. The tutorial exercises are appropriate for use either in self-training or in classroom settings.

MapInfo. MapInfo (Figure 8.14) was one of the first GIS programs to do desktop mapping. The vendor is MapInfo Corporation of Troy, New York. The software is well

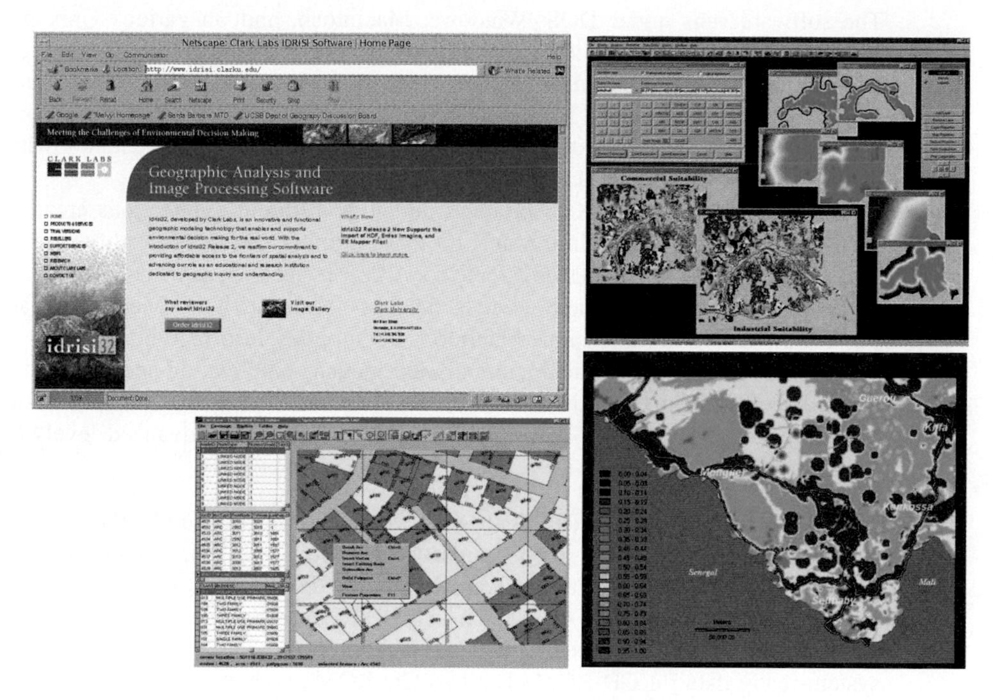

FIGURE 8.13: Left: The IDRISI project at `http://www.idrisi.clarku.edu`. Right side: Applications and examples of the Idrisi32 user interface taken from the Web site. (Courtesy of the IDRISI Project, Worcester, MA. Used with permission.)

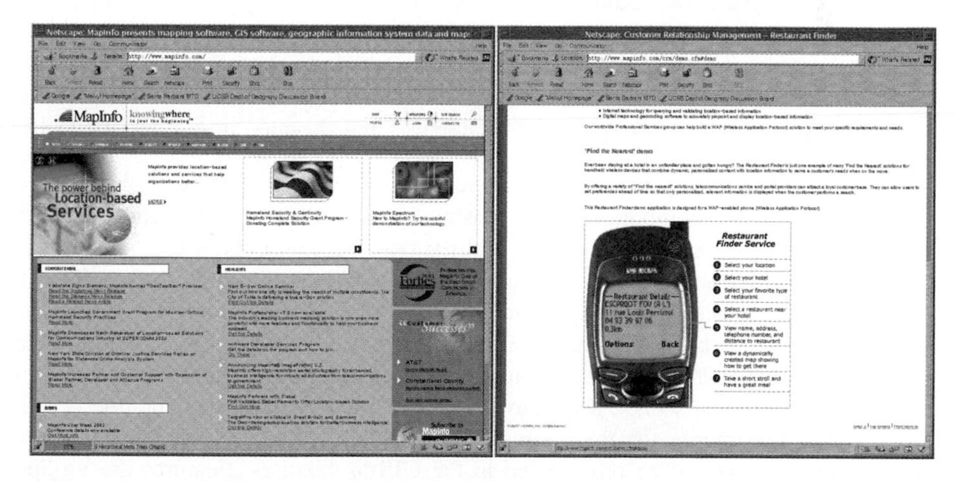

FIGURE 8.14: Left: The MapInfo home page at `http://www.mapinfo.com`. Right: MapInfo application, providing a mobile Internet search capability for restaurants using a cellular telephone. (Courtesy of MapInfo Corporation, Troy, NY. Used with permission.)

distributed and has many user groups and a broad variety of applications worldwide. The software runs under DOS, Windows, Macintosh, and on various Unix platforms. Although MapInfo's GIS retrieval and analysis functions are fewer than those of full-blown GIS packages, MapInfo includes a link to the Basic programming language via a language called MapBasic. This development environment permits the creation of customized "mapplications," extending MapInfo's built-in functionality and allowing use of a common graphical interface.

MapInfo has several GIS products aimed at different applications area, including MapInfo Professional, MapInfo MapX for programming GIS functionality, and specialist analytical modules such as MapXtreme for Web services, MapXSite for managing spatially enabled Web sites, and various database tools such as MapInfo Spatialware, Proviewer, and GIS Extension. MapInfo also supplies information products spanning geographic, economic, political, cultural, and industry application-specific content, each derived from leading worldwide sources to work the software. MapInfo also has an extensive training program, with classes at introductory and advanced levels for MapInfo and MapBasic.

Maptitude. Maptitude (Figure 8.15) is a GIS that works under the Windows operating system. The software is by Caliper Corporation, Newton, Massachusetts. Caliper has long been associated with the TransCAD and GIS-Plus GIS software packages. The latest version is Maptitude 4.5, which includes census data, a developer's toolkit and extended file support. The software comes with a considerable amount of geocoded and system-ready data on CD-ROM. The two CD-ROMs contain every street in the United States with the address information, state, county, zip codes, and census tracts as polygons with associated demographic data, and additional assorted U.S. and global data. Maptitude reads most standard PC file formats directly and can match each record against geographic data files using street address, zip code, and other features. Maptitude allows users to create and maintain geographic databases, analyze geographic relationships in data, and create highly professional map displays for presentations and reports.

Maptitude runs under Windows 3.1, Windows for Workgroups, Windows 95, or Windows NT, and with networks. The software uses the object linking and embedding of Windows, so that objects can be dragged and dropped into other applications.

GeoMedia. GeoMedia (Figure 8.16) is a widely distributed layer-based GIS with a tradition in computer-assisted design by the Intergraph Corporation of Huntsville, Alabama. The software runs on workstations, PCs, and under the Windows NT system. An extensive set of add-on modules allow users to configure GIS capability around their specific needs. The set of modules includes GeoMedia, GeoMedia Professional, Intelliwhere Ondemand (for mobile systems), GeoMedia Webmap, and GeoMedia WebMap Professional. There are extensions aimed at applications in land information, parcel management, public works, and transportation. The layered implementation permits efficient storage structures for the geometry and linkages to relational database records. Geographic elements are represented in the GIS as features. Features are grouped into the same categories as the maps on which they appear.

For the attribute data, GeoMedia incorporates use of the Oracle and SQL relational interface system, which facilitates client–server network communication to the relational DBMS so that multiple workstations communicate with the database server simultaneously. GeoMedia is fully integrated with Intergraph's traditional products, which

FIGURE 8.15: Upper left: The Caliper Corporation WWW home page at `http://www.caliper.com`. Right side: Sample applications and examples of the user interface. Bottom left: Example of a stepped statistical surface map generated by Maptitude. (Courtesy of Caliper Corp., Newton, MA. Used with Permission.)

include the MGE suite and tools for cartographic production. GeoMedia contains tools for building and maintaining topologically clean data without the processing and storage overhead of building and maintaining topology. In addition, it supports the open geodata interoperability specification and the spatial data transfer standard.

8.5.2 Selecting Software: Issues

Selecting the best GIS for use involves many other aspects than simply the technical capabilities of the software package. It could be argued that very little difference actually exists between GIS packages other than their user interfaces and their data structures. Conversely, many of the issues that determine how satisfied we are with the GIS we choose relate to how we acquire the software, how easily it installs itself on our computer, whether or not it is flexible enough to run on a given computer system, and how satisfied we are when the software is up and running.

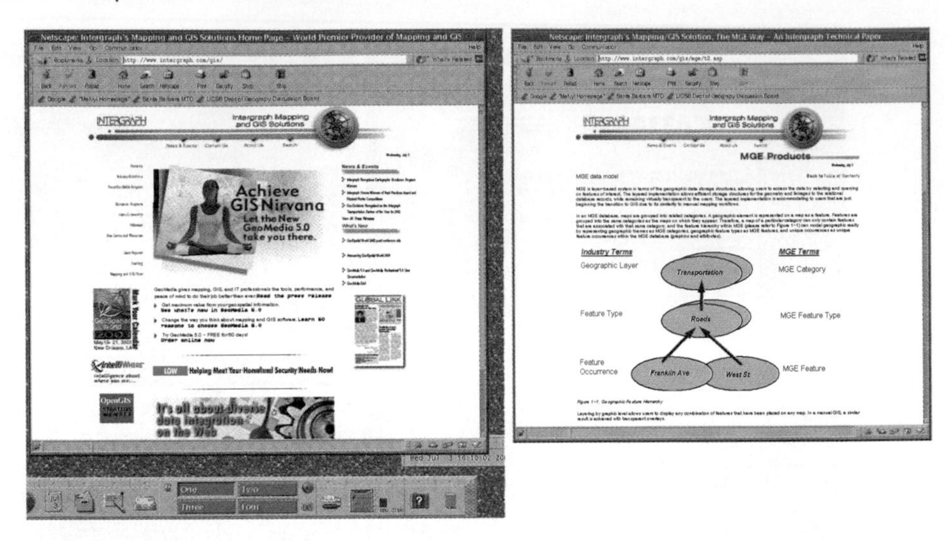

FIGURE 8.16: Left: The Intergraph home page at `http://www.intergraph.com/gis`. Right: Web page showing the basic data structure of Microstation MGE. (Courtesy of Intergraph Corp. Huntsville, AL. Used with permission.)

Obviously, cost is an important factor. Although the cost of basic GIS packages has fallen remarkably in recent years, cost can still be significant, especially when the hidden costs are taken into account. For example, GIS companies may charge not only a software purchase fee, but also include a maintenance fee, a fee for upgrades, a per call support cost, and sometimes other fees. Maintenance fees for workstation licenses, the sorts of licenses that would be used in a local area network configuration, can be a major proportion of the software cost. In addition, there is constant pressure to upgrade to new versions, usually by discontinuing support for older versions of the software. Especially if a large project is to be undertaken by the GIS user, this fact should be budgeted into the GIS software costs. Shareware and freeware, by contrast, may have less support infrastructure, but the software and update costs are zero.

Training is another important factor. Few GIS packages can be used by a novice right out of the box. The user may need help from a systems expert, may have special installation requirements, and may require the user to get some formal GIS training. Of course, this book can go a long way toward helping the user to understand GIS, but there is a great deal of straightforward technical information as well. Many GIS users take technical training from one of the GIS vendors or from other sources. These vary from one- or two-day workshops to entire college semester classes. They can also be rather expensive and time-consuming. Many GIS implementations, although well thought out and organized, fail for the lack of one or two people with the right technical expertise at the right time.

Once technical training ends, the real GIS use begins. At this stage, late on a Friday evening with a project deadline looming, the usual sole self-help mechanism is the GIS system manual. Again, these vary considerably in readability, comprehensiveness, and user-friendliness. Some are excellent, others poor. The user should ask to see documentation before making a major GIS purchase, as users will spend many hours poring over these pages. Best of all are online manuals, which can be searched, may be context-sensitive, have hypertext links, and will be available on a computer while the software is running in another window. This feature is worth extra expense, since it can speed the early learning and still serve as a reference later.

Regardless of the GIS's self-help capabilities, sooner or later almost all GIS users will eventually call a help line or interact in some way with the GIS vendor's technical support staff. In most cases this is done exclusively by telephone, but increasingly companies use fax, E-mail, and network conference groups as help facilities. Help lines can involve being placed on hold for long periods, or worse, waiting to be called back after leaving a phone number. E-mail is far better and gets around the time-zone problem of phone lines. When contacting a help line, a concise statement of the problem and a full set of information, usually including the serial number and date of purchase of the software, will greatly speed up your call. In general, using the reference manual or user guide until there is no other means of finding information is far preferable to calling a help line. Remember, if all else fails, *read the manual!*

Software maintenance can be another major consideration. For example, most software is updated by complete version upgrades, which require a new installation, or by "patches," a self-contained fix for a specific problem in the software. Maintenance is more of a consideration for large and networked systems, but every user needs to be concerned about too many large files and about how critical data are to be backed up in case of emergency. A GIS should also not be seen as a static entity, but rather one that will grow and evolve. A system that is big or powerful enough for a small prototype project today will probably not be able to deal with the follow-up project. Fortunately, as time passes the hardware becomes faster and faster, the disks get bigger and bigger, and the cost actually remains the same or falls. Conversely, the expertise required to install, maintain, and use the system is also important and should be planned for. GIS technicians typically get experienced enough to compete for better jobs very quickly. This should also be a part of the GIS cost plan.

Picking a GIS is obviously a complex and potentially confusing process. The most productive approach to the problem is to adopt the attitude of someone about to purchase a new car. First, the GIS user should assemble all the available details about the system requirements, the functional capabilities, the system constraints, and so on. The car buyer could, for example, determine a need for four doors, power steering, at least 14 cubic feet of luggage space, and front-wheel drive. Next, these should be matched against the systems available. Perhaps a trade-off is necessary between capabilities? Next comes the visit to the car dealership, followed by a test drive. Many demonstration versions of GIS packages are available to give a flavor of the system use before the purchase. Some demo versions can be downloaded free over the Internet or are given away at GIS conferences. One such demonstration version of ArcView is included with this book.

Finally, "You pays your money and you takes your choice." After the fact, however, the car will need to be maintained and perhaps repaired. One day it may be traded in for a new car. Every one of these issues should be considered. Although every vehicle will probably allow you to get home from the dealership, fundamental differences exist between a sports car and a Sports Utility Vehicle. Just so with GIS. To summarize: Before you choose, *research*, *select*, *test,* and *question*. Fortunately for the new GIS user, the early days of GIS failures are now over. Technically, today a GIS is much like a reliable automobile. Where and how you drive, however, is still entirely up to you!

8.6 STUDY GUIDE

8.6.1 Summary

CHAPTER 8: How to Pick a GIS

- GIS users need to be aware of different GIS software products during system selection and beyond.
- Informed choice is the best way to select the best GIS.
- GIS software has evolved very rapidly over its brief history.
- A historical GIS "snapshot" was the IGC survey conducted in 1979.
- In the 1979 survey, most GISs were sets of loosely linked FORTRAN programs performing spatial operations.
- Many early computer mapping programs had evolved GIS functionality by 1979.
- In the early 1980s, the spreadsheet was ported to the microcomputer, allowing "active" data.
- In the early 1980s, the relational DBMS evolved as the leading means for database management.
- Addition of a single integrated user interface and a degree of device independence led to the first true GISs.
- The second generation of GIS software used graphical user interfaces (GUIs) and the desktop/WIMP model.
- Unix workstations integrated GIS with the X-Windows GUI.
- As GUIs became part of the operating system, GISs began to use the operating system's GUI instead of their own.
- PCs integrated GIS with the variants of Windows and other OSs.
- GIS features are known as functional capabilities.
- Functional capabilities fall into the "critical six" categories.
- The critical six functional capabilities are data capture, storage, management, retrieval, analysis, and display.
- Some data capture functions are digitizing, scanning, mosaicing, editing, generalization, and topological cleaning.
- Storage functions are compression, metadata handling, control via macros or languages, and format support.
- Some data management functions are physical model support, the DBMS, address matching, masking, and cookie cutting.
- Some data retrieval functions are locating, selecting by attributes, buffering, map overlay, and map algebra.
- Some data analysis functions are interpolation, optimal path selection, geometric tests, and slope calculation.
- Some data display functions are desktop mapping, interactive modification of cartographic elements, and graphic file export.
- Many GIS functional capabilities are by-products of their particular data structure.
- Raster systems work best in forestry, photogrammetry, remote sensing, terrain analysis, and hydrology.
- Vector systems work best for land parcels, census data, precise positional data, and networks.

- **Eight GIS systems form the bulk of operational GIS in professional and educational environments.**
- **There are some significant differences among these "big eight" systems.**
- **A variety of issues should be considered in system selection:**
  - **cost**
  - **upgrades**
  - **network configuration support**
  - **training needs**
  - **ease of installation**
  - **maintenance**
  - **documentation and manuals**
  - **help-line and vendor support**
  - **means of making patches**
  - **workforce**
- **Selecting a GIS can be a complex and confusing process.**
- **The intelligent GIS consumer should research, select, test, and question systems before purchase.**

8.6.2 Study Questions

The Evolution of GIS Software

Make a timeline from about 1960 to today. Place on the timeline each of the packages mentioned in this chapter. How does the sequence of software packages relate to the history of GIS discussed in Chapter 1?

GIS and Operating Systems

Make a list of all the operating systems, mainframes, workstations, and microcomputers that can run the "big eight" GIS packages. Which are mentioned most frequently? Why?

GIS Software Capabilities

Make a word list of key functional capabilities structured under the headings of the "critical six." Score the functional capabilities by how essential they are for a GIS to qualify as a "true" GIS. Match the capabilities of one particular GIS against the list.

GIS Software and Data Structures

Review the Chapter 4 coverage of the different data structures for GIS. Classify each of the "big eight" by which data structure they support and whether or not they support data structure conversions. List some operational reasons why you might need to convert between data structures.

Choosing the Best GIS

Go through the "People in GIS" sections in this book and tally the mentions of specific GIS software packages. How does your list match up with the "big eight"?

Make a table of the "critical six" as columns and the "big eight" as rows. Fill in the table entries with observations from the text on the "big eight." Invent a scoring system and rank the "big eight" for their suitability in each of the applications areas listed in Section 8.4.

8.7 EXERCISES

1. *If you have access to more than one GIS, establish a common data set, such as a TIGER file or a DLG, read the file into the GIS system, and perform a simple retrieval or analysis operation such as a buffer or overlay. Take careful note of how long each step took, how many steps were necessary, and how useful the manuals and help systems were in troubleshooting. Place the two output maps together at the same size and scale. Are they identical? What might be the factors contributing to the differences?*
2. *Examine the manuals for two different GIS packages. Read the same section—for example, the section on digitizing lines—in each manual. Which is the better explanation? Why? Make a list of the features that you consider desirable in GIS documentation.*
3. *If you have the ability, install another operating system on your workstation or microcomputer, such as Linux and Windows. Alternatively, find two computers with different operating systems already installed. Do the same task—say, enter 50 numbers into a spreadsheet file—in each of the two operating systems. Time each process and make a chart showing how much total time you spent on each task. How much did the operating system help or hinder the task? How much system help was available in each system?*
4. *Using the Internet, the mail, or any other means available, make a comparative price list of the "big eight" and as many other GIS packages as you can find. Using the same functions checklist from the study questions, compute a "features per dollar" number for each GIS. Which is the best? Which is the worst? Why?*
5. *Follow the network conference group comp.infosystems.gis for one week and keep a tally of good and bad comments about the GIS systems that are discussed there. Would this be a good way to choose a GIS? Why or why not?*

8.8 REFERENCES

Brassel, K. E. (1977) "A survey of cartographic display software," *International Yearbook of Cartography*, vol. 17, pp. 60–76.

Day, D. L. (1981) "Geographic information systems: all that glitters is not gold." *Proceedings, Autocarto IV*, vol. 1, pp. 541–545.

Marble, D. F. (ed.) (1980) *Computer Software for Spatial Data Handling*. Ottawa: International Geographical Union, Commission on Geographical Data Sensing and Processing.

Tomlin, D. (1990) *Geographic Information Systems and Cartographic Modelling*. Upper Saddle River, NJ: Prentice Hall.

Tomlinson, R. F. and Boyle, A. R. (1981) "The state of development of systems for handling natural resources inventory data," *Cartographica*, vol. 18, no. 4, pp. 65–95.

8.9 KEY TERMS AND DEFINITIONS

active data: Data that can be reconfigured and recomputed in place. Spreadsheet term for data for attributes or records created by formulas within a spreadsheet.

address matching: Address matching means using a street address such as *123 Main Street* in conjunction with a digital map to place the street address onto the map in a known location. Address matching a mailing list, for example, would convert the mailing list to a map and allow the mapping of characteristics of the places on the list.

affine transformation: Any set of translation, rotation, and scaling operations in the two spatial directions of the plane. Affine transformations allow maps with different scales, orientations, and origins to be coregistered.

Autodesk Map: A GIS software package. See Section 8.5.

batch: Submission of a set of commands to the computer from a file rather than directly from the user as an interactive exchange.

big eight: The eight most popular GIS packages, established by the numbers of users, particularly among people getting started with GIS, at any given time.

buffer: A zone around a point, line, or area feature that is assumed to be spatially related to the feature.

CALFORM: An early computer mapping package for thematic mapping.

CAM (computer-assisted mapping): A map projection and outline plotting program for mainframe computers dating from the 1960s.

CGIS (Canadian Geographic Information System): An early national land inventory system in Canada that evolved into a full GIS.

clump: To aggregate spatially; to join features with similar characteristics into a single feature.

compression: Any technique that reduces the physical file size of data in a spatial or other data format.

cookie-cut: A spatial operation to exclude area outside a specific zone of interest. For example, a state outline map can be used to cut out pixels from a satellite image.

critical six: The GIS functional capabilities included in Dueker's GIS definition: map input, storage, management, retrieval, analysis, and display.

data exchange format: The specific physical data format in which exchange of data between similar GIS packages takes place.

data structure: The logical and physical means by which a map feature or an attribute is digitally encoded.

DBMS (database management system): Part of a GIS; the set of tools that allow the manipulation and use of files containing attribute data.

desktop mapping: The ability to generate easily a variety of map types, symbolization methods, and displays by manipulating the cartographic elements directly.

desktop metaphor: For a GUI, the physical analogy for the elements with which the user will interact. Many computer GUIs use the desktop as a metaphor, with the elements of a calendar, clock, files and file cabinets, and so on.

device independence: The ability of software to run with little difference from a user's perspective on any computer or on any specialized device, such as a printer or plotter.

dissolve: Eliminating a boundary formed by the edge or boundary of a feature that becomes unnecessary after data have been captured: for example, the edges of sheet maps.

Dueker's definition of GIS: "A special case of information systems where the database consists of observations on spatially distributed features, activities or events, which are definable in space as points, lines, or areas. A geographic information system manipulates data about these points, lines, and areas to retrieve data for ad hoc queries and analyses."

DXF: Autocad's digital file exchange format, a vector-mode industry standard format for graphic file exchange.

dynamic segmentation: GIS function that breaks a line into points at locations that have significance, and that can have their own attributes. For example, the line representing a highway can have a new node added every mile as a mile marker that can hold attributes about the traffic flow at that place.

edge matching: The GIS or digital map equivalent of matching paper maps along their edges. Features that continue over the edge must be "zipped" together and the edge dissolved. To edge-match, maps must be on the same projection, datum, ellipsoid, and scale and show features captured at the same equivalent scale.

entity by entity: Any data structure that specifies features one at a time, rather than as an entire layer.

FORTRAN: An early computer programming language, initially for converting mathematical formulas into computer instructions.

functional capability: One of the distinctive processes that a GIS is able to perform as a separate operation or as part of another operation.

functional definition: Definition of a system by what it does rather than what it is.

fuzzy tolerance: Linear distance within which points should be snapped together.

generalization: The process of moving from one map scale to a smaller (less detailed) scale, changing the form of features by simplification, and so on.

geometric test: A test to establish the spatial relationship between features. For example, a point feature can be given a point-in-polygon test to find if it is "contained" by an area.

GNU: Free Software Foundation organization that distributes software over the Internet.

GRASS: A GIS software package (see Section 8.5).

GUI (Graphical User Interface): The set of visual and mechanical tools through which a user interacts with a computer, usually consisting of windows, menus, icons, and pointers.

help line: A telephone service available to software users for verbal help from an expert.

import: The capability of a GIS to bring data in an external file and in a nonnative format for use within the GIS.

installation: The step necessary between delivery of GIS software and its first use, consisting of copying and decompressing files, data, registering licenses, and so on.

integrated software: Software that works together as part of a common user interface rather than software that consists of separate programs to be used in sequence.

local area network: An arrangement of computers into a cluster, with network linkages between computers but no external link. Usually, this allows sharing data and software licenses, or the use of a file server.

macro: A command language interface allowing a "program" to be written, edited, and then submitted to the GIS user interface.

map algebra: Tomlin's terminology for the arithmetic of map combination for coregistered layers with rasters of identical size and resolution.

map overlay: Placing multiple thematic maps in precise registration, with the same scale, projections, and extent, so that a compound view is possible.

Maptitude: A GIS software package (see Section 8.5).

mask: A map layer intended to eliminate or exclude areas not needed for mapping and analysis.

metadata: Data about data. Index-type information pertaining to the entire data set rather than the objects within the data set. Metadata usually includes the date, source, map projection, scale, resolution, accuracy, and reliability of the information, as well as data about the format and structure of the data set.

mosaic: The GIS or digital map equivalent of matching paper maps along their edges. Features that continue over the edge must be "zipped" together and the edge dissolved. To edge-match, maps must be on the same projection, datum, ellipsoid, and scale, and show features captured at the same equivalent scale. See also **edge matching**.

Motif: A graphical user interface standard that is common on Unix workstations.

multitask: The ability of a computer's operating system or a GIS to handle more than one process at once; for example, editing and running a command sequence while extracting data from the database and displaying a map.

node snap: Instructing the GIS software to make multiple nodes or points in a single node so that the features connected to the nodes match precisely, say at a boundary.

online manual: A digital version of a computer application manual available for searching and examination as required.

patch: A fix to a program or data set involving a sequence of data that are to be overwritten onto an older version.

proprietary format: A data format whose specification is a copyrighted property rather than public knowledge.

raster: A data structure for maps based on grid cells.

relational DBMS: A database management system based on the relational data model.

renumbering: Use of the DBMS to change the ordering or ranges of attributes. Also, especially in raster GISs, to change the numbers within grid cells into categories.

rubber sheeting: A statistical distortion of two map layers so that spatial coregistration is accomplished, usually at a set of common points.

sift: To eliminate features that are smaller than a minimum feature size.

spatial data transfer standard (SDTS): The formal standard specifying the organization and mechanism for the transfer of GIS data between dissimilar computer systems. Adopted as FIPS 173 in 1992, SDTS specifies terminology, feature types, and accuracy specifications as well as a formal file transfer method for any generic geographic data. Subsets for the standard for specific types of data—vector and raster, for example—are called profiles.

spreadsheet: A computer program that allows the user to enter numbers and text into a table with rows and columns, and then maintain and manipulate those numbers using the table structure.

SURFACE II: An early computer mapping package from the Kansas Geological Survey.

SYMAP: An early multipurpose computer mapping package.

topologically clean: The status of a digital vector map when all arcs that should be connected are connected at nodes with identical coordinates and the polygons formed by connected arcs have no duplicate, disconnected, or missing arcs.

Unix: A computer operating system that has been made workable on virtually every possible computer and has become the operating system of choice for workstations and science and engineering applications.

upward compatibility: The ability of software to move on to a new version with complete support for the data, scripts, functions, and so on, of earlier versions.

user interface: The physical means of communication between a person and a software program or operating system. At its most basic, this is the exchange of typed statements in English or a programlike set of commands.

vector: A map data structure using the point or node and the connecting segment as the basic building block for representing geographic features.

version: An update of software. Complete rewrites are usually assigned entirely new version numbers (e.g., Version 3), while fixes and minor improvements are given decimal increments (e.g., Version 3.1).

VisiCalc: A spreadsheet package for first-generation microcomputers. Supported data tables in flat files.

warping: See **rubber sheeting.**

WIMP: A GUI term reflecting the primary user interface tools available: windows, icons, menus, and pointers.

X-Windows: A public-domain GUI built on the Unix operating system and computer graphics capabilities, written and supported by the Massachusetts Institute of Technology and the basis of most workstation shareware on the Internet.

zip: See **mosaic.**

PEOPLE IN GIS

Assaf Anyamba Research Associate, The IDRISI Project

Assaf Anyamba is a research associate at the Clark University Laboratories for Computer Cartographic Technology and Geographical Analysis, the IDRISI Project, based in Worcester, Massachusetts. His elementary, secondary, high school, and undergraduate education was undertaken in Kenya under the British educational system. Assaf has a B.A. with a double major in geography and economics from Kenyatta University, Nairobi, and an M.A. in geography from Ohio University. He is a Ph.D. student and a

NASA Global Change Research Fellow at Clark, where he is working on a dissertation studying El Niño's impacts on Africa.

KC: Assaf, how did your GIS career begin?

AA: During my undergraduate years I had the opportunity to participate in the Koobi Fora Harvard University Field School Program in the Rift Valley region of Kenya. Later I spent two months at the Regional Center for Services in Surveying Mapping and Remote Sensing in Nairobi, annotating the Rift Valley Landsat Mosaic and worked as a research assistant on the Kenya Rift International Seismic Project with a team of scientists from Karlsruhe University Institute of Geophysics. The major aim of the project was to gain a better understanding of the deep structure of the rift valley system using digital geophysical techniques. This was my first introduction to digital "things." I went on to Ohio University, where I got my Master's degree in geography. My thesis was on the comparison of ecological variables with coarse resolution satellite data for ecological mapping over Kenya. While there, I worked one summer on a project involving electoral districting, my first experience of project-oriented GIS. I spent a summer at the NASA/Goddard Space Flight Center through a Universities Space Research Association graduate summer internship program working in the Global Inventory Monitoring and Modeling Systems (GIMMS) Lab. At Clark University I am pursuing a Ph.D., working on reconstructing the El Niño/southern oscillation from coarse resolution satellite time-series data for Africa. I have a NASA Graduate Fellowship in Global Change Research.

KC: When did you first hear about GIS?

AA: In 1988 from a professor in Kenya as "automated cartography" in a theoretical sense (we had no practical exposure) and from a UNEP brochure in 1989. I thought of GIS as "spatial database" organizing land resources information (land cover, land use, population, drainage, etc.).

KC: And how would you define GIS today?

AA: As computerized systems for input, archiving, and manipulating different forms of geographic data and for output of derivative products from the data to assist in providing answers to environmental questions and to highlight specific environmental problems or resources.

KC: What is your role at IDRISI?

AA: Primarily research. I am one in a group of project members that form the Change and Time Series Group. We are concerned with development and use of time-series analysis techniques to understand spatiotemporal change in environmental data. We want to apply these techniques to drought and famine early warning, food security issues, and climate variability. I also help with software testing, with the WWW site for the IDRISI Project, and in training.

KC: What do you see as developments in GIS that have made it a practical technology for use throughout the world, especially in developing nations?

AA: Perhaps the most revolutionary thing was "porting" the GIS engine to the PC platform. PCs are cheaply available worldwide and easy to use. They have made it possible for most developing countries to undertake GIS projects.

KC: What is important to someone just getting started in GIS?

AA: Education, education, education! There needs to be a revised curriculum at the university level in GIS, perhaps national standards. Most schools are training GIS software specialists rather than geographic information scientists. This may not be wrong but it does affect GIS implementation. Software specialists can be narrow-minded and lose track of the broad scientific view. Training of GIS specialists needs to be stressed and monitored, perhaps by a GIS standards consortium. GIS education should cover basic geodesy and computer science; and ecology, climatology, biology, demography, and so on, should have a GIS component attached to them.

KC: Thanks Assaf. (Used with permission.)

CHAPTER 9

GIS in Action

9.1 INTRODUCING GIS IN ACTION
9.2 CASE STUDY 1: GIS FIGHTS THE GYPSY MOTH
9.3 CASE STUDY 2: GIS AND ROAD ACCIDENTS IN CONNECTICUT
9.4 CASE STUDY 3: GIS AT THE WORLD TRADE CENTER AFTER SEPTEMBER 11, 2001
9.5 CASE STUDY 4: RESOURCE MANAGEMENT FOR CALIFORNIA'S COASTAL ISLANDS: THE CHANNEL ISLANDS GIS
9.6 CASE STUDY 5: USING GIS AND GPS TO MAP THE SLIDING ROCKS OF RACETRACK PLAYA
9.7 STUDY GUIDE
9.8 EXERCISES
9.9 REFERENCES
9.10 KEY TERMS AND DEFINITIONS

Let observation with extensive view
Survey mankind, from China to Peru.
—*Samuel Johnson*, The Vanity of Human Wishes *(1749)*

It is a map that moves
faster than real
but so slow;
only my watching proves
that island has being,
or that bay.
—*May Swenson*, The Cloud Mobile, *1958, 2nd verse*

"It's all fine you saying we're no longer in Kansas, but lets try to be a little more scientific about this..."

9.1 INTRODUCING GIS IN ACTION

As much as knowledge and understanding of the principles behind GIS are critical to getting started with GIS, the technology's true strength is and will always be in the power of its applications. In this chapter, five GIS case studies are presented. Each is unique in its own way, and the reader should pay attention to differences in data structures, software, procedures, and directions as the GIS systems we have discussed in theory now move out into the real world. What is also impressive is the extreme breadth and versatility of these applications. GIS is a tool that crosses disciplinary and professional boundaries with ease. Nevertheless, each field of expertise has an angle on GIS use and brings to the application a fresh set of approaches. The five applications cover oceans, rural, suburban, deserts, and urban areas, they encompass forestry, geology and ecology, public health concerns and insects, storms and runoff, and mysterious rocks, and also how GIS assisted at the site of the tragedy in the nation's worst terrorist attack. These five applications do not pretend to be comprehensive. Each has been contributed by the GIS experts in question as a summary of a broader-scale work that they have either completed or that remains in progress. Nevertheless, these applications are a perfect starting point from which to examine GIS in action.

9.2 CASE STUDY 1: GIS FIGHTS THE GYPSY MOTH

A Case Study of the Use of GIS to Understand Population Dynamics of the Gypsy Moth in Michigan

9.2.1 Contributors: Bryan Pijanowski and Stuart Gage, Michigan State University

The Entomology Spatial Analysis Laboratory in the Department of Entomology at Michigan State University is devoted to the spatial analysis of insect pests and the assessment of risk to Michigan's forests, among other projects. The laboratory is directed by Dr. Stuart Gage, who has conducted research on the spatial distribution of forest and crop pests for over 25 years. Dr. Bryan Pijanowski is an ecologist, and an associate in the laboratory. He has specialized in the use of GIS, such as Arc/Info and IDRISI, to model insect and human populations. The laboratory currently contains several Pentium computers, four Sun workstations, and a Silicon Graphics workstation for visualization of spatial data. The staff uses Arc/Info, IDRISI, ERDAS, ER-Mapper, and Atlas*GIS for research.

9.2.2 Background

The use of GIS to study the gypsy moth in Michigan provides an excellent example of the applicability of this tool in the biological sciences and for resource management. The gypsy moth (Figure 9.1) is an introduced forest pest that consumes the leaves and needles of nearly 300 woody plants. The insect was first discovered in the state 40 years ago, and outbreaks of the pest have been occurring in Michigan since the mid-1980s. Severe defoliation (i.e., loss of leaves) of oaks, aspens, and other tree species preferred by gypsy moth caterpillars has occurred throughout the northern Lower Peninsula, and

FIGURE 9.1: A gypsy moth caterpillar, shown against an oak leaf. (Photos in this section are courtesy of Dr. Bryan Pijanowski. Used with permission.)

populations continue to expand into southern Michigan and into the state's Upper Peninsula. Defoliation has increased from 2800 hectares in 1984 to over 280,000 hectares in 1992.

Unlike many native forest insects, the gypsy moth is a problem in both urban areas and forests. Multitudes of large, hairy caterpillars, abundant frass (fecal material), and loss of leaves on shade and ornamental trees create much annoyance for people in wooded residential and recreational areas. Management of the gypsy moth is carried out by aerial spraying of a biological insecticide called *Bacillus thuringiensis* (Bt) from helicopters or planes (Figure 9.2). This biological insecticide kills only moths and butterflies that eat the Bt from tree leaves and it degrades in the environment in a few days.

9.2.3 The Monitoring Program

In 1985, a statewide gypsy moth monitoring program was implemented to characterize this pest's population dynamics. Because the male is attracted to the female through the use of a pheromone that is emitted by the female, populations of male moths have traditionally been monitored through the use of pheromone-baited traps (Figure 9.3). A small pesticide strip is placed at the bottom of these traps to kill the moths once they enter. The statewide program entails the monitoring of 3000 pheromone traps placed in a grid-like design (Figure 9.4) with a 6-mile intertrapping distance. Several agencies have been involved in this monitoring effort, including the Michigan Department of Agriculture, Michigan Department of Natural Resources, the USDA–APHIS, Animal and Plant Health Inspection Service, and USDA Forest Service. Funding for the project has come from the Michigan Department of Agriculture.

Every year, these pheromone-baited traps are placed in designated locations in the spring. In the fall, trap tenders visit each location and record the number of moths contained in the trap. Trap catch data are recorded on specially designed forms and are

FIGURE 9.2: Helicopter with spray boom, spraying trees with the Bt biological pesticide to kill the gypsy moth.

FIGURE 9.3: A milk carton trap used by the spatial monitoring program for the gypsy moth across the state of Michigan.

sent to the state survey coordinator at Michigan State University for data entry and management. Trap locations are geocoded at Michigan State University's Entomology Spatial Analysis Laboratory by linking permanent site numbers to geographic coordinates. Data are then placed into a geographic information system for spatial analysis and association with other information, such as previous years' moth estimate, host distribution, and tree defoliation.

Once the data are brought into a GIS, the numbers of moths captured per year, initially represented as point data, are converted to a raster format using various interpolation methods. The most common interpolation method that is used is the inverse distance squared (IDW) method (we use a weighting of 2), which is available in both IDRISI and

FIGURE 9.4: Gypsy moth trap sample locations in Michigan.

Arc/Info. Interpolation of these trap data are conducted to obtain a raster layer of gypsy moth trap numbers at 1-km cell resolution. Once the raster data layer is developed for trapping results for each year, the data are reclassified into population-size categories For example, raster maps of moth counts of 1–25; 26–100; 101–200; 201–300; 300–400; and 400 or more moths are used frequently as starting maps for analysis and overlay with various other GIS layers, such as forest cover.

9.2.4 GIS Use and Data Analysis

The use of GIS to assess risk to Michigan's forests from the gypsy moth is one example of a case study that has special interest to both resource analysts and to biologists interested in studying the interaction of insects and forests. The main objective of this study, partly funded by the Michigan Department of Agriculture, is to determine areas where the most susceptible tree species, oak and aspen, may undergo defoliation. We approached this study by developing annual gypsy moth population-size category coverages (Figure 9.5). Annual high-risk population coverages, which we determine to be 400 moths or more, are created as simple binary maps; a "1" is coded as the presence of 400 or more moths and a "0" as locations of the state that contain fewer than 400 moths for that year. Susceptible forest data were obtained from a statewide 1-km resolution forest-type map (dominant tree species only) that was developed from a multitemporal analysis of AVHRR data and various forest cover maps from the Michigan Department of Natural Resources.

This forest cover map contains information for several classes of forests (e.g., oak–hickory; spruce–fir); we used GIS to extract only susceptible forest types (i.e., oak and aspen forests) from this database. To perform the final analysis, we coded all oak forests with a "2" and all aspen forest with a "1" and multiplied the high-risk population coverage with the susceptible forest coverage.

9.2.5 Summary

The Michigan Department of Agriculture has used these map series to help manage the state forests that are located in the western portion of the Lower Peninsula of Michigan. Analysis of these data could not be accomplished easily without the use of a geographic

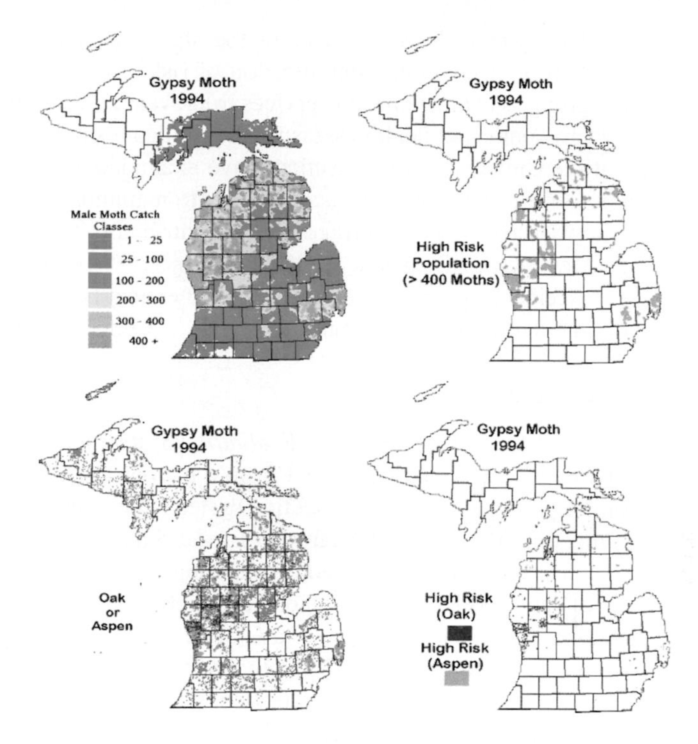

FIGURE 9.5: Upper left: Sample annual map for 1994 with seven density categories. Upper right: High-risk population map; area with greater than 400 moths trapped. Lower left: Susceptible forest types (i.e., oak and aspen forests) from the database. Lower right: Risk areas from an overlay of the data.

information system. Results of spatial analyses are generally easy to interpret and are thus powerful tools for use in resource management and policy development. Furthermore, because the results can be displayed as a colorful map, posting the maps and associated text on the World Wide Web has become a very effective communication device as well. The Entomology Spatial Analysis Laboratory in the Department of Entomology at Michigan State University maintains a World Wide Web site at `http://www.ent.msu.edu`.

9.3 CASE STUDY 2: GIS AND ROAD ACCIDENTS IN CONNECTICUT

Case Study of the Use of GIS to Inventory and Understand the Pattern of Traffic Accidents in Connecticut

9.3.1 Contributor: Ellen Cromley, University of Connecticut

Ellen Cromley is a medical geographer who studies geographical patterns of health and disease and the location and use of health services. Mapping has long been an important part of medical geography, and most mapping activity now uses GIS. Dr. Cromley has been involved in several public health GIS projects, including working with the Connecticut Department of Environmental Protection to develop a GIS for the Water Supplies Section of the Health Department, the unit that regulates public drinking water in the state; developing a GIS for statewide surveillance of Lyme disease and identification of risk areas; using GIS with a community planning group to evaluate health services to prevent HIV/AIDS; and compiling GIS databases for environmental health assessments. Graduate students in the Department of Geography at the University of Connecticut have examined exposure to electromagnetic fields from power transmission lines, accessibility to mammography services in the state, and Emergency services (911) coverage areas. Contributors to the Connecticut CODES GIS include Mary Kapp, Connecticut CODES Project Director; Brian Pope, Graduate Assistant, the Accident Records Section of the Connecticut Department of Transportation (ConnDOT); and the Connecticut Health Research and Education Foundation (CHREF).

9.3.2 Background

CODES stands for *Crash Outcome Data Evaluation System* and Connecticut is one of 20 U.S. states participating. CODES evolved from a national need to report on the benefits of regulations requiring automotive protection systems like seat belts and bicycle helmets. States are funded by the National Highway Traffic Safety Administration (NHTSA) to link motor vehicle crash data with medical outcome data to develop a better picture of the problem of motor vehicle injury and the effectiveness of protection systems. The linked database is the primary product of a CODES project, and a public-use version of the database is also required by NHTSA. In addition, CODES projects are allowed to develop state-specific products. The Connecticut CODES GIS is an example.

The Connecticut CODES Project links statewide automotive crash data from police accident reports for 1995 and 1996 (the two most recent complete years at the time the project began) coded by the Accident Records Section of ConnDOT to trauma registry, emergency department, and inpatient records maintained for the project by CHREF (an arm of the Connecticut Hospital Association), and mortality records maintained by the Vital Records Section of the state health department. The data include all collisions reported to police that occurred on Connecticut's state or federal roads in 1995 and 1996 and all collisions that occurred on local roads if the police report indicated a fatality or injury. There is one database for each year of collision data. The number of crashes is alarming: 72,672 involving 190,143 people in 1995 and 78,407 involving 202,792 people in 1996. This resulted in a linked accident-hospital database with 28,913 records for 1995 and 37,124 for 1996. All have spatial location of the accident as part of the record.

9.3.3 The Connecticut CODES GIS

The purpose of the Connecticut CODES GIS is to create a viewing environment for the linked motor vehicle crash records so that users can find collisions of interest and obtain data on their attributes and locations. CODES users can easily display, query, and map data. These capabilities are especially important for the public-use version of the databases, with which the GIS works. The CT CODES GIS is a combination of Microsoft Access databases and an ESRI ArcView application modified with Avenue scripts to create a GIS specifically for the project. Users can search the CODES databases by *WHAT* and by *WHERE*. In Access, users can perform detailed queries to identify *what* collisions are of interest, report them, and add them as a user-defined collision data layer in the GIS to see *where* the collisions occurred. In the GIS, users can find *where* a place of interest is, and then identify and report collision attributes to find out *what* kinds of collisions occurred in that place.

The Access databases contain related tables of collision attributes, traffic unit attributes (a traffic unit is a pedestrian or a combination of a vehicle and an operator), and involved person attributes including individual data on operators, pedestrians, and passengers. Every collision is assigned a unique identifier. Specially designed Access queries and reports allow users to find collisions of interest and either print reports or export a table of collision identifiers for the user-defined collisions to the GIS. DBase tables of collision, traffic unit, involved person, and user-defined collision attributes are exported from Access. They are automatically linked to ArcView point shape files of collision locations when the GIS user selects a database to view in the application (Figure 9.6).

Users open the GIS application by clicking a shortcut icon on the computer desktop. The application automatically adds data layers to the application and applies legends (Table 9.1). These data layers create the context for viewing the collision data. Users of CT CODES GIS can pan and zoom to locations of interest in a number of ways. They can enter a CODES_Id and the view zooms to display the collision location in the center of view at a scale of 1 : 24,000. Users can select from a tool menu of common map scales, click a point on the screen, and zoom to display the location where the user clicked in the center of the view at the selected scale. Users can select a type of place from the *PanTo* menu, scroll through the list of names for that type in the annotation

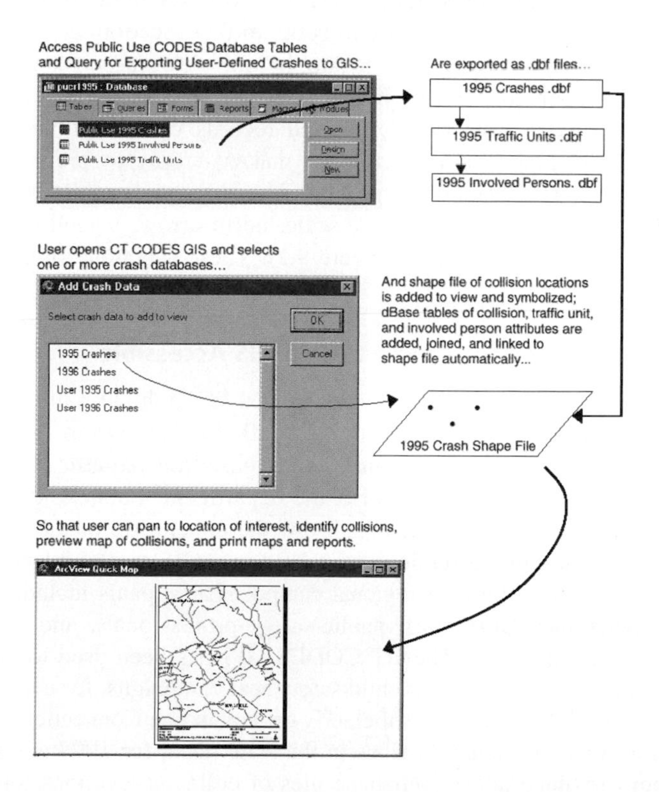

FIGURE 9.6: Process flow for queries in the Connecticut CODES GIS. (Courtesy of Ellen Cromley.)

TABLE 9.1: Data layers in CT CODES GIS

Data Layer and Source	User Action	Range of Map	Scale
Town names Connecticut DEP	Added when application is opened	1 to 125 000	400 000
Town index Connecticut DEP	Added when application is opened but not visible	1 to 125 000	400 000
Quadrangle index Connecticut DEP	Added when application is opened but not visible	1 to 125 000	400 000
Boundaries Connecticut DEP	Added when application is opened	1 to 6 000 35	0000
Roads ConnDOT	Added when application is opened		1 to 6 000 125 000
Annotation	Added when application is opened; user can turn off and on as needed		1 to 6 000

data layer, and zoom to display the location of the annotation the user selected. Finally, users can load the address-ranged street network database for a town of interest and enter an address or street intersection to zoom to that location.

Once locations of interest are in the view, users can graphically select collisions and open joined and linked tables of the collision, traffic unit, and involved person attributes from which selected records can be printed as reports or exported to files. The Map and Report menu makes it easy to preview and print color maps and reports of collisions in the view. Users can enter titles for maps and reports, but other elements of the layout like scale, north arrow, legend, and date printed are handled by the GIS application (Figure 9.7). These functions support data distribution and analysis.

9.3.4 Making Connecticut CODES GIS Accessible

The CT CODES GIS resides on a dedicated PC in the Department of Health in Hartford, Connecticut (CT). Health department staff use the system for their own analyses and they can extract data, maps, and reports based on requests. Individuals can also make appointments to use the system at the department. Public Use Access and CT CODES GIS User Guides are also available. At the conclusion of the first phase of the project, free hands-on training sessions were held in a GIS teaching laboratory at the University of Connecticut's Hartford regional campus. Participants included local health directors, EMS personnel, DOT staff, public-safety professionals, and public health researchers from around the state. The CT CODES GIS has been used to provide data for specific towns and regions, for local child safety seat campaigns, for evaluation of traffic calming devices by DOT, for studies of elderly drivers in one Connecticut county, and for research on fatal motor vehicle collisions in the state. Data for 1997 are now being added to the system. Attribute tables and shape files of collision locations for each of the 3 years of data available to date will be distributed for the entire state and for individual towns

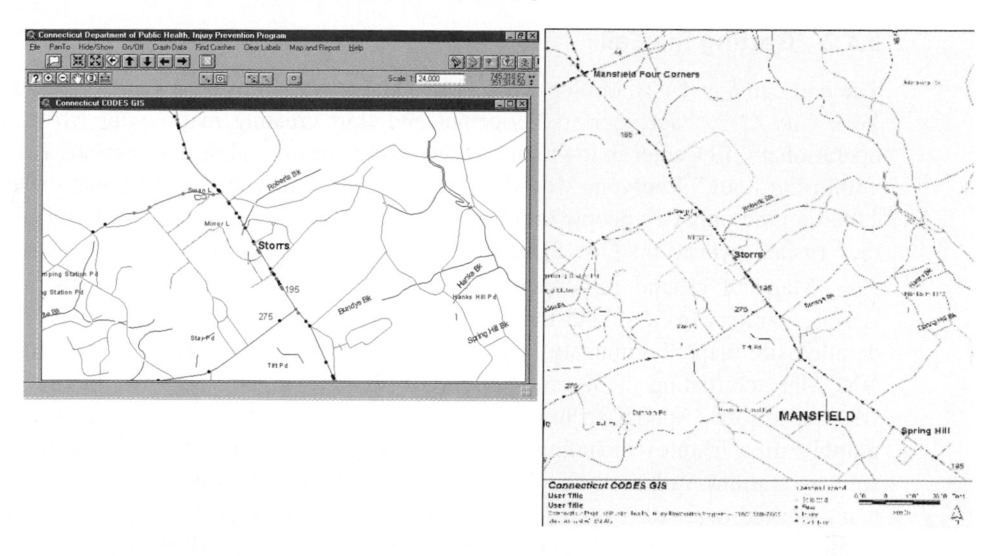

FIGURE 9.7: Map displays created by the Connecticut CODES GIS. Left: Screen query. Right: Automatically generated printed map. (Courtesy of Ellen Cromley.)

through the MAGIC (Map and Geographic Information Center, of the University of Connecticut libraries), which distributes digital geospatial data for the state through its Web site at `http://magic.lib.uconn.edu`.

9.4 CASE STUDY 3: GIS AT THE WORLD TRADE CENTER AFTER SEPTEMBER 11, 2001

How GIS Helped in the Rescue and Clean-Up Operations after the World's Worst Terrorist Attack

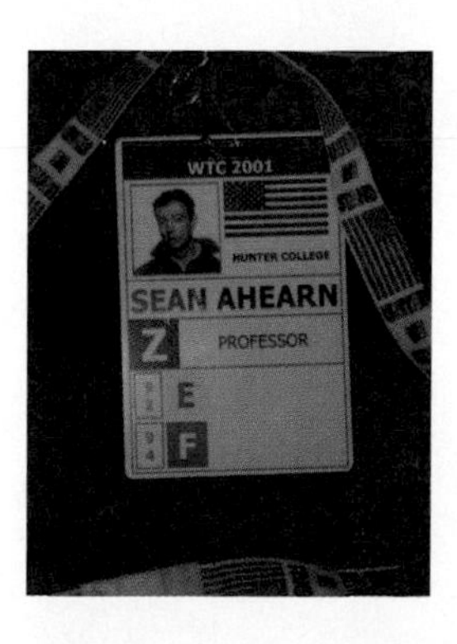

9.4.1 Contributor: Sean C. Ahearn, Hunter College–CUNY

September 11, 2001 saw the greatest peacetime tragedy of the recent era, the combined suicidal attack on the twin towers of New York City's world trade center (WTC) and the Pentagon by four hijacked planes. At Hunter College's CARSI (Center for the Analysis and Research of Spatial Information) laboratory, GIS was put to immediate and effective use in dealing with the aftermath. Fortunately, Geography Professor Sean Ahearn was ready and able to assist, having worked on the NYCMap, New York City's comprehensive GIS. The CARSI played a critical role, partly because the permanent New York City Emergency Operations Center had been located in the WTC complex, and was destroyed. This case study is dedicated to all of those who helped, but also to those who died, and especially to Geographer Robert LeBlanc of the University of New Hampshire who was on United Flight 175 on the way to a Geography conference in Santa Barbara when it was crashed into the WTC's south tower.

9.4.2 Getting the Call

The call came at 4 P.M. on September 11, 2001. It was from Alan Leidner, head of New York City GIS, "Get your staff together and start creating maps, your lab is the only operational GIS Center in town, everything else is destroyed or inaccessible, I'll be there within the hour." Everyone who didn't live in Manhattan had gone home except me. I called up the only two people on my staff from Manhattan, Ji Ding and Jeffery Bliss and they rushed over to the CARSI Lab at Hunter College. Leidner showed up soon after.

Maps of ground zero were needed for command and control of the operation. Rescue workers from around the country would be pouring in and they would all need detailed site maps. Fortunately New York City had recently created a "base-map" called NYCMap, consisting of 30 cm resolution orthophotography and planimetric map data with an absolute spatial accuracy of half a meter. NYCMap has over two dozen geographic map features including building outlines, curb lines, street centerlines, parks, subway stations, rails, towers, and so on. Over the next four hours, the Hunter College team worked to create a baseline set of maps using just the planimetric data (Figure 9.8), and maps showing the orthophotographs with planimetric overlays (Figure 9.9).

At about 10:30 P.M. on September 11, Leidner and Bliss headed downtown to the Emergency Operations Center (EOC) on 21st Street with armfuls of maps of the WTC site. The city virtually empty, the next morning at 7:00 A.M. the Hunter team piled three computers loaded with the NYCMap database into the back of a police car and sped down to the EOC. The data on these machines was to form the kernel of what would become a

FIGURE 9.8: Plot of the planimetric data from the NYCMap GIS. Figures in this section by Sean Ahearn. (Used with permission.)

FIGURE 9.9: Orthorectified photography overlain with vector data from the NYCMap. WTC area prior to 9/11/02.

twenty-four hour a day, seven days a week operation involving over fifty GIS professionals and lasting for over two months. The full range of mapping science technologies would be deployed: GIS, GPS, and remote sensing. Cartographic representation of data would prove to be critical in an environment in which the consumers of maps (such as firefighters and rescue workers) had never seen the likes of the data that we would be providing, including thermal imagery and Light Detection and Ranging (LIDAR).

9.4.3 Damage Assessment

Damage assessment and infrastructure status were the first tasks. Bruce Oswald from the New York State Office of Technology called the EOC on the morning of September 13 to discuss possible remote sensing instruments to be deployed. Since my background was in remote sensing I took the call. We decided on orthophotography with an accuracy of plus or minus 70 cm, a thermal sensing instrument and Bruce suggested LIDAR. We ended up by going with all three. Earth Data Holdings out of Maryland would fly the plane, man the instruments, and pass on the data to the CARSI Lab at Hunter College and the GIS team at the EOC. Earth Data would create an all-digital system with a turnaround time of less than 12 hours.

The first orthophotographs were shocking. Even the rescue workers who saw them were startled. This was the first synoptic view of the entire site and it was horrific (Figure 9.10), showing 16 acres of total destruction. The smoke from the fires still obscured a portion of the site confirming the need for the LIDAR system. LIDAR is an "active" remote sensing device, one that relies on its own energy for creating an image, in this case a laser with a wavelength of 0.9 micrometers. I was familiar with the technology because one of my classmates while I was a graduate student at the University of Wisconsin-Madison, Gordon McClean, did some of the early research on

FIGURE 9.10: Orthorectified photography from September 19th.

LIDAR. The systems had come a long way in 20 years and the new ones could fire over 10,000 pulses per second. More importantly their spatial accuracy was about 9 cm vertical and 30 cm horizontal. These incredible accuracies are made possible with the help of an on-board GPS that provides three-dimensional locational coordinates and an inertial navigation unit which provides angular orientation of the platform. The LIDAR system measures the time each pulse takes to hit an object on the ground and return to the sensor. By knowing the speed of light the travel time can be converted to a distance. Because we know the location of the sensor, the orientation of the platform and the scan angle of the sensor, we can derive a precise location of the object reflecting the LIDAR pulse. The result is a blanket of points of the "terrain" with precise X, Y, and Z coordinates (Figure 9.11). These points can be converted to a Triangulated Irregular Network (TIN) by fitting each set of three points with a triangular facet (Figure 9.12). The TIN is essentially a "wire frame" similar to those created in movie animations. It can be viewed in 3-D by adjusting the viewing azimuth and angle, and the sun azimuth and angle (Figure 9.13).

The first LIDAR image was captured on September 19, 2001. This was the first clear image of the site because LIDAR penetrates clouds and smoke (Figure 9.14).

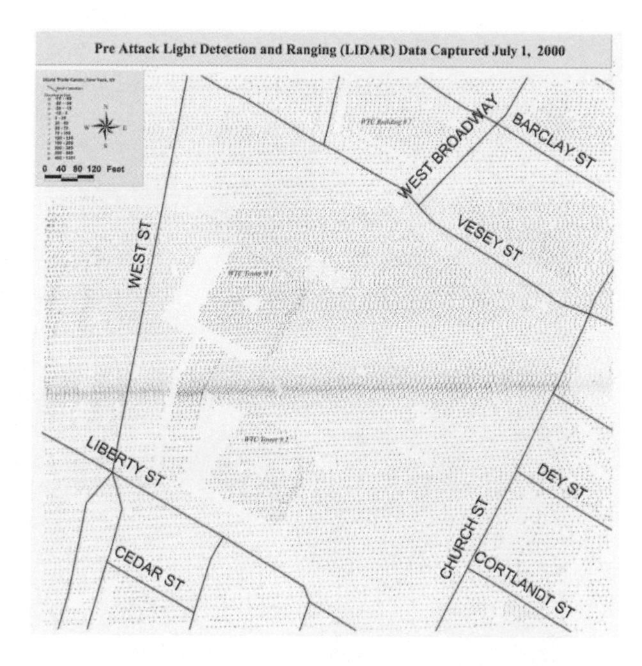

FIGURE 9.11: Point-based LIDAR height data collected before the attack.

FIGURE 9.12: Elevation data from a TIN derived from the LIDAR data in Figure 9.11.

FIGURE 9.13: Three dimensional GIS-based rendering of the LIDAR elevation data.

FIGURE 9.14: LIDAR was able to penetrate the pervasive smoke that hindered imaging.

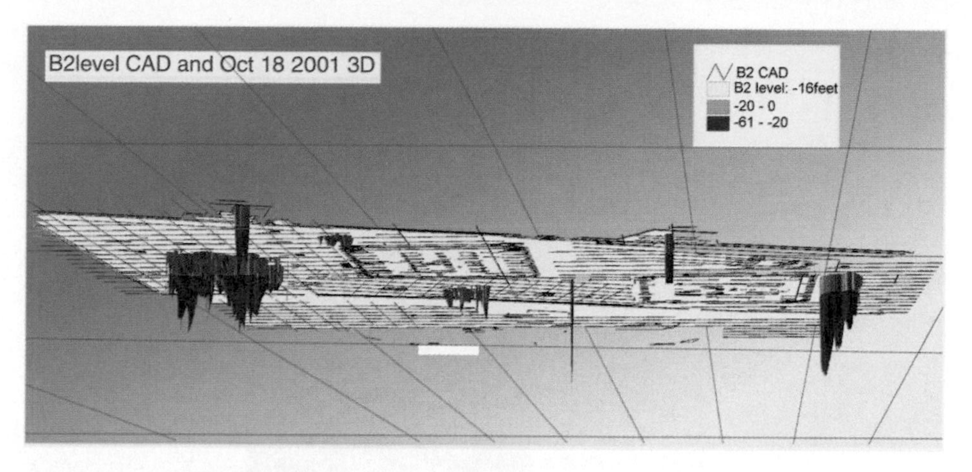

FIGURE 9.15: LIDAR elevations overlain with CAD-based building and room layout levels at the second level of basement, showing debris penetration.

The team at the CARSI Lab at Hunter College struggled to get the right cartographic representation for communicating the maximum amount of information while insuring that the image was interpretable by nongeographers. I still remember rolling out that first 50-by-60-inch LIDAR image of ground zero in front of a half dozen firemen and waiting for their response. After about two minutes of silence it "clicked" and they simultaneously began relating the image to their ground experience "oh my guys were working on that pile yesterday," "that's what's on the other side of that pile," and so on. The LIDAR proved to be a very valuable tool not only for damage assessment but also for getting a better understanding of the site. One of the real unknowns was the damage done to the underground infrastructure. By geometrically rectifying the CAD drawings to the NYCMap building outlines and incorporating the LIDAR TIN models the amount of surface penetration into the subsurface could be analyzed (Figure 9.15).

The thermal systems deployed were probably the weakest link in the remote sensing suite. The problem was that they were thermal videography systems that provided information on relative not absolute temperatures (Figure 9.16). It was good for seeing where the hotspots were but not good for knowing how hot something was. After three weeks a thermal system with on-board black body calibration was located and flown from a helicopter with GPS and inertial navigation systems after another week of bureaucratic approvals. Because it was a federal asset, owned by the Department of Energy, the request went from the City of New York to the State Office of Emergency to the Federal Emergency Management Agency, which made the request to the Department of Energy! By the time we got the data, it was too late as the fires had subsided and the threat to the underground Freon tanks had passed.

9.4.4 Infrastructure Status and the Role of Geographic Information

While the remote sensing efforts were underway the Emergency Mapping and Data Center (EMDC) under the leadership of Alan Leidner (Figure 9.17), had been set up on Pier 92 in the Hudson River, the new home of the Emergency Operations Center.

FIGURE 9.16: Thermal remote sensing data collected at the WTC on September 16th. Source: Roger Clark, USGS Open File report 01-0429.

Every part of the infrastructure was damaged and the GIS professionals at the pier, who consisted of a mix of Government, Industry, and Academia, were busy gathering information and creating maps of infrastructure status. These included utilities (such as gas, water, electric, telecommunications), building, transportation, and storage tanks (Figure 9.18). These maps were updated daily, posted on the Internet, passed on to city officials, and released to the news media. There were also a host of special use maps for a variety of local, state, and federal agencies.

One of the biggest problems was managing the building status database. While NYCMap had all of the building outlines delineated, these geographic features had yet to be tied to the attribute databases that described them. Additionally, the identifier used for the inspection of buildings was a street address, which was not unique! A building can span a whole block and can have one or more address for each block face. Fortunately through the foresight of Richard Steinberg of the Department of City Planning, buildings were assigned a unique Building Identification Number (BIN). In response to this need a project team was set up at the Pier to assign each of the building features of NYCMap for the area below Canal Street with a BIN.

The next problem was going from the paper inspection data to a map of building status. This process was fraught with all kinds of problems from a lack of consistency of the data collected, to data integrity, to the timeliness of the process. A team from IBM and Linkspoint Inc. stepped in and created a wireless hand-held inspection application.

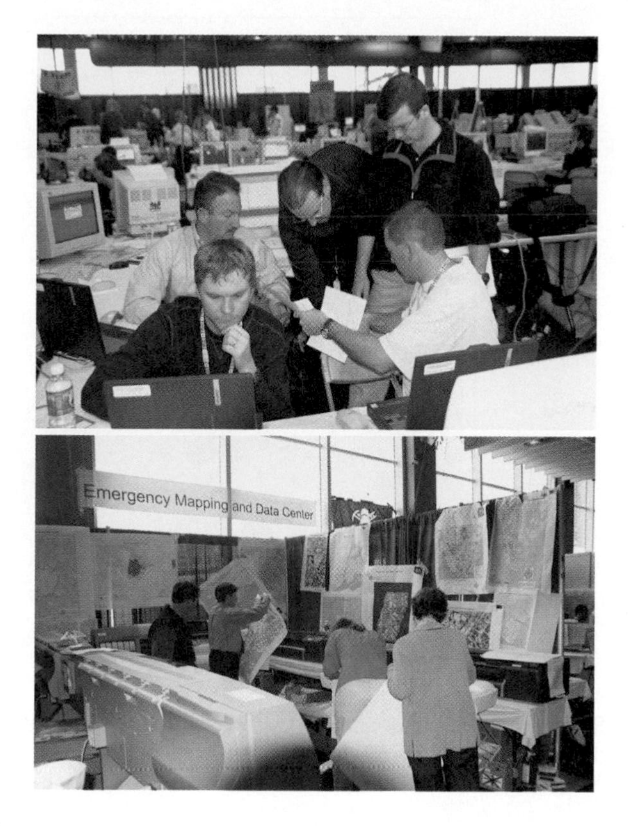

FIGURE 9.17: The Emergency Mapping and Data Center established at Pier 92 on the Hudson River.

The application enabled inspectors to access NYCMap wirelessly from a handheld iPAQ, click on a building to see the previous inspection (and make sure they were in the right place), fill out a new inspection, and then send it to the database. The Chief Inspector in the office could monitor which buildings were inspected that morning (or in the last minute) and create a map of building status at the stroke of the mouse. A process that up until then had taken three days and was fraught with problems had been reduced to a matter of minutes!

9.4.5 GPS Field Application

Firefighters' boots were literally melting as they stood scribbling down information on equipment and body parts recovered at ground zero and doing their best to estimate where they were on the 16-acre pile. A better methodology was needed to increase data integrity, obtain a more accurate location for the objects found, and reduce the time of collecting data in the severe environment confronted by the firemen. The solution was a ruggedized handheld computer with a barcode scanner, a Linkspoint GPS hardware attachment, and software. Extensive testing and the use of a specialized GPS antenna for the Linkspoint system in the second week after September 11 helped reduce the problems of multipathing due to the many pieces of metal on the site (Figure 9.19). The new process was as follows: The firemen would place the item in the bag, attach a barcode

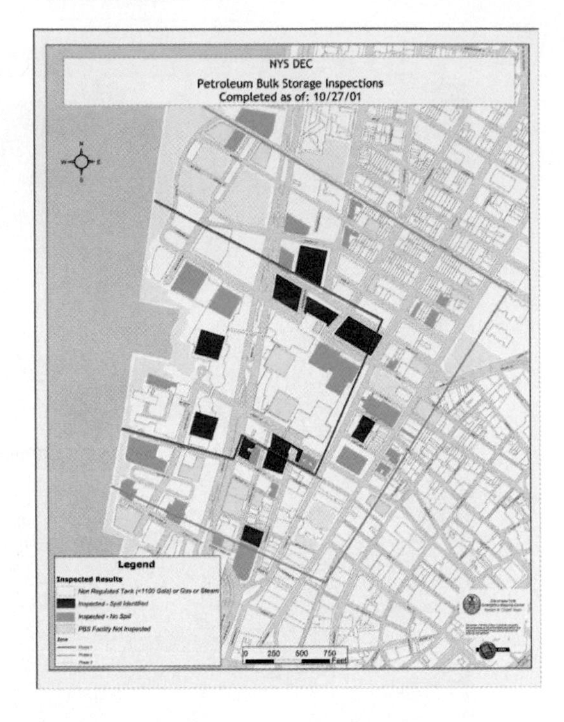

FIGURE 9.18: Infrastructure map products generated at the Emergency Mapping and Data Center for the New York City Department of Environmental Conservation. Locations and inspection conditions of gasoline storage tanks are shown.

to it, scan the barcode, select an item from a drop-down menu list, and the time and location was automatically captured. The technology reduced the whole process down to less than a minute, improved data accuracy and usability, and decreased firefighter exposure to the hazardous conditions they faced. The data would be mapped and used for victim identification and would provide information to loved ones on the location of victims bodies (Figure 9.20).

9.4.6 Lessons Learned

- New York City's GIS infrastructure played a critical role in the response to the WTC crisis.
- The federal government needs to provide options for available technologies (e.g., remote sensing).
- There were some severe data gaps:
  - The Census Bureau lists 1369 people as resident in the census tract containing the World Trade Center, but there was no daytime count of individuals who lived there.
  - Unique identifiers are essential for infrastructure management (such as the building IDs) but had to be added at the worst possible time.

FIGURE 9.19: Hand-held GPS data collection suffered from multipath signal deflections from the considerable amount of metal.

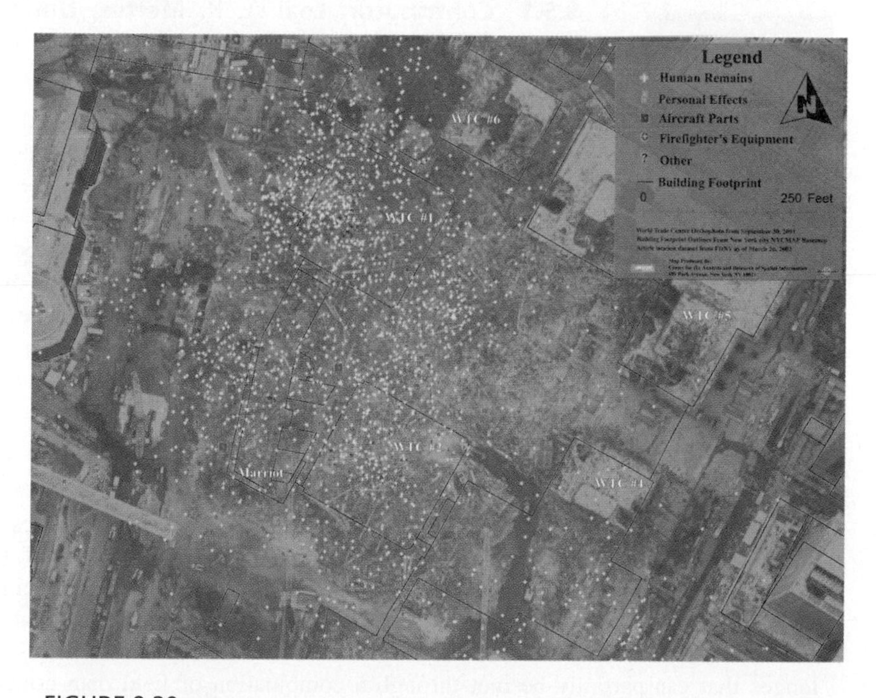

FIGURE 9.20: Map showing data collected from the mobile GPS and palmtop computer.

- Cities should make an effort to integrate their Geographic Information Systems into the rest of their management information system environments (i.e., connect spatial data to its attributes!).
- Cartographic standards for map production need to be established.
- Scenario development and exploration to anticipate data needs and modeling requirements should be planned in advance.
- Multiple levels of government involvement in scenario development are necessary to ensure cooperation in emergencies.
- Mobile access to GIS data is critical for "real-time" assessment and mapping of infrastructure
- Version management of data is a necessary tool for multiuser environments.

9.5 CASE STUDY 4: RESOURCE MANAGEMENT FOR CALIFORNIA'S COASTAL ISLANDS: THE CHANNEL ISLANDS GIS

Case Study Featuring a Collaborative GIS Helps in the Management of a Sensitive Coastline in the Santa Barbara Channel of California

9.5.1 Contributor: Leal A. K. Mertes, University of California, Santa Barbara

Dr. Leal Mertes investigates the processes responsible for creating wetlands and floodplains in large river systems and uses GIS and remote sensing in their analysis. The Channel Islands GIS (CIGIS) was created by undergraduate students, programmers with expertise in laboratory development for undergraduate courses, and GIS experts. The CIGIS is continuously being updated and used in teaching about the natural changes and ecosystems of the Channel Islands. Currently, new visualization interfaces are being created to enhance use of the database by managers and to provide educational materials to the general public. Contributors to the CIGIS include Ben Waltenberger, Ethan Inlander, Ceretha McKenzie, Amy L. Bortman, Melodee Hickman, John Dvorsky, and Olivia AuYeung.

9.5.2 Background

Viewing the coastal system of California as an ecosystem that includes both marine and terrestrial inputs and outputs allows managers to take account of the composition, structure, and function of the entire range of processes influencing the area's environmental health. Environmental management in the rapidly growing coastal areas of Southern California is controlled by a unique set of political and scientific challenges that can partially be met through a combination of field data collection, spatial modeling, and information technologies. In particular, digital databases, remote sensing data, and spatial analysis tools embedded in a GIS allow analysis of relations among the environmental variables. In addition to scientific challenges, coastal regions

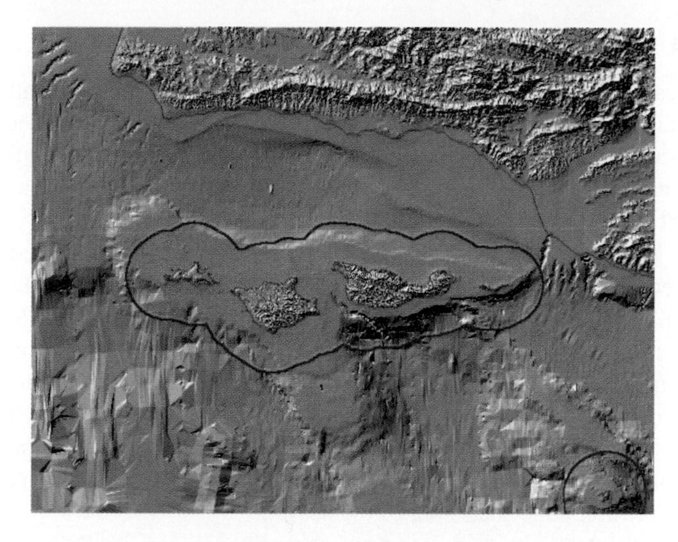

FIGURE 9.21: Topographic/bathymetric map of the area covered by the Channel Islands GIS. Area stretches from Point Conception to Ventura, California, west of Los Angeles. Image courtesy of Leal Mertes. (Used with permission.)

involve management by multiple agencies with distinct disciplinary and jurisdictional interests.

To meet the political and scientific challenges for management of the coastal region of central California, which includes the Channel Islands, the CIGIS was developed as a cooperative project among the University of California at Santa Barbara, the NOAA Channel Islands National Marine Sanctuary, the Channel Islands National Park, the Santa Cruz Island Reserve, the University of California Natural Reserve System, and the State of California Fish and Game Office of Oil Spill Prevention and Response. The area surrounded by buffers and covered by the database is shown by the topographic and bathymetric map (Figure 9.21).

9.5.3 CIGIS Users

As a resource management tool, the CIGIS provides information on flora and fauna (for example, kelp, sea grass, harbor seals, seabird colonies, shellfish), location of sensitive archeological sites, location and dimensions of sea caves, shipping lanes, oil platforms, bathymetry, geology, vegetation cover, soils, and topography. In response to the multiple roles it serves, CIGIS is customized to meet the needs of individual users. The master database is in Arc/Info and ArcView (Unix and PC version), while users receive versions with data relevant to the mission of the agency. For example, ArcView versions now reside at a field station and on a boat. Experience with multiple users shows that by promoting cooperation among agencies it is possible to create a database substantially more useful to the group as a whole than to have only emphasized the disciplinary or jurisdictional needs of an individual group. In addition, through pooling of resources, each agency has benefited from access to the entire database rather than only data explicitly related to their management mandate.

9.5.4 Using the GIS for Analysis

Success has been measured not by the size of the database, but rather by the new insight gained through spatial analysis of the environmental layers. Through analysis of a time series of Landsat remote sensing data (Figure 9.22) from 1972 to the present, Dr. Mertes and her collaborators have analyzed the characteristic patterns at the surface of the coastal waters of the Santa Barbara Channel.

More recently, shipboard measurements have been combined with water properties derived from advanced very high resolution radiometer (AVHRR) satellite data at a 1-km spatial resolution (Figure 9.22) During the El Niño storms of 1998, plumes were seen in these same positions from an airplane (Figure 9.23). In these images, the water color pattern in the ocean is dominated by the river plumes in the nearshore region that were generated by the Santa Ynez, Ventura, and Santa Clara Rivers. The surface pattern of the largest plume, which is the combined plume of the Ventura and Santa Clara Rivers is approximately 10 km long crossing the Santa Barbara Channel and is dispersed in different directions according to the season.

Digital terrain analysis of the watersheds of these three rivers shows that these rivers are the steepest and the largest of the rivers with input into these coastal waters and that the size of the river plume is correlated to storm size. To investigate the watershed characteristics associated with the smaller watersheds in the region, the surface expression of the plumes are under investigation. In addition, by analyzing properties of the watersheds using the cell-based modeling tools of the GIS, it is possible to compare features across the region in terms of watershed structure and the distribution of landcover types in different parts of the watershed. Watershed analysis was combined with analysis of the coastal geography (bathymetry and position of river mouths) to evaluate the potential impact of smaller rivers on the coastal waters (Figure 9.24).

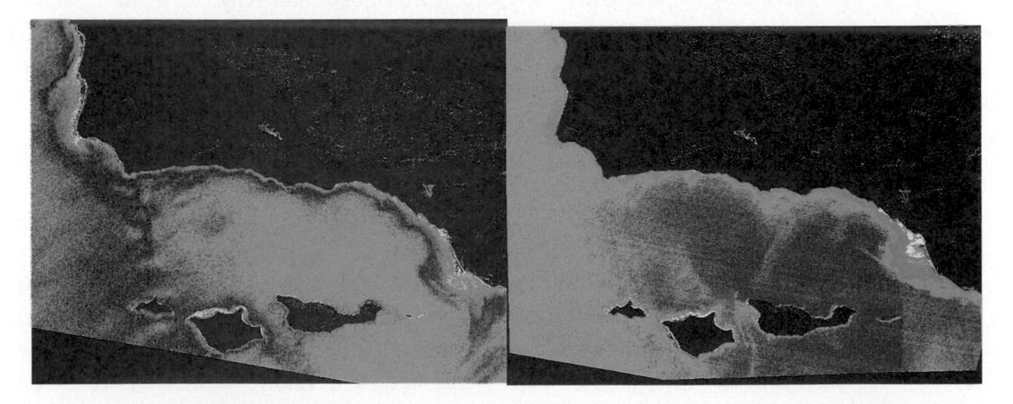

FIGURE 9.22: Sediment concentration from remote sensing of the Santa Barbara Channel. Ocean brightness indicates particulate concentrations, with brighter being higher. Left: Landsat 4 MSS image recorded 2/19/1983 after 28 cm of rain in the preceding month. Right: Landsat 5 TM image recorded 2/9/1994 after 10.5 cm of rain in the preceding month. Images courtesy of Leal Mertes. (Used with permission.)

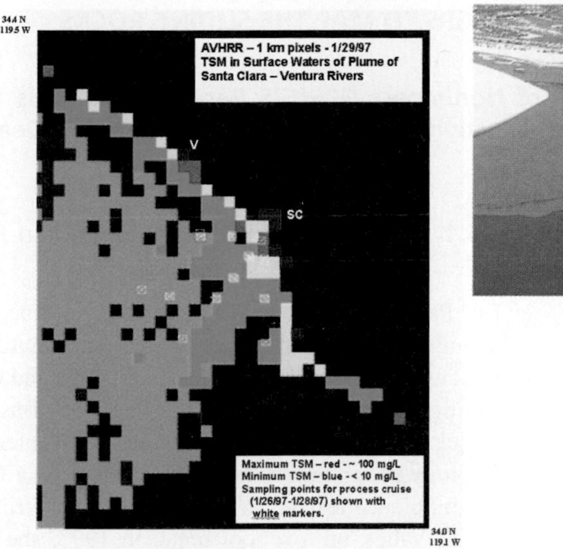

FIGURE 9.23: Left: AVHRR satellite image of the coastal region near the estuaries of the Ventura and Santa Clara Rivers during a typical storm event in 1997. Right: Oblique aerial photo taken of the Santa Clara River plume 2/10/1998, after El Niño-generated storms had resulted in 27 cm of precipitation in the preceding week. Images courtesy of Leal Mertes. (Used with permission.)

FIGURE 9.24: Streams on Santa Cruz Island, layer created using the DEM drainage functions in ArcInfo.

Conclusions from this research are that the greatest terrestrial influence on the Santa Barbara coastal waters is plumes, generated from the largest watersheds during and following significant (more than 3 cm of precipitation) winter storms. A second-order effect on the development of coastal plumes is the position of the watershed outlet with respect to the nearshore bathymetry and currents. Juxtaposition of smaller watersheds to shallow coastal waters may result in the appearance of anomalously large surface plumes.

9.6 CASE STUDY 5: USING GIS AND GPS TO MAP THE SLIDING ROCKS OF RACETRACK PLAYA

Case Study in the Use of Nonimpact GPS/GIS Research Methods to Map, Monitor, and Analyze the Activity of "Wandering Stones" in a Death Valley Wilderness Area

9.6.1 Contributor: Paula Messina, San José State University

Dr. Paula Messina investigates earth surface processes including the formation of desiccation features in desert regions, deflation of lacustrine sediments, and the unusual sliding rock phenomenon of the Racetrack Basin in Death Valley National Park. A graduate of Hunter College's Geology Program, she began her career as a high school Earth Science teacher in New York City. During a visit to Death Valley on a school break in 1993, she discovered the isolated Racetrack as a casual, but determined tourist. Upon observing the rocks—up to boulder in size—resting next to their enigmatic trails inscribed in the playa sediments, she was immediately captivated. She found herself returning repeatedly to try to understand how rocks could move along an almost perfectly flat dry lakebed, and realized the need for extensive spatial analysis of the area. As a graduate student of Keith Clarke, she pursued doctoral study that incorporated the use of Differential GPS, GIS, and terrain analysis techniques to explain the rocks' motions. She now teaches Geomorphology and GIS at San José State University, serves as a research associate for NASA Ames Research Center's Astrobiology Program, and still visits the Racetrack to monitor her rocks.

9.6.2 Background

The Racetrack Playa, at an elevation of 1131 meters, is a pluvial lake within the Panamint Range of Death Valley National Park, California. As the topographic low of the Racetrack Basin, rocks of various sizes tumble onto the dry lakebed from abutting cliffs and surrounding alluvial fans. Since the beginning of the twentieth century, prospectors, park rangers, and geologists have noted and described evidence of dynamic traction events that apparently occur once the rocks are deposited on the playa. Recessed furrows, up to two centimeters deep, suggest that rocks glide along the surface, a near-perfectly horizontal plane; in fact, rocks skid slightly *uphill* since most rocks are found on the ends of trails closer to the 5-centimeter-higher northern "shore." Trails are defined by lateral ridges, similar to scaled-down river levees, suggesting that the surface is saturated and pliant when the rocks move (Figure 9.25). Some trails exhibit evidence of splash marks, wakes, and bow waves, indicating that the rocks are propelled at speeds of about 2 meters per second, or even more. The longest trail, over 800 meters, is fairly straight, but others record extremely chaotic activity. The largest boulders are estimated to have masses up to 320 kilograms (Figure 9.26), and contrary to logic, their trails are by no means the shortest. To date, no one has witnessed the rocks in motion.

FIGURE 9.25: Two rocks, "Ellen" and "Bessie," apparently slid to the northwest, imprinting trails as evidence of their unusual activity. (Photograph by Paula Messina. Used with permission.)

FIGURE 9.26: Paula Messina stands next to "Karen," one of the largest boulders on the playa. The GPS antenna protrudes from Paula's backpack, where the receiver is carried during field mapping. (Photograph by Paula Messina. Used with permission.)

Scientific investigations of the phenomenon commenced in the 1950s, when several researchers visited the basin, described a few of the rocks, and mapped some of the trails. Survey chains, pacing, and compasses were their only tools. Maps of a few selected trails showed significant parallelism, suggesting that rocks may move while entrained in a cohesive wind-propelled ice sheet. Temperatures do indeed reach the freezing point on the Racetrack: The Playa has been blanketed with snow, and ice sheets up to 8 centimeters in thickness have been reported.

While some trails are parallel, most are not. Robert P. Sharp of Cal Tech tested the ice hypothesis in the1970s by constructing a "corral" composed of seven iron stakes encircling three rocks; he was surprised to see, upon a return visit, that although two rocks remained within the circular space, one rock had slid out beyond the stakes, leaving a trail as evidence of its route. Sharp, who assigned women's names to the rocks during his seven years of monitoring visits, concluded that ice is not a requirement for rocks to move, and that the wind alone, acting over a surface "lubricated" with wet clay, may provide enough force to set the rocks in motion.

If *some* trails are parallel, does that imply that ice moves only *some* rocks? Are there any trails preserved within the space *between* two highly parallel ones, and if so, are they congruent, too? Why were only *some* trails mapped, while others ignored? No doubt, part of the answer to this last question rests in Death Valley's isolation, extreme environmental conditions, and the time commitment required of traditional survey methods and the long journey to the Racetrack. The lack of complete trail network maps made it impossible to settle the "ice versus wind alone" argument. Research on the playa became more restricted with the passing of the 1994 Desert Protection Act, which designated the Racetrack as a wilderness area. After 1994, only noninvasive, nonpermanent instruments could be used for scientific investigations.

9.6.3 GPS and GIS to the Rescue

Having reached full operational capacity in April 1995, the global positioning system (GPS) provides surveyors with a new non-impact tool for constructing highly accurate maps in little time. Differential correction of GPS data can achieve centimeter-level accuracy for point and line features. The exact locations of all rocks and precise plans of all trails on the 667 hectare playa were captured by a field crew of two in July, 1996, requiring only ten days' time. Data were exported to ArcView GIS, and analyzed using a variety of spatial and statistical methods.

The resulting map (Figure 9.27) shows the point locations of every rock ($N = 162$) and polylines of trails. Measurements taken in the field indicated no correlation between the size, shape, or lithology of a rock and the length, or straightness of its resulting trail. Ice floes may explain parallel trails, or similar but divergent trails (should an ice sheet shatter while in motion), yet the complete trail network, as mapped in 1996, shows virtually no parallelism among trails, suggesting that ice may not play a role in this phenomenon. In fact, two highly convoluted trails (Figure 9.28) show a great degree of congruence while continually *converging*. It is impossible to describe these features using the ice floe model. And yet, while it seemed that the wind alone may be the mechanism responsible for this activity, the lack of obvious relations within the data set remained problematic. Why did some rocks move, independent of their weight or shape, while others remained motionless—in a single event?

9.6.4 Terrain Analysis

With no smoking gun there were still many questions left unanswered. The GIS did present some spatial patterns worth investigating. It seemed that straighter, longer trails were concentrated on the eastern margin of the playa, but those near the playa's center were sinuous or chaotic. Analysis of the surrounding terrain, using the USGS digital

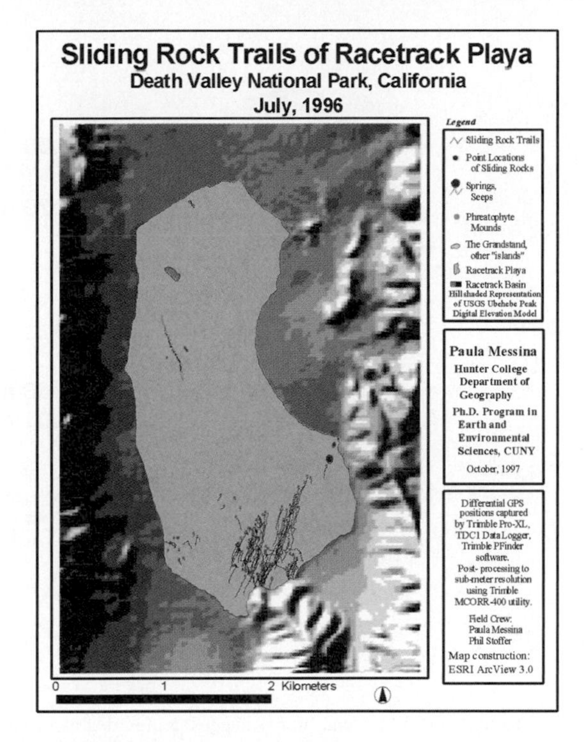

FIGURE 9.27: July 1996 map of the Racetrack's sliding rock network, as mapped by Trimble Pathfinder files, exported to ArcView. (Used with permission.)

elevation model (DEM), provided the clue that had remained hitherto elusive: The slope and aspect of the basin directs airflow along specific vectors (Figure 9.29). Direct measurements of the wind conducted with handheld anemometers revealed that wind speeds up to six times faster, and up to 50 degrees deviant in heading occurred synchronously at locations only 400 meters apart. Therefore, the nature of a trail has more to do with the location of the rock that inscribed it than to the physical characteristics of the rock itself. Follow-up visits to the Racetrack confirm these conclusions, since the rock that has slid farthest since the 1996 survey ("Diane") is the one that previously carved the longest trail (881 meters).

It was previously estimated by Sharp and Carey that rock sliding events occur with a frequency of every year or two. While this may have been the case in the 1970s, Messina's recent monitoring visits show a lower periodicity. For five years (1996–2001) most rocks remained in place. Sometime between February and May 2001, a storm propelled over 70 rocks to new locations, leaving fresh trails. The Racetrack may be thought of as a mosaic of microclimates, with different wind regimes in adjacent locations. A few days after a rain, when fine, saturated clays coat the surface, a "near-Teflon" state supports mobilization of Racetrack Playa's rocks by wind. *Where* they wander is a function of the conditions at their instantaneous locations (or perhaps of their whim!).

For more information and individual maps and photographs of the entire original data set, visit `http://geosun.sjsu.edu/paula/rtp`.

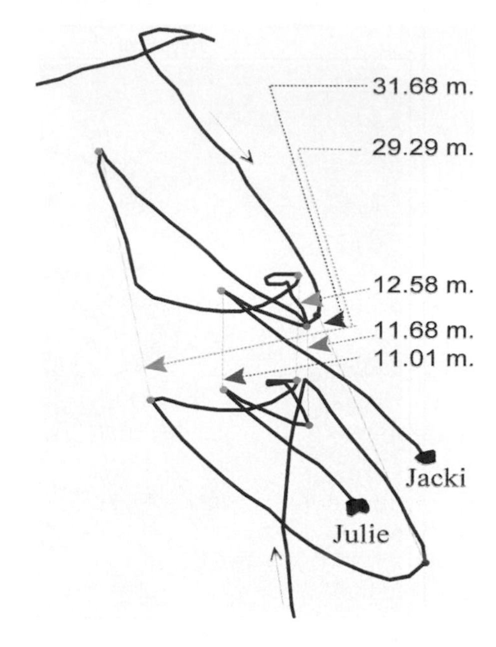

FIGURE 9.28: The trails of "Jacki" and "Julie" suggest a high degree of similar motion. However, although somewhat congruent, the rocks apparently converged during their calligraphic journeys. Trails such as these and the absence of ice-gouging features on the playa suggest that rocks are not driven lockstep within a sheet of ice. (Used with permission.)

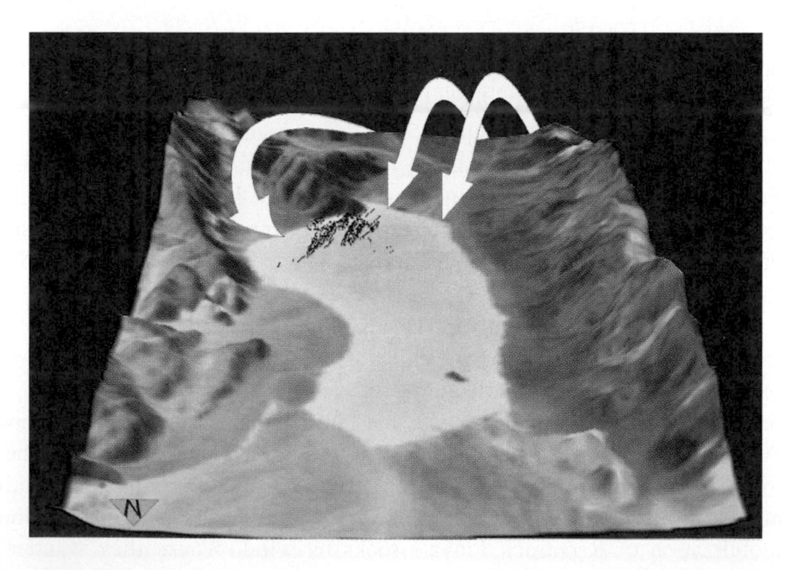

FIGURE 9.29: USGS aerial photograph of Racetrack Basin draped over a TIN based on the 7.5-minute Ubehebe Peak Quad DEM. The terrain channels air—generating turbulence resulting in erratic rock movement, and this explains the presence of convoluted trails in the playa's central region.

9.7 STUDY GUIDE

9.7.1 Summary

CHAPTER 9: GIS in Action

- **Use of GIS is best understood by examining case studies.**
- **Five case studies in this chapter cover desert, coastal, rural, suburban, and urban GIS applications.**

Case Study 1

- **In Michigan, GIS has been used by Michigan State University to monitor the spread of the gypsy moth.**
- **The gypsy moth has spread over the state from the north and east and defoliates trees.**
- **Information from the monitoring program, via a GIS in Arc/Info and IDRISI, is used to direct spraying trees with Bt.**
- **A statewide monitoring program uses milk carton traps in trees dispersed over a spatial grid.**
- **Data are aggregated annually in a central GIS; forms are entered and locations geocoded.**
- **Statewide gypsy moth infestations are interpolated using inverse distance squared weighting and are mapped.**
- **An overlay of tree species data is then used to map the trees at risk of defoliation and therefore to be sprayed.**
- **Results are used in resource management, policy development, and are posted for information on the WWW.**

Case Study 2

- **In Connecticut, a GIS has been developed to track automobile accidents.**
- **A key element in the GIS has been linking spatial data and maps showing locations of accidents with hospital information about injuries.**
- **A public access version of the GIS has been constructed for easy access to the information.**
- **Data include all collisions reported to police on state or federal roads in Connecticut in 1995 and 1996 and on local roads if the police report indicates a fatality or injury.**
- **GIS users can pan and zoom to locations of interest and select from a tool menu of common map scales.**
- **GIS users can select collisions and open tables to print reports or export these to database files.**
- **A Map and Report menu allow previewing and printing of color maps.**
- **ESRI's ArcView Avenue scripts allow users to enter titles for maps and reports, and to automate map layout.**
- **Training and workshops have produced system users, and the system is available to the public and community groups in the state capital.**
- **Additional data for more recent years are being added and analyses conducted.**

Case Study 3

- **Hunter College's CARSI laboratory came to the aid of New York City after the 9/11/01 destruction of the World Trade Center.**
- **Fortunately, New York City had a recently completed GIS-based digital map, the NYCMap with over two dozen mapped feature themes.**
- **The distaster relief effort was supported by a GIS operation involving over 50 GIS professionals that operated around the clock for two months.**
- **Data from remote sensing proved useful, including orthophotography, thermal imagery, and LIDAR.**
- **LIDAR allowed a TIN to be generated that helped relief workers in navigating and removing debris in the rescue and recovery efforts.**
- **Emergency crews used plotted GIS maps and new data in their work.**
- **Maps of infrastructure and building status were especially important.**
- **Each building had to be assigned a Building ID number in the GIS.**
- **A GPS-based field computing system proved effective in the recovery work.**
- **A series of lessons were learned that should be read by other communities around the country.**

Case Study 4

- **At the University of California, Santa Barbara, a GIS has been developed to provide data for environmental analysis of the Channel Islands in Southern California's Santa Barbara Channel.**
- **The GIS was built by collaborative efforts from several agencies, and included database assembly by both undergraduate and graduate students.**
- **Data layers assembled included information on topography, bathymetry, flora and fauna, archeological sites, sea caves, shipping lanes, oil platforms, geology, vegetation cover, and soils.**
- **The GIS has a master database and smaller, specialized subsets in use in different places and using different ESRI GIS products.**
- **The system has met operational, educational, and modeling needs.**
- **Analyses have included work incorporating remote sensing of sediment plumes, terrain analysis, and watershed modeling.**

Case Study 5

- **Studies of the mysterious Sliding Rocks of Racetrack Playa in Death Valley have been conducted using GIS and field GPS.**
- **The rocks sit on a flat playa, and have irregularly shaped trails indicating that the rocks have moved or slid up to 800 meters at a time.**
- **Research on the rocks dates back to the 1950s, with a leading theory being related to ice flow.**
- **Since the 1994 Desert Protection Act, intervention with the rocks has been illegal.**
- **A total of 162 rocks were inventoried and mapped over a period of 10 days in 1996 using differential GPS.**
- **Some of the mystery of the rocks' movement seems to be related to topographic forcing of winds. GIS mapping has shown how this could occur.**

- **Continued mapping has confirmed that the rock movement is wind driven and depends on the slick playa surface being wet from rain events.**

9.7.2 Study Questions

Case Study 1: Gypsy Moth

For this case study, make a flow diagram containing the following major boxes: data, data conversion, analysis, software, output. Work through the section, filling in the boxes with notes summarizing the project's stages.

What software packages were used in the study, and how well do the project's needs and characteristics suit the functional capabilities of the software?

Gypsy moths have been spreading slowly across the United States ever since a known accidental release of some moths over a century ago in Boston. The spatial process at work is called "spatial diffusion," the same process that occurs when smoke spreads in the air. What spatial analysis tools and quantitative measurements might suit the analysis of spatial diffusion?

Case Study 2: Traffic Accidents

For this case study, make a flow diagram containing the following major boxes: data, data conversion, analysis, software, output. Work through the section, filling in the boxes with notes summarizing the project's stages.

What software packages were used in the study, and how well does the project's needs and characteristics suit the functional capabilities of the software?

In this case study, an obviously important element is placing useful information into the hands of the public. What decisions might be made given the availability of the spatial information? What is added by having the linked hospital data? How can the GIS help in doing an analysis of the spatial patterns of accidents and injuries? Who might be the interested parties that could make productive use of the GIS?

How might this GIS application be said to be consistent with Chrisman's definition of GIS from Chapter 1?

Case Study 3: World Trade Center Aftermath and GIS

Which were the key layers in the WTC aftermath? Make a timeline of what data were used and when.

Call your local City Hall or County offices, or search their Web sites, and see if the NYCMap data equivalent exists for your own community. Is your city or community GIS-prepared for a major disaster?

Search the available literature about the 9/11 terrorist attacks at the World Trade Center, the Pentagon, and in Pennsylvania. How did the media cover the role that GIS played in the relief efforts?

Make a list of the barriers that might prevent the effective use of GIS in an emergency situation.

Write a short story about a "scenario" that could occur in your community. How might GIS help in planning and recovering from the situation? How might it prevent it from happening?

Case Study 4: Channel Islands GIS

This GIS contains a large suite of data layers with lnks to the environment. Make a list of the data layers. Which of them could be supplied from public-domain sources, and which would require additional data entry or acquisition of new data sets?

Several different user environments exist for the CIGIS, including the classroom and onboard research vessels at sea. Make a list of the user interface characteristics that would be best in each setting.

Make a flow diagram of the steps necessary for a GIS to translate digital topography, such as in a DEM, into the boundaries of watersheds.

Case Study 5: The Racetrack Playa and the Sliding Rocks

This GIS pivots on data collected in the field with GPS. Make a list of the data layers. Which of them could be supplied from public-domain sources, and which would require additional field data entry?

What might be some of the concerns if you had to conduct a GIS assembly operation entirely in the field? In Death Valley?

Suggest ways that the line information in the GIS about the rock trails might be described and summarized statistically.

How might DEM-based topography be used to prove that wind causes the rocks to move?

9.8 EXERCISES

1. *Use one of the sources of information in Chapter 1 to find out about a GIS application of interest to you. Find data on the application that may be publicly available on the Internet or elsewhere. Use the data input capabilities of your GIS to input the data used in the application. Make a concise listing of the problems you encounter, what data structure conversions are necessary, and a full listing of the suite of data formats that you had to deal with. If either the data or the software is not available, find a well-documented GIS applications project and try to surmise the information from the documentation and reports that accompany the study.*
2. *Choose a particularly simple applications study with a well-defined public-domain data set that uses your GIS software. Science dictates that experiments be repeatable. Try to repeat the application in exactly the same way as the original GIS expert. Are your results the same? If not, can you find where your results diverge from the reports or write-up that you are using as information?*
3. *If you are a student, ask your faculty or counselor about internships. Is a GIS company willing to give you a summer or part-time job (for college or school credit) for getting some real GIS hands-on experience? Find out about funding and awards to support internships. Remember, there is only so much that can be learned through books and manuals, even this one!*

9.9 REFERENCES

Alipui, I. (2002) "Dreams, ambition, enterprise." *American Congress on Surveying and Mapping Bulletin*, No. 195, January/February.

Center for History and New Media. (2001) The September 11th Digital Archive. `http://www.911digitalarchive.org`, George Mason University.

Mertes, L. A. K., Hickman, M., Waltenberger, B., Bortman, A. L., Inlander, E., McKenzie, C., and Dvorsky, J. (1998) "Synoptic views of sediment plumes and coastal geography of the Santa Barbara Basin, California," *Hydrological Processes*, Vol. 12, no. 6, pp. 967–979.

Messina, P. Stoffer, P., and Clarke, K. C. (1997) "From the XY Files: Death Valley's Wandering Rocks." *GPS World*, April 1997, vol. 8, no. 4, pp. 34–44.

Witter, J. A., Stoyenoff, J., and Sapio, F. (1992) "Impacts of the gypsy moth in Michigan," *Michigan Academician*, vol. XXV, pp. 67–90.

9.10 KEY TERMS AND DEFINITIONS

AVHRR (advanced very high resolution radiometer): An instrument on NOAA orbiting polar satellites that returns 1- and 4-km resolution data about the earth in four wavelengths. Used extensively for large-area land-cover and vegetation mapping and weather prediction.

base layer or map: A GIS data layer of reference information, such as topography, road network, or streams, to which all other layers are referenced geometrically.

biology: The study of living organisms and their vital processes.

defoliation: The removal of the healthy leaves of a plant or tree.

ecology: The science concerned with the interrelationship between organisms and their environments.

empowerment: Placing power in the hands of the citizen by providing effective and timely information.

entomology: The branch of zoology that deals with insects.

epidemiology: The science that deals with the incidence, distribution, and control of disease in a population.

GRID: The raster module of ESRI's Arc/Info GIS software.

gypsy moth: A tussock moth introduced into the United States in Boston about 1869. The early stage is a gray-brown mottled hairy caterpillar that defoliates trees.

Landsat: A U.S. government satellite program collecting data about the earth's surface in the visible and infrared parts of the spectrum. Two instruments, the multispectral scanner (79-meter resolution) and the thematic mapper (30-meter resolution), have been used. Landsat 7 is the next to be launched, for which the data will return to the public domain.

layer: A set of digital map features collectively (points, lines, and areas) with a common theme in coregistration with other layers. A feature of GIS and most CAD packages.

municipality: An administrative division of geographic space, usually for the purposes of election or service delivery.

Murphy's law: "Anything that can go wrong will go wrong." Long linked to the public use of computers.

parcel: A land surface partition recognized by law for the purpose of ownership.

pheromone: A hormonal substance excreted by an individual that elicits a response in the same species.

raster: A data structure for maps based on grid cells.

resource management: The intentional control or influence of environmental elements to accomplish particular goals.

TIGER: A map data format based on zero, one, and two cells, used by the U.S. Census Bureau in street-level mapping of the United States.

toxic release: Release of a toxic substance into the environment, such as the venting of a poisonous gas into the atmosphere.

unsupervised classification: The grouping of pixels by their numerical spectral characteristics without the intervention of direct human guidance.

UTM (universal transverse Mercator): A standardized coordinate system based on the metric system, and a division of the earth into sixty 6-degree-wide zones. Each zone is projected onto a transverse Mercator projection, and the coordinate origins are located systematically. Both civilian and military versions exist.

vector: A map data structure using the point or node and the connecting segment as the basic building block for representing geographic features.

CHAPTER 10

The Future of GIS

10.1 WHY SPECULATE?
10.2 FUTURE DATA
10.3 FUTURE HARDWARE
10.4 FUTURE SOFTWARE
10.5 SOME FUTURE ISSUES AND PROBLEMS
10.6 CONCLUSION
10.7 STUDY GUIDE
10.8 EXERCISES
10.9 REFERENCES
10.10 KEY TERMS AND DEFINITIONS

You cannot fight against the future. Time is on our side.
—*W. E. Gladstone, in a speech on the Reform Bill, 1866*

I have seen the future, and it works.
—*Lincoln Steffens.* Letters *1938. (On returning from a visit to the Soviet Union)*

Dum loquimur fugerit invida.
Aetas: carpe diem, quam minimum credula postero.
[While we're talking, envious time is fleeing.
Seize the day, put no trust in the future.]
—*Horace (65-8 B.C.),* Odes, *bk. 1,. no.11*

I think there is a world market for about five computers.
—*Thomas J. Watson,* IBM, *1958*

10.1 WHY SPECULATE?

The theme of this book has been an examination of the value that GIS brings to the workplace as a tool for understanding geographic distributions, and for describing and predicting what will happen to these distributions in the real world. History has shown just how powerful GIS can be as a new mechanism for managing information. From humble origins, a set of simple ideas, and some rather inefficient software, GIS has grown into a sophisticated, multibillion-dollar industry in only half the length of a human career. GIS's dual role as a mainstream technology for the management of geographic information and as an effective tool for the use of resources is no longer a promise, but a reality.

So why bother to discuss the future of GIS? Quite simply, why speculate? The future always seems to catch up with the present at an alarming rate. In only 3 years, for example, GIS and GPS technologies have met and merged in a seamless way without the slightest hitch. Why not just wait and let the technology deliver our dreams and speculations if they are realistic enough to come into being?

There are three good reasons to speculate. First, GIS use involves planning for the purchase of hardware and software. If a new product that is just around the corner can do more for less money, the GIS user stands to lose money and perhaps build obsolete systems if attention is not paid to market trends. Second, GIS has created a science, *geographic information science*, that has built itself as a field for study and research and which now is designing the systems of tomorrow. If GIS users, even new users, do not follow this research, they will miss information about what has been accomplished. Finally, GIS is expanding continuously into new fields and new areas of application, discovering new uses, and solving new problems.

In doing so, GIS is bringing fresh ideas into parallel sciences, oceanography, epidemiology, facilities management, disaster planning and relief, environmental management, forestry, geology, real estate—and the list goes on and on. Each new application brings GIS concepts and techniques to a new set of individuals and in turn is influenced by their own expert knowledge and needs. Many of these second-generation GIS users are not just scientists and professionals, but ordinary people. GIS can be used to defend or develop a community, to plan basic services, to educate, to win an election, to plan a major event, to avoid traffic jams—and again the list goes on. The inexpensive but powerful microcomputer has placed the basic tools of high technology into the hands of the educated citizen. Quite simply, things can never be the same again.

If speculation about the future is indeed valid, we should recognize two types. First, present trends and ideas already at the cutting edge of research can be extended into the future. This is the most certain form of speculation, for most of these predictions will actually come about, especially if the items are new products coming to market. The second type of speculation is really gazing into the crystal ball. Many of these ideas will be wrong, but what the heck, some will probably be right! We divide the discussion into three sections: first, an examination of where data will come from in the future; second, a look at hardware trends; and third, a discussion of software. The book will then conclude with a few of the broader issues and problems that GIS has brought to light and that we will have to grapple with in the future. If there is a GIS in your future, at some stage you will most likely be dealing with one or more of these issues. Just remember, you read it here first!

10.2 FUTURE DATA

10.2.1 Data Are No Longer the Problem

The blood of a GIS is the digital map data that runs through its software veins and hardware body. The future holds immense promise for new types of data, more complete data, higher-resolution data, and more timely data. Once the major obstacle to GIS development, data have now become GIS's greatest opportunity. Some of the types and sources of GIS data have already been described in detail in this book. The years ahead will bring us even more new types of data, and vast revisions of the existing types. As such, this summary of future data can be only a glimpse of what is still to come.

First, it should be stressed yet again that the entire mechanism for GIS data delivery has been revolutionized by the Internet and by the search tools built upon the structure of the World Wide Web. Most public-domain data, most shareware and freeware, and an increasingly large proportion of commercially produced GIS data use the Internet in place of computer tapes, diskettes, the U.S. Postal Service, and the so-called sneaker-net (i.e., hand delivery). This single trend has had, and will continue to have, the most impact on the field of GIS. Rarely does a new GIS project have to begin by digitizing or scanning geographic base maps. Instead, the majority of GIS work now involves bringing into the system a base layer of public-domain data and enriching it by capturing new layers pertinent to a particular GIS problem.

10.2.2 GIS and GPS

Another critical step in data provision has been the ability, using the global positioning system (GPS), to go directly to the field to collect data rather than relying completely on maps. The GPS has also improved mapping significantly, because the geodetic control once only marginally available to mapping projects is now as easy as pushing a GPS receiver button and doing a differential correction to submeter accuracy. So precise is this new mechanism for data collection that existing GIS maps of cities, parks, and other areas will have to be revisited for field verification. The ability to register a map quickly to a given map geometry (projection, ellipsoid, and datum) means that GIS layers can quickly and efficiently be brought into registration for overlay and comparative analysis. The field of GIS has greatly benefited, and the GIS-to-GPS link is now such that many GPS receivers and their data loggers can write data directly into GIS formats or include satellite images, air photos, or regular photographs directly in the field.

The flexibility of this system, when integrated with in-vehicle navigation systems that also use inertial navigation and stored digital street maps, has evolved a technology that is becoming standard equipment in public and private vehicles. Never again will the driver have to stop to ask the way to a destination! Now moving into large-scale production, these systems have already been incorporated into a car's dashboard (Figure 10.1). The rapid generation of street, highway, and city maps resulting from the growth of these systems—data that are by definition of great locational accuracy—is greatly benefitting GIS. Although the data have so far been digitized almost exclusively by private companies, competition has led to a data price war in recent years, and costs have fallen remarkably. Hand-held receivers with map displays can now be purchased for under a hundreds dollars, and are finding uses in hunting, hiking, travel, and driving. GPS has also found use in fleet vehicles such as the trucking and moving industries, and in the delivery business. In each case, the common element is the need for moving around a street network efficiently.

FIGURE 10.1: An in-vehicle dashboard mounted navigation system, installed in a production model car. GPS equipment is mounted on the roof and in the trunk. Image by Rich's Car Tunes of Watertown, MA. `www.cartunes.com`. (Used with permission.)

10.2.3 GIS and Image Maps

Another significant new data source now exists owing to the arrival of *digital orthophotoquads*. Digital orthophotoquads are geometrically corrected air photos with some cartographic annotation. Their historical use has been as sources of information for the U.S. Department of Agriculture. Recently, however, these data have been made available by the USGS on CD-ROMs in digital format in quarter-quadrangles; that is, one-fourth of a 1 : 24,000 7.5-minute quadrangle as one data set, with an equivalent scale of 1 : 12,000 and a ground resolution of 1 meter.

These astonishingly accurate data have already become a new national base layer, eclipsing the role of the current digital line graph data. Rather than being vector data, though, the raster nature of this layer and the fact that it is monochrome have resulted in its use as a background image for GIS, over which field and existing geocoded data are assembled. The primary function of the orthophoto will be to assure the same type of layer-to-layer registration discussed in the case of GPS above. Over the next few years, the coverage (Figure 10.2) will expand to cover the entire United States, and a 10-year revisit will assure that city and other maps can be updated as required. In addition, new raster images of the entire United States, *digital raster graphics*, have also been made available in CD-ROM format.

The digital raster graphic (DRG) is a scanned image of a U.S. Geological Survey (USGS) topographic map, including all the information on the map edge, or "collar." The image inside the map neat line is georeferenced to the surface of the earth. The USGS is producing DRGs of the 1 : 24,000-, 1 : 24,000/1 : 25,000-, 1 : 63,360- (Alaska), 1 : 100,000-, and 1 : 250,000-scale topographic map series. These maps make excellent

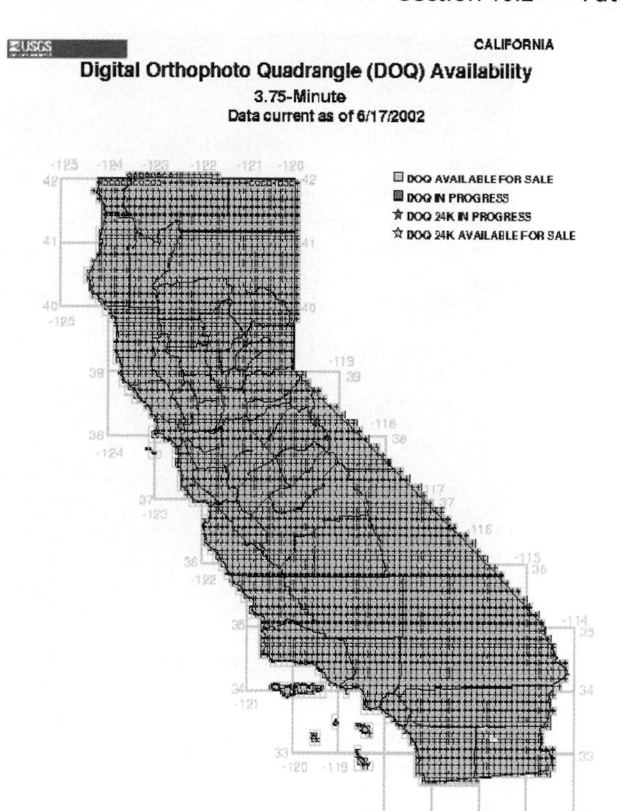

FIGURE 10.2: Status graphic for digital orthophoto map production in California. Note that the entire state is finished (compare with older editions of this book). Source: USGS. For other states see: `http://mcmcweb.er.usgs.gov/status/doq_stat.html`.

starting points for GIS projects, and they often contain many features that can be extracted for use, such as contour lines and building footprints. With these projects largely complete, the major mapping agencies, led by the United States Geological Survey are beginning work on the National Map, an extraordinarily ambitious project that hopes to build a national map coverage from contributed and existing data and imagery, to serve the information seamlessly, and to update the map with a very rapid turn-around time, on the order of days. A prototype has already been produced for the State of Delaware (see: `http://www.datamil.udel.edu/nationalmappilot`) (Figure 10.3).

10.2.4 GIS and Remote Sensing

An additional increasingly high-resolution source of map data is that coming from aircraft and spacecraft in the form of remote sensing data. New spacecraft with the next generation of space instruments will provide an extremely rich set of both new and existing forms of data. Among the new programs are NASA's Earth Observation System (EOS), consisting of a huge variety of new instruments for mapping that will continue the NOAA polar

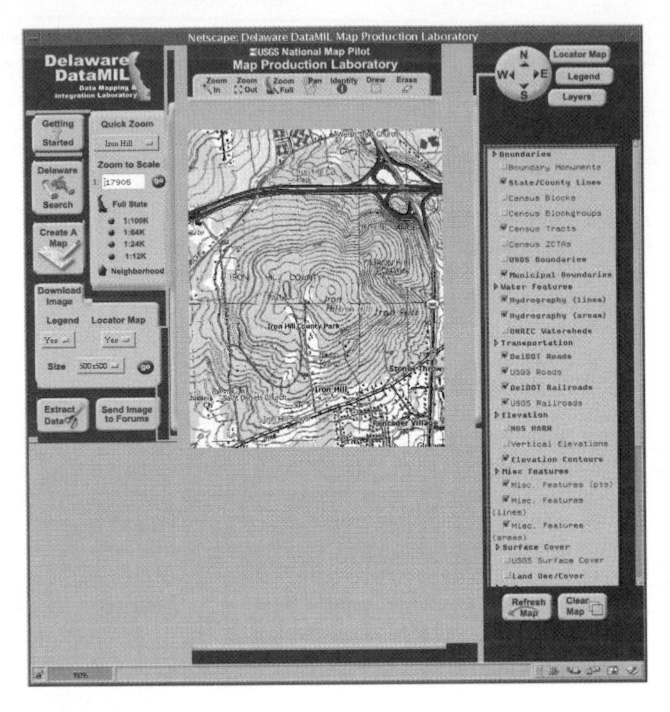

FIGURE 10.3: Delaware prototype of the National Map, a pilot project being undertaken by the USGS and collaborators. See: `http://nationalmap.usgs.gov/nmpilots.html`.

orbiting programs and Landsat type data flows (Figure 10.4). NASA's Terra satellite, launched in 1999, has already begun to set the flow of Earth Science Enterprise data into the NASA databases. The IKONOS commercial satellite returns high-resolution data at about a 1-meter ground resolution.

In addition, Landsat 7, also launched in 1999, has better spatial resolution as well as continuity with previous Landsat satellites. Several new commercial satellites, as well as a new generation of French SPOT satellites, will ensure that the diversity of instruments will increase. Similarly, the shuttle-carried radar mapping capabilities of SIR (shuttle imaging radar), as well as the Canadian RADARSAT, the European Space Agency's ERS, and the Japanese JERS, all promise nighttime and weather-invariant terrain mapping capabilities. The highly successful Shuttle Radar Topographic Mapping mission of Spring 2000 returned highly detailed topographic data and radar images for much of the world. As of this edition's writing, the first data sets are being released for use.

Finally, the release of previously top-secret government spy satellite data, from the CORONA, LANYARD, and ARGON programs during the 1960s and 1970s, has allowed a significant amount of historical high-resolution imagery, much of it covering the United States, to be used for new mapping purposes (Figure 10.5). Evident after the release from the "black" world of intelligence is the fact that this program and its successors have contributed significantly to the U.S. national mapping program, perhaps implying a higher degree of fidelity in these data than might have been imagined. As a historical record, these data are often able to show the "before" image necessary to understand the "after" of the present-day information.

FIGURE 10.4: NASA's Terra satellite carries an array of new earth science instruments. Among them is the moderate resolution imaging spectrometer (MODIS), returning data suitable for broad scale land cover mapping. Example shows work by a team at Boston University on mapping vegetation productivity. Source: http://earthobservatory.nasa.gov/Newsroom/EVI_LAI_FPAR.

FIGURE 10.5: CORONA spy satellite image acquired in 1966 for the Goleta, CA, area. These images were used extensively in revisions of the 1 : 24,000 USGS maps.

Coupled with this plethora of new systems is a completely new infrastructure for data access, searching, and distribution. NASA's EOS program uses the EOSDIS, a program designed to make most satellite and other data, especially those of interest to scientists studying global change, available over the Internet. The USGS distributes land-process data via the EROS Data Center in Sioux Falls, South Dakota, for EOS and many other programs, including the United Nations' GRID program. Even the CORONA data are distributed in this way, ensuring open and inexpensive public access to this map data. Landsat data are broadly available at reasonable costs, with up to 15-meter resolution. Most EOS data are also publically available on the Internet.

The successful launch of Landsat 7 moved satellite data from the U.S. government back into the public domain, as the Landsat program was originally intended back in

1972. The 1992 Land Remote Sensing Act reversed the limitations of the 1984 Landsat Commercialization Act, finally admitting that the commercialization efforts for Landsat did not produce the desired effect. Already, this act has led to some major reductions in the price of satellite data; as a result, remotely sensed information will find its way back into the GIS mainstream, especially in the form of integrated GIS databases and GPS ground observations.

Another major switch in policy will be the return to a continuous data stream. With commercialization, Landsat was moved over to a program that collected data only when a customer ordered it. As a result, much of the world remained uncovered, and searching back in time for data to show changes was impossible. Continuous coverage will allow far more images showing and contrasting changes, especially in the environment. The multiagency Pathfinder program has attempted to demonstrate this capability, generating a U.S. coverage for three decades using historical multispectral scanner data. As a project of "data mining" or searching existing data to extract products of value, another successful effort has been the AVHRR-based land-cover and vegetation index mapping conducted by scientists at the EROS data center and now released regularly on CD-ROM. In the future, this data set is planned for global coverage and periodic release, a massive boost to global-scale GIS use.

Clearly, remotely sensed data are highly structured around the raster data format. As much more data become available in this format, the demand upon software that converts between raster and vector data will increase, as will intelligent software for correcting lines and boundaries that come from pixel-based images. If this software becomes powerful and inexpensive, the possibility of having it work directly on the orbiting spacecraft becomes attractive, since the resulting vectors use far less data storage to either store or transmit to earth, allowing more efficient use of the orbit time. If the existing digital map could also be loaded, the spacecraft need only send back to earth revisions reflecting construction, natural changes, and so on. The prospects for automatic up-to-date maps seem bright.

10.2.5 GIS and Data Exchange

The final prospect for GIS data discussed in this chapter is one of exchange. As GIS becomes more widespread, the various map-generating and map-using communities will need to trade data more than ever before. Already, nautical charts and world maps have needed to be standardized, edge-matched, and cross-checked across national and even continental borders. This implies that there is a need to build a formal structure for data exchanges, and the several new standards for data transfer have already had a major impact on this issue. Standard transfer formats mean, for example, that a ship sailing into foreign national waters can download the latest navigation chart for immediate use.

Many sets of standards for data have emerged. In the United States, map data have evolved the Spatial Data Transfer Standard (SDTS), now formalized as the FIPS 173 (a FIPS is a Federal Information Processing Standard). The year 2000 Census in the United States was the first full-scale mapping effort to generate all of its digital map data in the SDTS format. Other agencies, such as the USGS, have already converted many files, such as the Digital Line Graphs, into FIPS 173 format and structure. Internationally, NATO has produced the DIGEST standard, the International Hydrographic Organization has produced the DX-90 standard, and other nations have established their own data transfer standards. Industries such as television, computer software, and communications have seen standards take on a critical role—and even critically influence technologies, such as videotape formats. Standards will have a great impact on the future of GIS. With formal, explicitly defined formats for features, open exchange will be easy and data will no longer be a constraint to GIS use. As the world becomes more and more a single global market, early elimination of the data transfer barrier will assure the future of GIS for many years to come.

A critical element of data exchange is simply finding out who has data that already exist about a geographic area. Those who have GIS data may be willing to share not necessarily all the data but at least the metadata that give information about data coverage, accuracy, timeliness, and availability. Standards have now been developed by the U.S. Federal Geographic Data Committee that specify how data can be indexed for effective search. Prototype systems for coding data, and Web and other computer-based tools for searching and browsing for data, have given rise to the concept of a digital map library. Such a library allows searching, and then allows the user to access a public or other Web location that can provide data for downloading. As data become more and available, these metadata systems will become increasingly useful for sorting through the huge quantity of available digital map data.

Libraries of digital map and GIS-usable data are already finding their way onto the Internet. ESRI has established its Geography Network (see Chapter 8), and other prototype systems exist such as the Alexandria digital library and Microsoft's Terraserver. Some continental and global level data sets are now finding their way into the public domain, many from the United Nations GRID program and the USGS EROS data center. A global level map project is the Global Map being generated cooperatively by the International Steering Committee for Global Mapping (`http://www.iscgm.org`). National data at 1 : 1,000,000 scale for eight themes are now available for 11 countries, and the remainder of the world is taken from the existing Digital Chart of the World and its revision, termed VMAP-0.

10.2.6 GIS and Location-Based Services

Location-based services (LBSs) are computer-based services that exploit information about *where* a user is located in geographic space. Location-based services take advantage of GPS, but also may rely on E911, an initiative of the Federal Communications Commission that requires wireless telephone carriers to pinpoint a caller's telephone number to emergency dispatchers. This may use the location of the telephone itself with respect to the nearest cellular transmitters, solved by signal triangulation. E911 is the most widely used location-based service in the United States, although manufacturers of cellular telephones are also incorporating GPS chips into new cellular telephones. It has been estimated that LBS will account for $40 billion in revenue by 2006. The power of LBS means that the Internet can also be made location oriented. For example, an Internet search can be made such that the "hits" are returned to the user in order of the distance that they are from the user's current location. Many such Web-based services already exist, often using map providers like MapQuest to provide maps and directions along with the geographic search capacity.

Users of LBS so far seem to be either vehicle-based, where the GPS and computer are resident in the car and are used to query geographically ordered information, or mobile. Mobile users are usually either working on a personal digital assistant that contains a cellular phone connection to the Internet and a GPS card (which is often an add-on feature, and comes on a PCMCIA or other card), or they are using the fairly limited interactive communications capabilities of a cellular telephone. Early uses of the systems have included automotive roadside assistance, emergency and collision notification, stolen vehicle tracking, on-demand navigation assistance, traffic alerts, and vehicle diagnostics. For example, On-Star LBS is available in 32 models of General Motors cars, and the TeleAid system is standard on all Mercedes cars sold in the United States. Other systems include tracking devices for children and house-arrest prisoners, pet location devices, and others.

In essence, LBS uses selected subsets of GIS functionality, but delivers them to the user on demand. Most applications will be in navigation, route finding, and space-constrained search. For example, the Web site for MapInfo (see Chapter 8), features an example of searching for nearby French restaurants in an unfamiliar city. One unresolved issue with LBS is how "open" the geographic information will be. For example, consumer goods could be made to report back their location and condition to their manufacturers or sellers. While this has some positive implications for avoiding breakdowns, the privacy issues involved are obvious. Sales calls to cellular telephone owners who walk close to a particular store or restaurant may be informative, but are likely to be very intrusive. Similarly, cellular calls could be used to locate people in time and space. This may be relevant for criminals or terrorists, but if the information were broadly available some significant abuses would be possible. Nevertheless, interest in LBS by some very mainstream businesses is high, and we are likely to see them in broad use in the near future.

10.3 FUTURE HARDWARE

Hardware for GIS has gone through at least four revolutions in the last decade: the workstation, network, microcomputer, and mobility revolutions. Each one of these has already had a profound impact on computer hardware and will influence the future of GIS significantly.

10.3.1 The Workstation Revolution

The first of these—the workstation revolution—has given GIS an operating platform that has all of the necessary power and storage to work with massive databases. In the space of just a few years, the capability of a $15,000 workstation has gone from megabytes to gigabytes of storage, while increasing the size of RAM beyond 64 megabytes and the processor speed well above and beyond the capabilities of most mainframe computers. Examples are Sun's Sunblade, DECs DecStation, and Silicon Graphics workstations. Along with the expansion of the workstation has been the spread of Unix, the TCP/IP communications protocol, and graphical user interfaces such as Sun's OpenLook, Motif, and MIT's X-Windows. The more powerful systems of the future and the falling price of workstations seem to make this the preferred GIS work environment for large-scale projects, although Windows, Macintoshes, Linux, and even DOS remain for low-end systems, small projects, and for education.

10.3.2 The Network Revolution

The network capabilities built into workstations have broadened to include many other types of computers, including microcomputers. Many computers are now connected to the Internet and can use network search tools such as Windows Explorer and Netscape to "surf" the World Wide Web (WWW). Already, the Internet has become a primary means for data exchange and information search and retrieval. Many GIS packages, including Arc/Info, GRASS, and IDRISI, have support services on the Internet's network conference groups. The national spatial data infrastructure, a linked distributed database of public GIS information with common metadata, is being built upon the capabilities of the Internet and the WWW.

Many commercial GISs have now developed modules that allow entire GISs to be Web-enabled, including ESRI's Internet Map Server, MapInfo's MapXtreme, and Intergraph's GeoMedia WebMap. This means that the GIS can be searched, queried, or analyzed over the Web and the results displayed locally on a client using software tools such as Java and a standard browser. GIS is behind many of the map display tools now proliferating on the Internet, including the Web serving of public information in many communities and cities around the country. Full GIS functionality is rarely delivered over the Internet, and these systems usually feature simplified user interfaces and simplified data searching and map construction. If complete functionality were deliverable, then the GIS user need not "own" the GIS software, or even the data, and could simply pay for their use over the network when desired. Some Web-based educational systems already use this approach, such as ESRI's virtual campus.

10.3.3 The Microcomputer Revolution

The microcomputer has matured and increased in power significantly, making this platform widely distributed, relatively inexpensive (especially when compared with the other components of a GIS), and easily capable of running many GIS packages. Here, the Intel Pentium chip, the CD-ROM drive, and simple graphical user interfaces such as Microsoft Windows, Linux, and others have led the way. While the first and even the second generation of microcomputers were at best only modestly suitable for GIS applications, present-day systems have crossed the size and power threshold and become useful professional and educational GIS platforms. Estimates are that in 1999 annual sales of microcomputers were $185 billion worldwide from $95 billion in 1994. This translates to an installed base in 2000 of approximately 100 million microcomputers with the capability of running GIS software, a number possibly far exceeded by today. The implication of this revolution has been largely one of broad distribution—GIS can now go almost anywhere a microcomputer can go.

10.3.4 The Mobility Revolution

The fourth major technological revolution represented by microcomputers has been the trend toward mobility. Here, driving forces have been the laptop, portable, subportable, and even palm-top computer; the PCMCIA and USB interface allowing easily transferable data storage and interoperability of devices; and the mobile communications and GPS technology that now accompany them. The fact that I am writing this section of the book on a notebook computer on a train in a tunnel beneath Park Avenue in New York City is remarkable enough. However, the linkage that allows a GPS unit to compute a position, download it to a portable computer, receive by modem and mobile phone differential corrections to the GPS location, and then write these data directly into a GIS format, and to do all this so simultaneously that the points appear as if by magic in real time on the portable computer's GIS map display, was literally beyond belief only a few years ago.

Some GIS vendors now offer limited versions of their GIS for use on highly portable devices, such as the Palm Pilot and Compaq iPAQ. Among these are MapInfo and ESRI, with the ArcPad software. When these devices are coupled with a GPS card, often available as a plug-in on a PCMCIA card, they become completely mobile GIS systems in their own right. They often make inexpensive alternatives to the dashboard navigation systems available from auto manufacturers, and can be brought hiking and touring. When more integrated with the mobile Internet for data and information retrieval based on location, these systems are indeed likely to become a new killer-app appliance, broadly used in vehicles and elsewhere. Certainly, many commercial producers are rushing similar products onto the consumer market to take advantage of this growing area of business.

Added to the continued miniaturization of computer and communications equipment, personal mobility of GIS hardware has reached and gone far smaller than the field-portable minimum level. Some universities are now conducting research into "wearable" computers; that is, computers that are so small they can be integrated into clothing or objects such as eyeglasses (Figure 10.6). Along with these new capabilities come the terms *ubiquitous computing* (go anywhere, remain connected to the Internet via cellular telephone) and *augmented reality*, in which the GIS data view can be superimposed on the "real" view by direct entry into the human vision field. These are prototypes now, but are apparently already in use in some professions. At the least, a "see-through" eyeglass display seems safer for automobile navigation than a dashboard navigation system.

10.3.5 The Impact of the Revolutions

Extending these concepts into the future gives us the following four observations. First, the workstation and all its characteristics will continue to dominate the GIS workplace as the primary tool for advanced applications, but will become immensely more powerful. This will entail more local disk, perhaps workstations capable of terabytes of storage locally, and more distributed and shared data resources, with file servers acting as the data libraries or depositories for GIS projects.

Similarly, as the amount of random access memory (RAM) available approaches the gigabyte range, many processes now performed as input/output or file manipulation operations will be possible to do inside the workstation RAM in real time, making even computationally complex and sophisticated operations very fast, perhaps interactive, and certainly fast enough to allow use of the new techniques of scientific visualization. The dominance of the Unix/Motif/OpenLook/X-Windows environment looks certain, as does a shift toward programming GIS in new systems, languages, and environments. The move toward visual programming tools, object-oriented programming, expert systems, and so on has already started to deliver new and more user-friendly GIS systems; at the same time, the high-end systems are likely to acquire new and even more powerful capabilities. The new computing method most likely to have a major impact on GIS is the move toward parallel processing, which, once in place within the high-end workstations (and already in effect today), will allow real-time processing of imagery in new ways, promising immense speed-up in processing.

The role of the network is another simple extension of today's environment. Already we have prototype systems in place of future systems. NASA's EOSDIS, the National Spatial Data Clearinghouse, and the entire WWW are testimonials to the rapid growth, acceptance, and exploitation of the Internet as the primary future tool for the searching, distribution, and distributed storage of spatial data. Yet the Internet can deliver far more

FIGURE 10.6: Wearable GIS system with eyeglass display, GPS (antenna on shoulder), and full GIS functionality. For information, see: `http://dg.statlab.iastate.edu/dg`. System worn by Andrea Nuernberger. (Photograph by Susan Baumgart.)

than data and metadata (data about data). It can deliver information, advice, and assistance, often tailored to a specific environment or GIS package.

The Internet can offer formal means for the dissemination of ideas and research, much as today we depend on the printed page in books and journals. It can also deliver labor; in other words, it can remove the GIS analyst almost entirely from the traditional workplace. Also of significance is the fact that the Internet can deliver shareware, meaning that the new user can experiment with a free or inexpensive GIS before making a purchasing decision. Finally, and most important to the academic world, the Internet can deliver both real-time and programmed university education, in the form of multimedia and hypertext "virtual" classrooms free from the restraints of national boundaries and geographic separation. Again, the democratization of the GIS field offers some exciting prospects for a future information-based economy.

Both the power and the increased flexibility of the microcomputer have been pivotal not in increasing the power of GIS applications, for this has been the domain of the workstation, but in penetrating new fields of GIS application and in the domain of GIS education. New fields to GIS in the 1990s were archeology, forestry, epidemiology, emergency management, real estate, marketing, and a host of others. In every instance, the first steps in these areas were taken by new users in a microcomputer environment.

Obviously, improvements in microcomputer user-friendliness have been critical, especially the move to Windows-based graphical user interfaces (GUIs). The acceptance of GISs, which are necessarily complex and often counterintuitive to the newly initiated, has really dated only from the widespread use of these windows-based GUIs. In addition, the movement away from smaller hard disks to CD-ROMs, PCMCIA cards, and tape backup has helped. Another important step has been the large price decreases in devices for basic graphic input, including small digitizing tablets and scanners, and for output, such as color printers and pen plotters. Even a small office can now become fully equipped for GIS for only a few thousand dollars, a far cry from setup costs in the hundreds of thousands of dollars in the early 1980s.

Education has benefited significantly from the low cost of hardware, because the budget for hardware in colleges, universities, and schools is usually small and under constant threat. The trend toward the microcomputer classroom with a networked server running shared software licenses is broad enough that this configuration is now common in many high schools, and even there GIS has entered the curriculum in some places. As geography moves back into the curriculum in high schools, GIS will lead the way, bringing forth a new generation of GIS-literate students for the information economy.

Increased mobility has also generated many new GIS uses. Here, however, it is the coalescence of mobile technologies, communications, navigation, and data processing that has been pivotal. Obviously, the exciting new data capture prospects of GPS have been very important; however, the migration of software and hardware for image processing and remote sensing into the mobile environment offers many exciting prospects.

10.3.6 Speculations about Hardware

Finally, what about the speculative future? Some of the trends on the edges of computer science and engineering have real prospects for GIS application. Among these are stereo and head-mounted displays; input and output devices that are worn; parallel and self-maintaining fault-tolerant computers, and above all, mass storage and computing power much greater and faster than that available today.

A vision of a future GIS system might be a pocket-held integrated GIS, GPS, and image-processing computer capable of real-time mapping on a display worn as a pair of stereo sunglasses. Data capture would consist of walking around and looking at objects, and speaking their names and attributes into an expert-system-based interpreter that encodes and structures the data and transmits them immediately to a central network-accessible storage location. This implies that a single person, or even an unmanned vehicle or pilotless aircraft, could move around gathering data while any interested person displays and analyzes the information in real time in his or her office or home. Perhaps a nationwide set of mobile data collectors could roam the countryside, constantly field-checking and updating the digital maps being used by virtually every automated system, from power supply to emergency vehicles to the U.S. postal system. Such mobile systems were clearly of great value in the World Trade Center relief effort case study covered in Chapter 9. Rather than being the "big brother" of fiction, universal and open access to the information generated would ensure that the public good is being served.

Another future prospect is that of the data analyst becoming a data explorer, delving into three-dimensional realistic visualizations of the data, seeking out patterns and structure instead of the user of the simple statistical analysis of today. The human mind is capable of some amazing parallel processing of its own and can easily seek out structures that computers and even some scientists often miss. Similarly, the same systems could manage the very systems they support, perhaps allowing for integrated modeling and prediction of future "what-if" scenarios.

Regardless of the actual hardware used, there is little doubt that the tools and devices required for GIS work will become commonplace in the very near future: perhaps never cheap enough to come free with a fill-up at the local gas station, but undoubtedly cheap enough that the likelihood of GIS hardware being a limiting factor in the GIS future is minimal.

10.4 FUTURE SOFTWARE

A review of GIS software trends of recent years is in order if we are to speculate in a similar way about where GIS software is going in the future. Several themes suggest themselves.

10.4.1 Software Trends

The first major trend over the last few years has been in operating systems. In the 1970s, complex mainframe operating systems predominated, and system interaction was limited both by the inflexibility of the user interface and by the nature of the early time sharing of systems. The first minicomputer operating systems were little better, with the exception of Unix, a simple and much abbreviated set of instructions for doing file and systems management that has proven very flexible and long lived. Today, operating systems can "multitask," working on two problems at once, with ease. Microcomputer operating systems now also have this capability.

Early systems were somewhat poor at user interaction, yet the revolutionary Apple Macintosh system, followed by the various flavors of Windows and X-Windows, led to a significant improvement in user simplicity and comprehensiveness. Standardization was an additional unseen improvement: that is, every application could use a standard and commonly understood set of menus instead of making its own flavor. Most recently, operating systems that run on multiple platforms have flourished, including Unix. The ability to divorce standard operations such as printing and digitizer communication from the GIS led to some major improvements. Similarly, commonly accepted industry standard formats and languages, such as PostScript, led to another level of standardization, this time for hardware devices such as printers and plotters.

10.4.2 The User Interface and WIMPs

The computer era has seen radical changes in the very nature of both the computer and GIS user interfaces. Early systems used only the screen and the keyboard to communicate to the user. Systems now have these same functions, but also a mouse, pointing devices such as a track ball or light pen, multiple windows on the screen, sound, animation, and many other options.

Most significant has been the rise of the WIMP (windows, icons, menus, and pointers) interface. *Windows* are multiple simultaneous screens on a single display, usually serving different tasks and fully under user control. When inactive, windows can be closed and kept visible as *icons*, or icons can be attached to tasks and used to activate them—programs, for example. *Menus* can take a variety of forms. Many user interfaces place a set of menus along a bar at the top of the screen, controlling more and more specific tasks as one goes from left to right. Menus are often "nested," that is, a selection reveals another menu level and even more selections. Menus can "pop up" from a space or window, or can be "pulled" from other menus of messages. *Pointers* are devices for communicating location on the screen and in windows, and they most commonly take the form of a mouse or a track ball.

Central to the GUIs of recent years has been a *metaphor*. The metaphor most commonly used has been the desktop; that is, the screen of the computer is designed to resemble the top of a desk, and the icons and other elements are allowed to rest on it, awaiting use. Some operating systems have gone beyond the constraints of this suite of interactions, and many operating systems now allow input from voice, touch screen, and even direct input from GPS receivers and other recording devices, such as digital cameras and videocams.

The map itself is a useful metaphor, and a future GIS can easily be imagined in which the map and its elements, such as the scale and the legend, are used to manage and manipulate the data. This is already what a GIS does, but the user-interactive element would be a new addition to the system. Several systems already use icons as elements of a process or transformation model to track sequences of operations. For example, the spatial modeler in Idrisi32 allows a set of tasks to be constructed as a *flow diagram*, and then the model can be activated directly from the diagram when complete. A flow could consist of selecting an image, running a geometric correction routine, classifying the data, selecting a category, doing a map overlay, and then printing the result. The entire sequence of operations can then be copied, subsetted, or further manipulated just by treating the flow diagram as a graphic, with the elements of the graphic "standing" for the data sets, GIS operations, and cartographic transformations they represent in both a virtual and a real sense (Figure 10.7).

This is clearly a taste of the future of GIS. The GUIs will probably allow the user to specify tasks independent of the data, in the abstract. Possible alternative metaphors are the English language, a symbolic language such as Dana Tomlin's Map Algebra, or pictorial languages. As most GIS operations contain maps, they have been used as a metaphor. Max Egenhofer has attempted a GIS-metaphor user interface for direct use of GIS operations, called the Geographer's Desktop (Figure 10.8). In this GUI, the data layers are represented in the standard "layer cake" graphic, and operations are done by clicking and dragging. It is highly likely that the next generation of GIS will incorporate some or all of these features, making them considerably easier to use.

10.4.3 The End of the Raster versus Vector Debate

Another major software trend has been a massive change in the distinctions between systems based on their data structures. As we have seen in earlier chapters, quite often the process of geocoding, or sometimes a particular GIS process such as map overlay, leaves an "imprint" on the data that remains as one of the restrictions on data use and flexibility. The last few years have seen almost every GIS package become capable of supporting both raster and vector data structures, and in some cases many others besides.

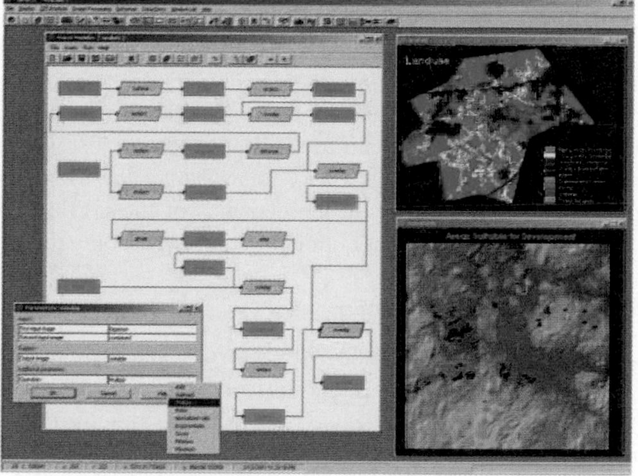

FIGURE 10.7: The spatial modeling interface in Idrisi32. The flow of information processing and algorithms applied to generate the output maps shown in the windows is symbolized by the flow diagram on the right. In this case, the user interface becomes the process. Image courtesy of Clark Labs, Worcester, Massachusetts. (Used with permission.)

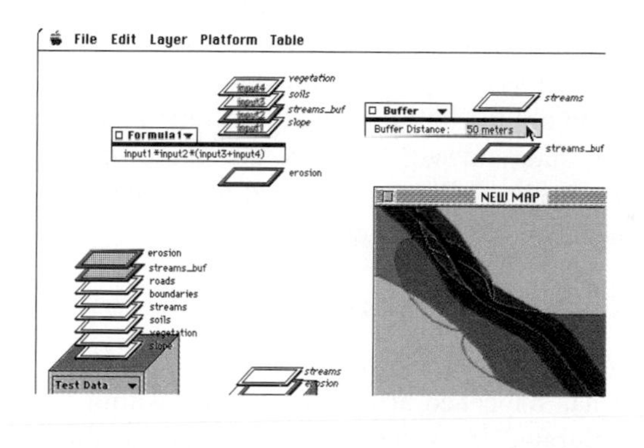

FIGURE 10.8: The Geographer's Desktop by Max Egenhofer, Jim Richards, and Tom Bruns at NCGIA, University of Maine. GIS data layers and operations are represented as icons that can be clicked on, dragged, and so on.

This has become the sort of single superflexible data structure that many sought to develop in the early days of GIS research. Instead of one structure winning out, GIS developers have realized that each structure has its strengths and weaknesses, in particular for analytical operations. Systems can take advantage of the strengths of a particular structure for a particular operation—map overlay or edge detection, for example. The disadvantage is that the transformation between data structures often entails significant error in and of itself and can lead to some serious problems in GIS analysis. Nevertheless, if done carefully, the raster/vector dichotomy can be eliminated.

In the future, GIS software is likely to have incorporated the strengths of the various structures and should be capable of intelligently converting data between structures without the intervention of the GIS user. This means that some of the principles of what is happening may be "hidden" from the user. Self-configuring GIS software does not seem too far-fetched. In addition, the spatial data transfer standard has allowed data to be encoded along with the necessary information to move easily between structures. A GIS could simply read a standard file header, establish just what is stored in the file, and then reconfigure the data as necessary for whatever the user demands. In time, also, an intelligent GIS could learn about the demands of the GIS's own user and hold data in suitable structures for the most commonly performed operations and analyses.

10.4.4 Object-Oriented GIS

Another major development in the software world has been languages, and now databases, that support "objects," called *object-oriented systems*. Geographic features map very closely onto objects. Object-oriented programming systems (OOPSs) allow the definition of standard "classes" that contain all the properties of an object. As a simple example, an object class could be a point containing the latitude and longitude of the point, a feature code for the point such as "Radar Beacon," and any necessary text describing the object. If we wish to create another point feature, this can be done simply by cloning the original with all its class information, a process called *inheritance*.

In addition, we can encode the fact that points often have data conversion or analysis constraints. For example, the centroid of a set of points is itself a point and can inherit a point's properties. This approach has allowed the development of entire GIS packages, and is seen as a way of building far more intelligent GIS systems in the future. While the OOPS is not the tool for all GIS operations or systems, it is indeed a powerful way of modeling data and will influence the future of GIS software significantly. The first GIS based entirely on the object model is the Smallworld GIS (see the Web site at `http://www.smallworld-us.com`).

10.4.5 Distributed Databases

A major transition within the GIS industry has been the movement toward distributed databases. This has happened at two levels, first within a local area network; data and software have migrated from individual hard disks to file servers, computers dedicated solely to disk storage and moving information over the local network to the client workstations or sometimes microcomputers. This is a direct equivalent of the transformations made possible when the availability of printed books was revolutionized by the advent of public libraries.

Library users need not worry about getting the latest information, specifics of book ordering from publishers, and so on. They can use the library as an information delivery service. The price to the user is a security system of some kind and the loss of "ownership" of the data or software on the server. The ability of computers to make almost unlimited immediate copies of files without loss from the original source has changed the library model somewhat. Quite clearly, though, a distributed data system can lead to a large-scale reduction in storage duplication.

Second, connection to the Internet has made it possible to have distributed databases on a massive scale, across national boundaries and even across major hardware and software barriers. Thus it is possible to let the USGS or the Census Bureau maintain a library of data and to download the data sets of interest only when they are needed. This arrangement is ideal but leans heavily on the ability to locate and transfer data on demand. Various network search tools such as WAIS, Netscape, Archie, Gopher, and Mosaic have made this metadata accessibility possible, leading to some major breakthroughs in Internet-wide distributed databases. Threats to this situation would be privatization of the Internet, implementation of a pay-per-use system for data retrieval, or taking public data out of this broadly accessible distribution system.

The Internet supplies far more to GIS users than data. It delivers software, research papers, advice, shared knowledge, and the routine contact necessary for efficient operation of a GIS. Increasingly, GIS companies and shareware services are using the Internet as the primary means by which support is delivered. A GIS user can send e-mail questions to an expert anywhere in the world. Use of the File Transfer Protocol (FTP) in "anonymous" mode allows downloading of software fixes (called *bridges* and *patches*), and even some tailor-made debugging and testing. Remote log-ins allow an expert to get onto a sick computer and cure software ills without leaving the office. In the future, this sort of service will grow to become the major means of GIS software user support.

As GIS systems have grown, so has that part of the GIS industry that acts as a supplier of data. Many companies work to update, enhance, or correct all sorts of existing data and many also generate new data from scratch. These services have acted to provide data in a broad variety of common GIS formats and offer subscription services for regular data updates, after a new release, for example. Recent trends have shown that these services have become more willing to help on a project-by-project basis.

New GIS projects especially often require digitizing and scanning even before basic operations can begin. The data services conduct turnkey operations, handing over to the GIS staff a complete data set for use. As the costs of data supply fall, and the distribution mechanisms such as CD-ROM become more widespread, the cost of GIS data is likely to plummet. This GIS data price war should result in very low cost data in the future, at least for basic cartographic data. This is exactly the model that has been followed for paper maps, at least in the United States, where they are very low cost and sometimes even free. The data services will turn increasingly to custom services and data enhancement as a means to survive and prosper.

10.4.6 GIS User Needs

Another issue of interest to the future of GIS is how the industry will continue to develop. Obviously, GIS users have broadened into two types: the large organization-wide projects with huge databases and often specific missions; and the small, usually one-person operations run by a jack-of-all-trades. Although GIS can serve both sets of users, the specifics of hardware, software, and the computing environment mean that different GISs suit each world.

At the organizational level, labor can be divided. One staff member can take care of data maintenance and software updates, one education and training, one data analysis, and so on. In the one-person shop, all of these tasks are the responsibility of one individual, often a GIS pioneer who is the key person in championing the use of GIS to begin with and who is also the computer expert, systems administrator, hardware engineer, and coffee-maker. Small users will probably not be able to add significant amounts of new data, with the exception of field data collection with GPS. They will be more reliant on public-domain data, and the data will probably be less up to date and at a coarser scale.

It is at this level that the GIS use is closest to the domain expertise. Getting the GIS as far into the field as possible is often a key to the success of a system. Field operatives can use the GIS quickly to make ordinary but informed decisions about the use of resources on a day-to-day basis; that is, where the payoff is greatest. Sophisticated analytical operations may not even be necessary at this level, and using the GIS as a graphic inventory and map production system is more than sufficient for success.

Large systems, by contrast, can maintain up-to-date and detailed information, and can use it in its full GIS context, performing the roles of inventory, analysis, decision making, and management. Here, also, better information means better use of resources. Clearly, the GIS industry must continue to exploit both types of environment. Often, this means taking large systems and packaging them small, or taking lessons learned by advanced users and translating them for the general user.

Finally, the GIS users themselves have become a sort of self-help facility. Most major software packages or regional-interest organizations using GIS have user groups, often with special conferences, workshops, newsletters, and Internet discussion groups. This is an excellent grass-roots level for GIS to flourish, one that GIS vendors have discovered. As GIS packages become more complex but also more user friendly, these user groups will converge on some common principles for GIS use. These principles should be, and are, shared with all users. Often, a good idea in one software environment can lead to productive duplication in another.

10.4.7 GIS Software Research

Some of the future expectations for GIS software are the results of research now under way, and as such are also somewhat predictable. For example, for some time, scholars in GIS have been interested in the impact on GIS of supporting geographic and attribute data from many time periods. Obviously, the digital map in a GIS is "time stamped" at the time the data were created. In the real world, however, data become out of date and must be revised, or new data sets are released to replace the old.

Some data have very short duration—weather forecasts or shipping notices, for example, and revision and update quickly become a major part of the GIS maintenance. In most cases, GIS data are simply given an additional attribute of the date the data were created, even though often the date of the data and the date of entry into the GIS are not always the same. The implications on the design of the GIS to facilitate use, automatic update, for instance, or automatic selection of the most up-to-date version of every feature are now being integrated into the GIS's functions.

Another trend that today fills research journals is use of the more recent object-oriented programming systems and database managers as the tools with which to construct GISs. One such system, using the SmallTalk object-oriented system, is already on the market, and many others are now switching to the programming language C++ and Java, or contain object-supportive languages. The advantage of object-oriented systems is that the features within the GIS can be described in advance, categorized by types, and that actual data represent an "instance" of one of these types or "classes" of object.

This advance knowledge of types allows operations and algorithms to be stored with the objects. The objects become a "hidden layer" for which the user need not perform many of the operations one performs as routine in a regular GIS. For example, an object-oriented GIS can know in advance the steps necessary for, and outcomes of, map overlay along with any data conversions necessary for its performance. Disadvantages of object-oriented systems are that they are often memory and computationally intensive and that their sophistication is unnecessary for most of the basic GIS operations.

Some GIS research has focused on the user interface with GIS. Most GIS systems have evolved from command-line and macro control to an interactive menu system. As the GUI improves, GISs can improve, too. One suggested improvement has been to incorporate natural language interfaces, in which the user communicates in English with the system. Others have suggested that GISs incorporate the "fuzzy" characteristics of English as well. Such a system could be asked to show a buffer containing features that are "near" something rather than within 2.6 kilometers of it.

More advanced user interfaces could be icon driven, as in the Geographer's Desktop, and could use a symbolic manipulation language such as the Idrisi, ERDAS Imagine, ER-Mapper, Khoros and Stella modeling systems, in which the user plans out operations by drawing a highly stylized flow diagram and then makes the process operational to carry out the tasks. Even more sophisticated interfaces are obviously possible, and we have yet to even start work on effective use of interfaces for multimedia, interactive, and animated GIS systems.

10.4.8 GIS Interoperability

Another area of concentrated GIS research is that of *interoperability*. An effort is currently under way to standardize and publish a set of specifications for GIS functions and capabilities, allowing a standard language and a higher degree of mobility among systems. This effort, termed OPEN/GIS, is an attempt to repeat the success that an open description of the user interface had for GUIs, an effort known as Open/Systems, which gave us OpenLook and Motif. Such a specification, when openly published, allows vendors to develop products along a common line and toward common goals while maintaining the individuality of their own software package. The payback from this effort will be that GIS software will run in a manner that is totally unaltered from the user's perspective, on virtually any computer and under any operating system.

The last, and a major trend as far as interoperability is concerned, is the arrival of the U.S. federal standards for spatial data, the spatial data transfer standard (FIPS 173). This standard means that data that comply with the standard will be able to move directly into a GIS with all the stored characteristics, topology, attributes, and graphics fully intact. For the first time, identical data sets can migrate between GIS software packages without losing the resolution, accuracy, or descriptive poignancy necessary for rigorous GIS analysis. This effort is already close to realization, and most GIS vendors have declared their intent to support the standard in the very near future.

In this section we have pointed to some likely and some less likely possible futures for GIS. Perhaps the least emphasized future has been the most obvious one, and that is yours. As the world moves into the information age, information management will become the leading, driving sector of the economy. Few information management systems have the integrative and analytical power of a GIS.

As GIS becomes mainstream, the GIS package will become yet another basic requirement of using a computer, and the software will become so ubiquitous that it will be available either bundled on a computer on purchase, or shrink-wrapped at the local computer store. When GIS reaches this stage and when even the advanced, let alone the basic, GIS operations become standard operating parts of decision making on a daily basis, GIS will be a part of every person's life, known or unknown. A GIS will affect how we live, eat, travel, communicate, manage our finances, work, and even how and when we have fun.

10.5 SOME FUTURE ISSUES AND PROBLEMS

Assuming that GIS is now only a few years away from this degree of permeation into the economy, if it is not already there, it is a good idea to finish this chapter, and indeed the entire book, with a glimpse at the issues and problems we are likely to face with the future of GIS. How well we as a user community react to the challenges of the issues will play a major role in the future of GIS. As a person now introduced to the possibilities, it is you, the reader, who will have to deal with these issues at a practical level.

10.5.1 Privacy

An issue that raises itself again and again as GIS databases become more and more widespread is that of personal privacy. We very often take our right to privacy for granted, yet all the time, by the use of telephones, credit cards, mail order, and the like, we are constantly revealing to other people what can be personal property. Facts we consider of the greatest privacy—our personal income, information about the family, our health record, and employment history—are all tucked away in somebody's database. GIS offers the integration of these data through their common geography. Although it is to the public benefit, for example, to build a link between environmental pollution and health, the more local and individual the link, the more the issue of personal privacy arises. Even the federal census, with its highly general information about groups of individuals, has strict restrictions on availability of information that can identify specific people, holding such data private for over 70 years before releasing it.

Whole sectors of the economy now rely on linking data from individuals, such as magazine subscriptions and purchases by mail, with demographic and other information by district, such as census tract or zip code. A personal credit history can be amazingly revealing about an individual, and data are often bought and sold as a side benefit of computerized ordering and mailing systems. Just assembling every item of information about an individual, once an extremely difficult task, is now considerably easier.

Who draws the lines? A whole new area of subinterest in GIS is in GIS and the law. As GIS becomes used in lawsuits, in lobbying, voting district delineation, and, as always, in mapping of property, the legal profession will come increasingly to use GIS as a tool, and then by extension to challenge the means by which data are collected and transformed, analyses are conducted, and conclusions are drawn.

This will force GIS analysts to become somewhat more explicit in their methods and more accountable in their operations. GIS software, for example, should keep a log of the functions used, commands given, menu choices selected, and somehow attach this "data lineage log" to the data sets themselves. It is well known that regular statistics can be used to support many viewpoints, and even maps can be manipulated to show different points of view (see, for example, Mark Monmonier's [1991] book *How to Lie with Maps*). GIS offers the mapping and analysis processes full accountability, and this must be stressed in the future if GIS is not to become yet another courtroom gimmick, like computer graphics, as far as the law is concerned.

10.5.2 Data Ownership

There are two philosophies about GIS data ownership. At the one extreme, the federal government produces and distributes digital data in common formats at the marginal cost of distribution, the "cost of fulfilling user requests." This means that the cost of producing the data should not enter into the pricing of the data. The logic here is that, because the federal government has already created the data at the public's expense, it cannot charge a second time for data to the same people when they request copies for their own use. The computer networks have made the dissemination cost for the user effectively zero, so that data are usually available for setting up and using a GIS free or at least for only a very modest price.

At the opposite extreme lie the groups (and nations) who believe that GIS data are a commodity, a product to be protected by copyright and patent and sold only at a profit. The argument for this view is that when the market demands a data set, the profit motive will generate the data, and the profit will draw in competitive data producers, who will eventually drive down the cost. In a few cases, this has happened, but rarely is the profit motive capable of generating data in complete, systematic, and standardized coverages that are regularly maintained. There is a great deal of motive to produce a data set that may sell many times, but little motive to map a corner of the country with little demand and poor existing digital maps. Extended to the international context, neither will there be a motive to map for GIS the poorest and most needy nations, especially in Africa and South America.

Most nations have evolved some combination of these two approaches. The United States uses the federal government data, especially the DLGs, DEMs, and TIGER files, as a base, but adds other, more detailed and up-to-date information by geocoding new maps or buying the data from private companies. The companies sell their data based on its timeliness, accuracy, completeness, and so on, but originally most of the data derives from one or another of the free federal data sets.

This two-way relationship between government and business has generally served GIS well, although it should be clearly noted that without the free federal data, the entire system collapses. The private mechanism can rarely produce data for every small planning office and project, and rarely can the small office afford the high cost of such data. As always, most people will continue to work with the "least cost solution." For GIS, this usually means a microcomputer, inexpensive or public-domain software, and free federal data.

10.5.3 Scientific Visualization

A critical issue for the future of GIS is the degree to which the systems become integrated with those new parts of computer graphics and cartography most suitable for GIS applications. The entire field of scientific visualization is an example. *Scientific visualization* seeks to use the processing power of the human mind, coupled with the imaging and display capabilities of sophisticated computer graphics systems, to seek out empirical patterns and relationships visible in data but beyond the powers of detection using standard statistical and descriptive methods.

Key to the issue of visualization is the ability to model very large and complex data sets and to seek the inherent interrelationships by visual processing alone or with the assistance of standard empirical and modeling methods. Obviously, GIS is the provider of such data sets. GIS data are complex, and the use of maps to begin with already implies that a visual processing mechanism is being used. GIS should move toward full integration with the tools and techniques of scientific visualization and has much to gain by doing so. This would greatly enhance the analysis and modeling component of GIS use, and in a way that is inherently compatible with a GIS and the tools in the GIS toolbox.

Many GIS data are also inherently three dimensional, such as atmospheric and ocean concentrations of chemicals, topography, or abstract statistical distributions such as crime rates and population densities over space. New software allows the user of a GIS not only to map and analyze three-dimensional distributions, but also to model and display them in new ways. Among the cartographic methods now familiar to GIS and to automated cartographic system users are simulated hill-shading, illuminated contour, gridded perspective and realistic perspective views, and stepped statistical surfaces.

Even simple maps, such as weather maps, now use sophisticated hypsometric coloring with interwoven hill-shading. In addition, new types of display, such as stereo screens with shutters and head-mounted displays, along with the new types of three-dimensional input devices, gloves, track balls, and three-dimensional digitizers, have expanded the suite of interaction means for the GIS user remarkably. Many people who deal with image registration and digitizing work with anaglyphic (red and green) stereo and use soft-copy or computer screen photogrammetry to take measurements.

Animation has added another dimension to display and is now commonplace. What was once highly innovative, such as the *LA—the Movie* animated Landsat image sequence, is now commonplace during the weather forecast of the evening television news. Usually, weather satellite data such as GOES is animated and the perspective changed to simulate a flyby.

The possibilities of animated and interactive cartography, the sort we now see as interactive kiosk-type displays at hotels, airports, and supermarkets, are remarkable, and will strongly influence the future of GIS, especially as the computing power and tools necessary for animation become cheaper and more widespread. Animation has a particular role to play in showing time sequences in GIS applications. Just as it is hard to see exactly what happened during a particular play in a sports contest without slow-motion viewing of film or videotape, so GIS users can compress long time sequences or view short time sequences to reveal geographic patterns that were not visible in other ways.

10.5.4 New Focus

As GIS moves into the future, changes are inevitable, for GIS is a science and a technology based on change. Nevertheless, there are broad movements within science toward topics or challenges that are national or international areas of new emphasis. A few trends are already obvious; fortunately, GIS has a role to play in each of them.

First, science has become increasingly focused on issues of global importance. The earth as a whole system is now a valid way at which to approach issues of global climate change such as global warming and the ozone hole; global circulation, such as the patterns and flows within the earth's oceans and atmosphere; and the global scale of the impact of people on the environment. The new global nature of the world economy, the increasingly strong efforts to solve the world's problems with global legislative bodies, such as the World Bank and the United Nations, and the coming into being of methods and tools for approaching these problems with hard data have all led toward a new global science.

GIS has an immense amount to offer this global science. Global distributions need mapping, global mapping needs map projections, and the understanding of flows and circulations are based on an understanding of spatial processes. Even global data collection efforts for GIS are now under way, and organizations use GIS to attack global problems such as crop-yield estimation and famine prediction.

Moreover, GIS has also been at the forefront of a new approach to science. More and more the traditional boundaries between disciplines in the sciences and the social sciences have disappeared, although there are many who fail to recognize it and even resist this trend. Most major research is now conducted by teams, with representatives from a host of different but interrelated sciences working together on a problem. GIS is a natural tool for this sort of work environment because it is able to integrate data from a variety of contexts and sources and seek out interrelationships based on geography, the mapping of distributions, and visualization. In his 1992 book, *Mapping the Next Millennium*, Steven Hall pursued the theme of mapping as a basis for visualization and integrative context throughout the sciences, including astronomy, biology, geology, and physics.

While Hall's definition of mapping is very broad, his book highlights the benefits of an across-the-board mapping approach in the sciences. Just as no scientist is literate without calculus, matrix algebra, and statistics, the methods and principles of GIS are likely to become essential tools in the scientist's toolbox, at least as an integral part of one's educational background, well into the future. With this book, you too have gotten started with GIS, and by now, if you have followed the assignments, questions, and projects, you are already on your way toward getting a jump-start on this new scientific approach.

10.6 CONCLUSION

To summarize, we have covered in this chapter some speculations and some predictions about the future of GIS. Clearly evident is that new issues will arise and will have to be resolved as GIS matures into a mainstream science rather than a new approach. In the preceding section we discussed these in detail. Here we focus on the shorter-term problems that are being confronted as this book goes to press.

First, the impact of standards has become clear. The spatial data transfer standard has already led the way in confronting such issues as accuracy, file transfer, terminology, and definitions. The same standards groups in the United States, led by the federal Geographic Data Committee, have now set forth metadata standards. Other standards, especially for the WWW are now being formulated, such as GeoVRML and GML.

A good analogy to this effort is that of the catalog systems used by libraries. If each library is allowed to design and use its own system, there is little or no possibility for libraries to work together. For example, imagine an interlibrary loan request from one library that files its books by size to one that files its books by the color of the cover! Metadata standards are to digital cartographic and geographic data what the Library of Congress catalog system is to a library. Standardization is essential, not necessarily for a single library but certainly for a network of libraries. As GIS enters the era of the Internet and the World Wide Web, such standardization is the foundation upon which all later work will build. The free exchange of data will lead to massive savings in duplication and in incompatible and unmatched direct data equivalents.

Related to the metadata issue is the issue of archiving. While maps were and still will be maintained and archived in map libraries, who will archive and preserve for history the digital maps of the GIS world? The answer may be the map libraries themselves, which are moving rapidly into the digital era. A map library is already often an essential first step in the search for digital data and will become more focal as the generations of modifications that maps used to go through, and now the data go through, make many data redundant. Who can tell when GIS data will be necessary for historical research? Perhaps the legal profession will be the driving force for effective data archiving, just as the law requires effective archiving when maps become legal documents, such as a property deed or a courtroom exhibit.

Another ongoing movement is the internationalization of GIS. Major GIS activities are taking place in organizations such as the United Nations, the European Union, and in foreign countries, including the United Kingdom, Germany, Mexico, and Japan. The demands of foreign-language use on GIS activity are sometimes complex, especially Arabic, Chinese, and Japanese versions of software. In each nation, also, digital maps have a political context, and the data-sharing and data access standards used in the United States are not always evident. For example, environmental or health agencies, much less private citizens, may not even have access to data created by other branches of their own government.

Yet another issue of immediate concern is that of education and training in this new and expanding field. Many colleges and universities, both inside and outside the United States, now offer comprehensive programs in GIS. As we have seen in this book, it is unnecessary in GIS to make the traditional distinction between pure and applied versions of the learning experience. Most of the concepts are virtually meaningless without a hands-on learning environment, just as a program of lectures on the breaststroke will never teach you to swim. Most educational programs blend science and technology and pure and applied geographic information science.

In the future we are likely to see the substance of GIS, hopefully retaining this holistic educational nature, integrated into the curriculum at community colleges and high schools. Part of the beauty of GIS is that the actual curriculum can be in biology, earth science, geography, environmental science, or in many other areas. Only a few of the concepts in this book really require more than high school math, and the potential for GIS use by the millions of high school students who could benefit from the experience is phenomenal. GIS could lead to a new integrative multi- and interdisciplinary approach to science that will truly succeed only when future generations of scientists have used this approach throughout their precollege education.

An issue that raises its head regularly in GIS is that of the certification of training. When cartography underwent its own revolution in computerization, there was a call for national certification, perhaps administered by the various professional organizations in the field. Such a program would have a standardized curriculum and system of testing so that employers and others would be able to have realistic expectations. Instead, what has happened is that no organization has stepped forward to certify GIS training, and certainly not GIS education.

As a result, what experience you get from a GIS course, commercial or institutional, will vary according to the instructor's perceptions of the discipline. This has led to a certain degree of anarchy and a great deal of opportunism; nevertheless, this plethora of approaches has served to keep the field of GIS in a state of high energy and relevance. The closest yet to a national GIS curriculum has been that of the National Center for Geographic Information and Analysis (Kemp and Goodchild, 1991). This curriculum, although taught across many college campuses, has no textbook and is structured around the quarter system taught at the NCGIA's lead university. The revised core curriculum in GIS is a suite of interconnected Web pages, each representing a module for GIS learning. This is likely to be a highly successful format for GIS learning in the future (see: `http://www.ncgia.org`).

Finally, a word about GIS experience. As a new GIS user, your first experience will probably be in the "push-button" category. You will carry out a program of analysis or processing already prescribed by your predecessors or instructor. It is hoped that this book will help you gain the substance needed to understand more deeply what you are doing. A second group of those experiencing GIS for the first time are those who have been hired as a GIS expert, with perhaps only a little experience and are then thrust deep into a new project with only vendors' manuals for help. In this case, it is hardest of all to see the forest for the trees, especially when deadlines loom. Here too this text can help, by adding supplementary background to the often minimalist explanations and references in manuals. In either case, you are well on your way to mastering the basics of geographic information science, those principles necessary to build a GIS expert. Nevertheless, you will probably spend many hours perplexed and frustrated at first. If so, do not hesitate to fall back on the sources of help covered in this book. At the very least, the reassurance that you are not the only person in the world facing a seemingly impossible GIS problem is great.

If you have reached this page, either working alone or with a class, and have grasped the concepts behind GIS, there are two paths forward. First and foremost, there is no substitute for grappling with the problems and issues raised in this book. Many GIS packages are inexpensive, are shareware or freeware, and may even be available at your public library or school. Dive in head first, and use the knowledge that *Getting Started with Geographic Information Systems* has given you to master the technology. If you have followed along with the assignments in this book or have used the parallel manual for ArcView, you are already most of the way to being a GIS specialist.

At this point you are ready for the next step. The title of this book was chosen deliberately. It is a first guidebook, a tutorial to get you started on GIS with the necessary background information to avoid major mistakes. Quite simply put, there is much more. Most of the next steps you should take were dealt with in Chapter 1. From now on, you must be your own guide. In preparation for this step, take a look at the Code of ethics for GIS professionals prepared by the Urban and Regional Information Systems Association at `http://www.urisa.org/ethics/code_of_ethics.htm`. As a professional entrusted with spatial information, you will be confronted with both positive and negative uses of this powerful technology, and indeed, you must make your own choices.

As you move onward with GIS, or even if you have used this book to find your way out of a one-time-only GIS problem, always bear in mind the power that GIS can bring to bear. Many are the problems and ills of society and this world that GIS can help with. Most of all, GIS is a tool for a sustainable human future because it offers the promise of managing what resources we already possess with the highest level of efficiency. Indeed, GIS can help us live better with the resources we have rather than always demanding more.

Finally, GIS offers the promise of reducing waste, improving living standards, eliminating disease, helping manage risk and disasters, understanding global change, and even advancing the principles of democracy. In the new era of the Information Age, it is you, the intelligent GIS user and analyst, who can use this tool productively, or perhaps just as easily waste its immense capabilities. To paraphrase Sir Arthur Conan Doyle, now you know my methods—now go out and apply them!

10.7 STUDY GUIDE

10.7.1 Summary

CHAPTER 10: The Future of GIS

Why Speculate? (10.1)

- **The theme of *Getting Started with Geographic Information Systems* has been GIS's place in understanding geographic distributions and their mapping and prediction in the real world.**
- **Speculating on GIS's future is valuable because of**
  - **Planning for the purchase of hardware and software**
  - ***Geographic information science*, a new science that is used to design future information systems**
  - **Expansion into new fields and application areas**

Future Data (10.2)

- **Acquiring data for a new GIS is no longer a major problem.**
- **GPS has become a major source of new GIS data and comes increasingly from integrated GPS/GIS systems.**
- **Digital map images such as scanned maps and air photos are often used as a background image for cross-layer registration and update.**
- **Remote sensing will become an important source of GIS data as the cost of data falls and new sorts of data arrive.**
- **A new industry based on GIS, location-based services, is emerging as GIS, mobile communications, the Internet, and GPS have been brought together.**
- **Data exchange will become more common and has been facilitated by exchange standards.**

Future Hardware (10.3)

- **Advanced GIS work has been influenced significantly by the workstation.**
- **GIS has quickly incorporated distributed systems and databases.**
- **The microcomputer has allowed GIS to be applied to new fields and has improved GIS education.**
- **The mobility of portable GIS and GPS systems has revolutionized GIS use.**

Future Software (10.4)

- **Improvements in the user interface have substantially altered GIS's "look and feel."**
- **Basic data differences such as raster versus vector have disappeared as GISs have become more flexible.**
- **Object-oriented programming and databases are likely to improve GIS.**
- **Many GIS databases are now distributed over local or wide area networks.**
- **GIS user needs are both for small one-person systems and large multiperson systems.**
- **GIS software research is active and continues to build new developments.**
- **GIS will become increasingly interoperable as concepts, user interfaces, and functions become more standardized.**

Some Future Issues and Problems (10.5)

- **Privacy will become a critical issue for GIS as use expands to legal applications.**
- **Data ownership will remain critical to GIS, with a delicate balance between public and private GIS data.**
- **GIS research is threatened by a lack of funding and should be protected by the GIS community.**
- **Scientific visualization and computer graphics will be increasingly integrated with GIS capabilities, especially animated and interactive maps.**
- **GIS is critical to new scientific trends, especially multidisciplinary research, global systems study, and the scientist as advocate.**
- **GIS education has not been standardized, but has produced a great variety of learning environments and teaching materials.**

Conclusion (10.6)

- **The future of GIS is in your hands, and as a GIS analyst you have the capability to use or misuse the power that GIS brings to the world's problems.**

10.7.2 Study Questions

Future Data

In studies from the 1980s, data digitizing and assembly costs have been thought to account for anywhere from 60% to 80% of the total costs of setting up a GIS. Why will this change in the future? What new data sources are available now that were not available then? What data distribution systems might be used in data assembly now?

How do GPS data find their way into a GIS? What use would a large number of GPS differentially corrected points be to a GIS project?

What function might a geometrically correct air photo play in the development of a GIS project on environmental remediation? Under what circumstances might satellite remote sensing data be preferable to an air photo?

What is the Spatial Data Transfer Standard? Use the World Wide Web or a library to find out more about it. Suggest how the SDTS might lead to a more successful GIS project in the near future.

Future Software

Draw a generic diagram for a new user of any GIS package with which you are familiar so as to instruct that user on the characteristics of the user interface of the GIS. Suggest three ways that the user interface could be improved.

What is interoperability? How might full GIS interoperability assist the GIS professional's day-to-day work?

Future Issues and Problems

How might GIS be used to impinge on the individual's right to privacy? What GIS applications are closest to these scenarios today?

What are the elements of (a) hypertext, (b) multimedia, and (c) scientific visualization, and how are they likely to find their way into a GIS project in the future?

10.8 EXERCISES

1. *Investigate for your GIS, or for a GIS package about which you can find data, what the GIS license agreement says about redistribution of the GIS software and what rights you have as the creator of a GIS data set using the software. Create a table characterizing and contrasting various data and software ownership policies on a spectrum from freeware and public domain to copyrighted and restricted or proprietary. What mechanisms of enforcement of ownership are available for GIS packages, and how might these make using a GIS more difficult?*
2. *If possible, download from the Internet both a satellite image or an air photo for a district and a vector map coverage. Overlay the two digital maps, and examine carefully the discrepancies between features on the two data sources. What sources of error account for the differences? How might this and other layers be rectified so that they can be used together in overlay analysis?*
3. *Make a menu flow diagram of the path taken through a set of menus to perform a simple task. If your GIS does not support menus, choose a simple GUI like that of Windows or OS/2's presentation manager. Are any of the steps unnecessary? How might the flow be improved to speed up the process from the menus?*
4. *At your local or school library, examine as many textbooks as you can that cover the field of GIS. Using their table of contents information, compile a list of topics that constitute the "core" knowledge areas of GIS. About what areas is there disagreement? How specialized does GIS knowledge need to be for a particular field—for example, real estate, marketing, forestry, or geology?*
5. *How big would a GIS need to be, in terms of both software and hardware, to be used effectively to hold data for the whole world at the resolution of AVHRR instruments? What layers might a small "global awareness" GIS need to serve as an effective GIS learning tool in high schools? Choose an appropriate resolution for such a project so that the entire data set could fit on a CD-ROM.*
6. *Draw up a list of ways that you as a student or professional could prepare yourself for the future of GIS or the changes in GIS that the future will bring. Rank the tasks by short- and long-term priority. Select one task from the top of the short-term list and do it!*

10.9 REFERENCES

Clarke, K. C. (1995) *Analytical and Computer Cartography*, 2nd ed. Upper Saddle River, NJ: Prentice Hall.

Egenhofer, M. J. and Richards, J. R. (1993) "Exploratory access to geographic data based on the map-overlay metaphor," *Journal of Visual Languages and Computing*, vol. 4, no. 2, pp. 105–125.
Freidhoff, R. M. and Benzon, W. (1989) *Visualization: The Second Computer Revolution*. New York: H. A. Abrams.
Hall, S. S. (1992) *Mapping the Next Millennium: The Discovery of New Geographies*. New York: Random House.
Kemp, K. K. and Goodchild, M. F. (1991) "Developing a curriculum in geographic information systems: The National Center for Geographic Information and Analysis core curriculum project," *Journal of Geography in Higher Education*, vol. 15, no. 2, pp. 121–132.
Krol, E. (1992) *The Whole Internet: Users Guide and Catalog*. Sebastopol, CA: O'Reilly and Associates.
Monmonier, M. (1991) *How to Lie with Maps*. Chicago: University of Chicago Press.
Peterson, M. P. (1995) *Interactive and Animated Cartography*. Upper Saddle River, NJ: Prentice Hall.

10.10 KEY TERMS AND DEFINITIONS

anonymous FTP: Ability to connect with another site on the Internet without a user account, and to download files. FTP is the File Transfer Protocol, available on a large number of computer systems. Log-in is usually under the name "anonymous" with the password set to your own user ID on the Internet.

AVHRR (advanced very high resolution radiometer): An instrument on the NOAA orbiting polar satellites that returns 1- and 4-km resolution data about the earth in four wavelengths. Used extensively for large-area land-cover and vegetation mapping and weather prediction.

background image: A satellite image or air photo that serves as a backdrop for display and registration purposes only, rather than as a layer for analysis with the GIS.

base layer or map: A GIS data layer of reference information, such as topography, road network, or streams, to which all other layers are referenced geometrically.

CD-ROM: Compact disk read-only memory; a hardware storage device capable of making extensive data and software available for distribution on removable CDs or as offline storage for a microcomputer.

CORONA: Originally a secret satellite-based remote sensing system carrying the early generation Keyhole cameras for the period 1962–1974. Very high resolution monochrome data from CORONA, some covering the United States, are now available online through the USGS's EROS data center.

data mining: Revisiting existing data to explore for new relationships using new and more powerful tools for analysis and display.

DIGEST: A U.S. military and NATO data transfer standard, best known as the format for the digital chart of the world, which is in the Defense Mapping Agency's vector product format (DIGEST-A).

digital orthophoto quad (DOQ): One element of a national mapping effort to cover the lower 48 United States at a 1-meter ground resolution with monochrome air photos in digital format with a 1 : 12,000 equivalent ground extent. Collections of DOQs are distributed compressed on CD-ROM.

distributed network: A network-connected set of locations, each storing one element of a system. A distributed GIS may have the GIS software running on a workstation but use data dispersed at many computer storage locations over a local or wide area network.

download: To move a file across a network for eventual residence locally.

DX-90: A data transfer standard in use among members of the International Hydrographic Organization, primarily to assure standardization and free exchange of digital nautical charts.

EOS (NASA's Earth Observation System): A multisatellite 15-year program to increase the available data about the earth's land, water, clouds, ice, and air.

EOSDIS: The information distribution, dissemination, and storage section of the EOS program, which eventually will provide full network access to the data collected by all the various EOS instruments and programs.

file server: A computer whose primary function is to store data and make them available on a network as part of a distributed system.

FIPS 173: The federal information processing standard maintained by the USGS and the National Institute of Standards and Technology, which specified a standard organization and mechanism for the transfer of GIS data between dissimilar computer systems. FIPS 173 specifies terminology, features types, and accuracy specifications, as well as a formal file transfer method.

fix: A solution to a software problem or bug. Usually, a section of a computer program or a file to be overwritten to correct the problem, called a "patch."

freeware: Software and data made available on the networks to any user at zero cost. Immense amounts of freeware, including entire operating systems and GIS packages, are available on the Internet.

geographic information science: The scientific use and study of methods and tools for the capture, storage, distribution, analysis, display, and exploitation of geocoded information.

GRID: A United Nations program to assemble, use, and disseminate data sets of global extent of use to United Nations and other agencies.

GUI (graphical user interface): The set of visual and mechanical tools through which a user interacts with a computer, consisting of windows, menus, icons, and pointers.

hypertext: Textual information in which direct links can be made between related text through "hot links," where pointing to a highlighted term moves the user to the text context for that term in the same or a different document.

interoperability: The extent to which users, software, and data can move between computer environments without change or retraining. In a fully interoperable GIS, the user interface will look and feel the same in two different environments (say, a microcomputer and a Unix workstation), and the same set of functions will have the same effect on the same data.

in-vehicle navigation system: A navigation aid allowing the driver of a vehicle, pilot of a plane, or navigator of a boat direct assistance during operation. Combinations of GPS, on-board digital maps, GIS functions such as routing, and voice information are common in these systems. Most use outside aids. Those using the sensed motion of the vehicle are called "inertial."

Landsat: A U.S. government satellite program collecting data about the earth's surface in the visible and infrared parts of the spectrum. Two instruments, the multispectral scanner (79-meter resolution) and the thematic mapper (30-meter resolution), have been used. Landsat 7 is the next to be launched, for which the data will return to the public domain.

metadata: Data about data. Index-type information pertaining to the entire data set rather than the objects within the data set. Metadata usually includes the date, source, map projection, scale, resolution, accuracy, and reliability of the information, as well as data about the format and structure of the data set.

metaphor: For a GUI, the physical analogy for the elements with which the user will interact. Many computer GUIs use the desktop as a metaphor, with the elements of a calendar, clock, files and file cabinets, and so on.

microcomputer: A stand-alone computer containing at least a microchip, a keyboard, a display, and some sort of memory, running under an operating system. Typically, microcomputers have separate elements, and extensions such as network links, graphical pointers, and storage are added to a core system.

multimedia: The use of multiple simultaneous means of communication in a single "document," normally including sound, graphics, animation, and hypertext.

multispectral scanner: An instrument carried on the Landsat series of satellites capable of capturing four parts of the spectrum simultaneously for pixels 79 meters on a side.

multitask: The ability of a computer's operating system or GIS to handle more than one process at once: for example, editing and running a command sequence while extracting data from the database and displaying a map.

National Spatial Data Clearinghouse: A World Wide Web resource that serves as a cross-reference point for the distributed database of all U.S. government public-domain and other geographic information.

national spatial data infrastructure: The set of base geographic data necessary for effective operation of the federal government and its suppliers, made accessible as a distributed database.

network conference group (also network news group, usenet group, List): An Internet resource on which users with similar interests can share broadcast exchanges of information. Several major GIS groups exist, including `comp.infosystems.gis`; also called GIS-L.

object oriented: Computer programming languages and databases that support "objects." Objects are standard "classes" that contain all the properties of an object. As a simple example, an object class could be a point containing the latitude and longitude of the point, a feature code for the point, such as "radar beacon," and any necessary text to describe the object.

Open/GIS: An active effort to assure interoperability among GIS software packages by specifying a standard set of functions and a common user interface.

OpenLook: A set of GUI specifications agreed upon by several software vendors for an interoperable graphical user interface. Sun's Open Windows is an example of an implementation of OpenLook.

operating system: The suite of software programs and utilities necessary for the control and use of a computer, including as a minimum the management of files and the use of the computer's processor.

OS/2: A microcomputer operating system produced by the IBM Corporation.

parallel processing: A computer or workstation configured with multiple microcircuits, each functioning as a separate computer but usable in tandem. Either each computer can handle a separate task or each computer can hold data to which a uniform process is applied.

patch: A "fix" to a program or data set involving a sequence of data that are to be overwritten onto an older version.

Pathfinder: A U.S. government prototype effort to use data mining of older Landsat and other data to provide complete GIS coverages of the United States at different time periods.

PCMCIA: A credit card–like device interface for microcomputers and other devices, such as GPS receivers, that meets the standards of the Personal Computer Memory Card International Association. PCMCIA cards can act as memory, connectors to disk drives, and links to other types of devices; they can perform many other functions, and are interoperable across computers.

public domain: Information that has been made available to the general public and is distributed and redistributed without copyright or patent.

radar mapping system: An active form of remote sensing in which a radar beam is transmitted to earth and the reflected signals are detected and stored. These systems have the distinct advantage of being operable at night, through clouds, and through vegetation and therefore are used extensively for mapping in the tropics and for mapping terrain.

remote login: The ability of a computer user to log directly into another computer through a network connection or a modem.

remotely sensed data: Data collected by a sensor that is not in direct contact with the area being mapped. Active remote sensing involves transmitting a beam that is detected after reflection; passive remote sensing simply measures light from the sun being reflected by objects being sensed. Similar instruments for remote sensing can operate from aircraft or satellites.

scientific visualization: Use of the human visual processing system assisted by computer graphics, as a means for the direct analysis and interpretation of information.

shareware: Data or software placed in the public domain for distribution, whose use or support involves the payment of a (usually token) fee to the author.

spatial data transfer standard (SDTS): The formal standard specifying the organization and mechanism for the transfer of GIS data between dissimilar computer systems. Adopted as FIPS 173 in 1992, SDTS specifies terminology, feature types, and accuracy specifications as well as a formal file transfer method for any generic geographic data. Subsets for the standard for specific types of data, vector, and raster, for example, are called "profiles."

SPOT (Systeme Proprietaire pour l'Observation de la Terre): A French remote sensing satellite system with 10- and 20-meter resolution and stereo capability.

TCP/IP: A network communications protocol that forms the basis of most communications on the Internet.

Unix: A computer operating system that has been made workable on virtually every possible computer and has become the operating system of choice for workstations and science and engineering applications.

user interface: The physical means of communication between a person and a software program or operating system. At its most basic, this is the exchange of typed statements in English or a programlike set of commands.

WIMP: A GUI term reflecting the primary user interface tools available: windows, icons, menus, and pointers.

workstation: A computing device that includes, as a minimum, a microprocessor, input and output devices, a display, and hardware and software for connecting to a network. Workstations are designed to be used together on local area networks, and to share data, software, and so on.

World Wide Web (WWW or W3): A distributed database of information connected by the Internet and special-purpose software for browsing, searching, and downloading.

X-Windows: A public-domain GUI built on the Unix operating system and computer graphics capabilities, written and supported by the Massachusetts Institute of Technology and the basis of most workstation shareware on the Internet.

PEOPLE IN GIS

Michael F. Goodchild, Executive Director, National Center for Geographic Information and Analysis

Mike Goodchild is professor of geography at the University of California, Santa Barbara, where he teaches and does research in GIS. He is the executive director of the National Center for Geographic Information and Analysis, a consortium of three universities doing basic research in geographic information and analysis established in 1988 by the National Science Foundation. Most interested in data quality and the problems associated with GIS data accuracy, Dr. Goodchild, a member of the National Academy of Sciences, is one of the world's leading GIS scholars, and his work has been widely published in the United States and throughout the world. This interview was conducted over the Internet.

KC: Would you explain how you got started with GIS?

MG: I grew up in the UK, and studied Physics at Cambridge University. I spent a lot of my spare time exploring caves in England and Wales. In 1965 I joined a new research team at McMaster University in Canada doing scientific research on caves. We spent our summers in the Canadian Rockies, and the winters in the lab. In 1969 I graduated with a PhD in geography. Along the way I gained an interest in the application of statistical and computing techniques to geographical research, and that's what eventually led me to GIS.

KC: How important was your science background for GIS?

MG: Very important, because it gave me a rigorous foundation of mathematics, particularly geometry.

KC: When did you first hear about GIS?

MG: One of my first PhD students went to Ottawa to work at Environment Canada and in 1972 linked me up with the people at the Canada Geographic Information System.

KC: And what were your early impressions of GIS?

MG: Early GIS was about computer mapping and not very much about analysis. Everything was very crude, particularly the computing hardware. I remember struggling for a long time to figure out how one could print a symbol at a given coordinate on a sheet of paper using a line printer and a simple FORTRAN format.

KC: And how would you define GIS today?

MG: It seems that we now use the term to describe almost any kind of computer processing of geographic data, from simple archiving and storage to sophisticated analysis and modeling. We even use it to describe the data. It's become a very broad and comprehensive term for a field that is actually very diverse.

KC: What role do you see GIS as having in the future?

MG: I think we will see more and more applications of GIS in our daily lives, from navigation systems in cars to maps and images of our local communities accessed through our television cable systems. I think we've hardly begun to see the power of geography as a way of finding information and understanding our surroundings.

KC: What developments would you personally like to see in the field of GIS?

MG: I think two developments are badly needed in GIS—better ways of finding information, and systems that are easier to use. The first requires development of new network applications that go far beyond the capabilities of our current browsers such as Netscape and Mosaic. The second requires a rethinking of GIS functionality and user interfaces to hide much of the technical detail of system operation from the user, just as the automatic starter allowed automobile designers to hide early manual features, such as chokes and spark timers, from drivers.

KC: Do you have suggestions for those seeking an education or career in GIS?

MG: It's very important to find the right balance between an education, which gives an understanding of the underlying theory and concepts, and training in the use of specific packages. Both are necessary, because theory and concepts can be very dry without the hands-on experience that GIS is really all about, but if there's no theory to provide context, the best training in the world can be useless in the long term. The basic sciences are very important for GIS, but above all I think the key element that a GIS professional needs is an understanding of how the bits and bytes in the database are related to real phenomena on the earth's surface—in other words, the meaning of the data. That kind of understanding can be found in a number of disciplines, and perhaps geography comes closest to a comprehensive perspective.

KC: Thank you very much, Mike. (Used with permission.)

GLOSSARY

What's in a name? That which we call a rose
By any other name would smell as sweet.
—Shakespeare, Romeo and Juliet, *Act II, Scene II*

Glossary: A collection of glosses; a list with
explanations of abstruse, antiquated, dialectical,
or technical terms; a partial dictionary
—Oxford English Dictionary

Nope, "HEIMER." Doctor Livingheimer. The chap you want is in the next valley over. In my opinion, young fellah, you need a GIS.

absolute location: A location in geographic space given with respect to a known origin and standard measurement system, such as a coordinate system.

academic research: New learning derived from the activity of university and other scholars.

accuracy: The validity of data measured with respect to an independent source of higher reliability and precision.

active data: Data that can be reconfigured and recomputed in place. Spreadsheet term for data for attributes or records created by formulas within a spreadsheet.

ad hoc: For the particular case at hand.

address matching: Using a street address such as *123 Main Street* in conjunction with a digital map to place a street address onto the map in a known location. Address matching a mailing list, for example, would convert the mailing list to a map and allow the mapping of characteristics of the places on the list.

address range: The range from the highest to the lowest street number on one side of a street, on one block.

adjacency: The topological property of sharing a common boundary or being in immediate proximity.

affine transformation: Any set of translation, rotation, and scaling operations in the two spatial directions of the plane. Affine transformations allow maps with different scales, orientations, and origins to be coregistered.

analog: A representation where a feature or object is represented by another tangible medium. For example, a section of the earth can be represented in analog by a paper map, or atoms can be represented by ping-pong balls.

analysis: The stage in science when measurements are sorted, tested, and examined visually for patterns and predictability.

anonymous FTP: Ability to connect with another site on the Internet without a user account, and to download files. FTP is the file transfer protocol, available on a large number of computer systems. Log-in is usually under the name "anonymous" with the password set to your own user ID on the Internet.

arc: (1) A line represented as a set of sequential points. (2) A line that begins and ends at a topologically significant location, represented as a set of sequential points.

arc/node: An early name for the vector GIS data structure.

area: A two-dimensional (area) feature represented by a line that closes on itself to form a boundary.

area feature: A geographic feature recorded on a map as a sequence of locations or lines that, taken together, trace out an enclosed area or ring that represents the feature. Example: lake shoreline.

area qualitative map: A type of map that shows the existence of a geographic class within areas on the map. Colors, patterns, and shades are generally used. Examples are geology, soils, and land-use maps.

array: A physical data structure for grids. Arrays are part of most computer programming languages and can be used for storing and manipulating raster data.

ASCII (American Standard Code for Information Interchange): A standard that maps commonly used characters such as the alphabet onto one-byte-long sequences of bits.

attribute: A characteristic of a feature that contains a measurement or value for the feature. Attributes can be labels, categories, or numbers; they can be dates, standardized values, or field or other measurements. An item for which data are collected and organized. A column in a table or data file.

Autocad: A leading CAD program by Autodesk, often interfaced with GIS packages and used for digitizing, especially floor plans and engineering graphics.

AUTOCARTO (International Symposium on Automated Cartography): A sequence of computer cartography and GIS conferences.

AVHRR (advanced very high resolution radiometer): An instrument on the NOAA orbiting polar satellites that returns 1- and 4-km resolution data about the earth in four wavelengths. Used extensively for large-area land-cover and vegetation mapping and weather prediction.

azimuthal: A map projection in which the globe is projected directly onto a flat surface. Only one "side" of the globe can be shown at a time.

background image: A satellite image or air photo that serves as a backdrop for display and registration purposes only, rather than as a layer for analysis with the GIS.

base layer or map: A GIS data layer of reference information, such as topography, road network, or streams, to which all other layers are referenced geometrically.

batch: Submission of a set of commands to the computer from a file rather than directly from the user as an interactive exchange.

bearing: An angular direction given in degrees from zero, as north, clockwise to 360.

bell curve: A common term for the normal distribution.

big eight: The eight most popular GIS packages, established by the numbers of users, particularly among people getting started with GIS, at any given time.

biology: The study of living organisms and their vital processes.

bit: The smallest storable unit within a computer's memory, with only on and off states, codable with one binary digit.

block face: One side of a street on one block that is between two street intersections.

border: The area between the neat line and the edge of the medium or display area on which a map is being displayed. Occasionally, information can be placed within the border, but this area is usually left blank.

bounding rectangle: The rectangle defined by a single feature or a collection of geographical features in coordinate space, and determined by the minimum and maximum coordinates in each of the two directions.

browse: A method of search involving repeated examination of records until a suitable one is found.

buffer: A zone around a point, line, or area feature that is assumed to be spatially related to the feature.

byte: Eight consecutive bits.

CAD (computer-aided design): Computer software used in producing technical and design-type drawings.

CALFORM: An early computer mapping package for thematic mapping.

CAM (computer-assisted mapping): A map projection and outline plotting program for mainframe computers dating from the 1960s.

cartographic convention: The accepted cartographic practice. For example, water is usually cyan or light blue on a world map.

cartographic elements: The primitive component part out of which a map is assembled, such as the neat line, legend, scale, titles, figure, and so on.

cartographic spaghetti: A loose data structure for vector data, with order as the only identifying property of the features.

cartography: (1) The science that deals with the principles, construction, and use of maps. (2) The science, art, and technology of making, using, and studying maps.

CD-ROM: Compact disk read-only memory, a hardware storage device capable of making extensive data and software available for distribution on removable CDs or as offline storage for a microcomputer.

centroid: A point location at the center of a feature used to represent that feature.

CGIS (Canadian Geographic Information System): An early national land inventory system in Canada that evolved into a full GIS.

choropleth map: A map that shows numerical data (but not simply "counts") for a group of regions by (1) classifying the data into classes, and (2) shading each class on the map.

clarity: The property of visual representation using the absolute minimum amount of symbolism necessary for the map user to understand map content without error.

clump: To aggregate spatially; to join together features with similar characteristics into a single feature.

color balance: The achievement of visual harmony between colors on a map, primarily by avoiding colors that show simultaneous contrast when adjacent to each other.

color table: The part of the header record in a digital image file that stores specifications of colors based on simple index values, which are then stored in the data part of the image file.

complaint: To register a concern or problem with an agency, in this case by telephone from a person to a complaint hotline or service.

compression: Any technique that reduces the physical file size of data in a spatial or other data format.

compromise: A map projection that is neither area preserving nor shape preserving. An example is the Robinson projection.

compute: A data management command that uses the numerical values of one or more attributes to calculate the value of a new attribute created by the command.

compute command: In a database manager, a command allowing basic arithmetic on attributes or combinations of attributes, such as summation, multiplication, and subtraction.

computer mapping: Producing maps using the computer as the primary or only tool.

computer memory: A sequence of nonrandom bytes that are nonrecoverable after a computer has been turned off and on again.

conformal: A type of map projection that preserves the local shape of features on maps. On a conformal projection, lines on the graticule meet at right angles, as they do on a globe.

conic: A type of map projection involving projecting part of the earth onto a cone-shaped surface that is then cut and unrolled to make it flat.

connectivity: The topological property of sharing a common link, such as a line connecting two points in a network.

context-sensitive help: A component of a user interface that can reveal to the user information that assists with the current status of other elements of the user interface.

continuity: The geographic property of features or measurements that gives measurements at all locations in space. Topography and air pressure are examples.

contour interval: The vertical difference in measurement units such as meters or feet between successive contour lines on a contour map.

contour map: An isoline map of topographic elevations.

converge: The eventual agreement of measurements on a single value.

cookie-cut: A spatial operation to exclude area outside a specific zone of interest. For example, a state outline map can be used to cut out pixels from a satellite image.

coordinate pair: An easting and northing in any coordinate system, absolute or relative. Together, these two values, usually termed (x, y), describe a location in two-dimensional geographic space.

coordinate system: A system with all the necessary components to locate a position in two- or three-dimensional space; that is, an origin, a type of unit distance, and two axes.

CORONA: Originally a secret satellite-based remote sensing system carrying the early generation Keyhole cameras for the period 1960 to 1972. Very high resolution monochrome data from CORONA, some covering the United States, are now available online through the USGS's EROS data center.

credits: A cartographic element in which the sources, authorship, and ownership of the map and the map attributes are cited, often including a date or reference.

critical six: The GIS functional capabilities included in Dueker's GIS definition, map input, storage, management, retrieval, analysis, and display.

cylindrical: A type of map projection involving projecting part of the earth onto a cylinder-shaped surface that is then cut and unrolled to make it flat.

data: A set of measurements or other values, such as text for at least one attribute and at least one record.

data analysis: The process of using organized data to test scientific hypotheses.

data definition language: The part of the database management system that allows the user to set up a new database, to specify how many attributes there will be, what the types and lengths or numerical ranges of each attribute will be, and how much editing the user is allowed to do.

data dictionary: (1) A catalog of all the attributes for a data set, along with all the constraints placed on the attribute values during the data definition phase. It can include the range and type of values, category lists, legal and missing values, and the legal width of the field. (2) The part of a database containing information about the files, records, and attributes rather than just the data.

data entry: The process of entering numbers into a computer, usually attribute data. Although most data are entered by hand, or acquired through networks, from CD-ROMs, and so on, field data can come from a global positioning system receiver, from data loggers, and even by typing at the keyboard.

data-entry module: The part of a database manager that allows the user to enter or edit records in a database. The module will normally allow both entry and modification of values, and will enforce the constraints placed on the data by the data definition.

data exchange: The exchange of data between similar GIS packages among groups with a common interest.

data exchange format: The specific physical data format in which exchange of data between similar GIS packages takes place.

data extremes: The highest and lowest values of an attribute, found by selecting the first and last records after sorting.

data format: A specification of a physical data structure for a feature or record.

data mining: Revisiting existing data to explore for new relationships using new and more powerful tools for analysis and display.

data model: A logical means of organization of data for use in an information system.

data retrieval: The ability of a database management system to get back from computer memory records stored there previously.

data structure: The logical and physical means by which a map feature or an attribute is digitally encoded.

data transfer: The exchange of data between noncommunicating computer systems and different GIS software packages.

database: (1) A collection of data organized in a systematic way to provide access on demand. (2) Any collection of data accessible by computer. (3) The body of data that can be used in a database management system. A GIS has both a map and an attribute database.

database manager: A computer program or set of programs allowing a user to define the structure and organization of a database, to enter and maintain records in the database, to perform sorting, data reorganization, and searching, and to generate useful products such as reports and graphs.

datum: A base reference level for the third dimension of elevation for the earth's surface. A datum can depend on the ellipsoid, the earth model, and the definition of sea level.

DBMS (database management system): Part of a GIS, the tools that allow the manipulation and use of files containing attribute data.

decennial census: The effort required by the U.S. Constitution that every 10 years all people in the nation be counted and their residences located.

decimal: A counting system based on 10.

default: The value of a parameter or a selection provided for the user by the GIS without user modification.

defoliation: The removal of the healthy leaves of a plant or tree.

Delaunay triangulation: An optimal partitioning of the space around a set of irregular points into nonoverlapping triangles and their edges.

DEM (digital elevation model): A raster-format gridded array of elevations.

DEP (New York City Department of Environmental Protection): The city's government agency for the environment.

dependent variable: The variable on the left of the equals sign in a formula model, whose values are determined by the values of the other variables and constants.

design loop: The iterative process in which a GIS map is created, examined for design, improved, and then replotted from the modified map definition until the user is satisfied that a good design has been reached.

desktop mapping: The ability to generate with ease a variety of map types, symbolization methods, and displays by manipulating the cartographic elements directly.

desktop metaphor: For a graphical user interface, the physical analogy for the elements with which the user will interact. Many computer GUIs use the desktop as a metaphor, with the elements of a calendar, clock, files and file cabinets, and so on.

device independence: The ability of software to run with little difference from a user's perspective on any computer or on any specialized device, such as a printer or plotter.

difference of means: A statistical test to determine whether or not two samples differ from each other statistically.

DIGEST: A U.S. military and NATO data transfer standard, best known as the format for the Digital Map of the World, which is in the Defense Mapping Agency's vector product format (DIGEST-A).

digital elevation model: A data format for digital topography, containing an array of terrain elevation measurements.

digital orthophoto quad (DOQ): One element of a national mapping effort to cover the lower 48 United States at a 1-meter ground resolution with monochrome air photos in digital format with a 1 : 12,000 equivalent ground extent. Collections of DOQs are distributed compressed on CD-ROM.

digitizing: Also called semiautomated digitizing. The process in which geocoding takes place manually; a map is placed on a flat tablet and a person traces out the map features using a cursor. The locations of features on the map are sent back to the computer every time the operator of the digitizing tablet presses a button.

digitizing tablet: A device for geocoding by semiautomated digitizing. A digitizing tablet looks like a drafting table but is sensitized so that as a map is traced with a cursor on the tablet, the locations are identified, converted to numbers, and sent to the computer.

dimensionality: The property of geographic features by which they are capable of being broken down into elements made up of points, lines, and areas. This corresponds to features being zero-, one-, and two-dimensional. A drill hole is a point, a stream is a line, and a forest is an area, for example.

disease host: The animal or plant that plays the role of nourishing or sheltering a disease. For Lyme disease, this is usually the mouse or the deer.

disease transmission: Passing a disease from one individual to another.

dissolve: Eliminating a boundary formed by the edge or boundary of a feature that becomes unnecessary after data have been captured; for example, the edges of sheet maps.

distortion: The space distortion of a map projection, consisting of warping of direction, area, and scale across the extent of the map.

distributed network: A network-connected set of locations each storing one element of a system. A distributed GIS may have the GIS software running on a workstation but use data dispersed at many computer storage locations over a local or wide area network.

DLG: A vector format used by the USGS for encoding the lines on large-scale digital maps.

dot map: A map type that uses a dot symbol to show the presence of a feature, relying on a visual scatter to show spatial pattern. Most often used where point features are the GIS data, but dots can be scattered at random throughout areas.

double digitized: The same feature captured by digitizing twice.

download: To move a file across a network for eventual residence locally.

dropout: The loss of data due to scanning at coarser resolution than the map features to be captured. Features smaller than half the size of a pixel can disappear entirely.

drum scanner: A map input device in which the map is attached to a drum that is rotated under a scanner while illuminated by a light beam or laser. Reflected light from the map is then measured by the scanner and recorded as numbers.

Dueker's definition of GIS: "A special case of information systems where the database consists of observations on spatially distributed features, activities or events, which are definable in space as points, lines, or areas. A geographic information system manipulates data about these points, lines, and areas to retrieve data for ad hoc queries and analyses."

DX-90: A data transfer standard in use among the members of the International Hydrographic Organization, primarily to assure standardization and free exchange of digital nautical charts.

DXF: Autocad's digital file exchange format, a vector-mode industry standard format for graphic file exchange.

dynamic segmentation: GIS function that breaks a line into points at locations that have significance, and that can have their own attributes. For example, the line representing a highway can have a new node added every mile as a mile marker that can hold attributes about the traffic flow at that place.

easting: The distance of a point in the units of the coordinate system east of the origin for that system.

ecology: The science concerned with the interrelationship between organisms and their environments.

edge matching: The GIS or digital map equivalent of matching paper maps along their edges. Features that continue over the edge must be "zipped" together and the edge dissolved. To edge-match, maps must be on the same projection, datum, ellipsoid, and scale and show features captured at the same equivalent scale.

editing: The modification and updating of both map and attribute data, generally using a software capability of the GIS.

editor: A computer program for the viewing and modification of files.

elevation: The vertical height above a datum, in units such as meters or feet.

empowerment: Placing power in the hands of the citizen by providing effective and timely information.

encapsulated PostScript: A version of the PostScript language that allows digital images to be included and stored for later display.

end node: The last point in an arc that connects to another arc.

entity by entity: Any data structure that specifies features one at a time, rather than as an entire layer.

entomology: The branch of zoology that deals with insects.

enumeration map: A map designed to show one census enumerator the geographic extent and address ranges within a district.

environmental assessment: Using measurement and analysis to show the collective consideration of the local environment and its threats to individuals.

environmental violation: Breaking a law related to the environment, such as illegal storage of toxic substances.

EOS (NASA's Earth Observation System): A multisatellite 15-year program to increase drastically the amount of data available about the earth's land, water, clouds, ice, and air.

EOSDIS: The information distribution, dissemination, and storage section of the EOS program, which eventually will provide full network access to the data collected by all the various instruments and programs.

epidemiology: The science that deals with the incidence, distribution, and control of disease in a population.

equal area: A type of map projection that preserves the area of features on maps. On an equal-area projection, a small circle on the map would have the same area as on a globe with the same representative fraction. See also **equivalent**.

equatorial radius: The distance from the geometric center of the earth to the surface, usually averaged to a single value for a sphere.

equirectangular: A map projection that maps angles directly to eastings and northings. A cylindrical projection, made secant by scaling the height-to-width ratio. The nonsecant or equatorial version is called the Plate Carree. Credited to Marinus of Tyre, about A.D. 100.

equivalent: A type of map projection that preserves the area of features on maps. On an equal-area projection, a small circle on the map would have the same area as on a globe with the same representative fraction. See also **equal area**.

error band: The width of a margin plus and minus one standard error of estimation, as measured about the mean.

expected error: One standard deviation in the units of measure.

export: The capability of a GIS to write data out into an external file and into a nonnative format for use outside the GIS or in another GIS.

FAQ: A list of frequently asked questions, usually posted on a network newsgroup or conference group to save new users the trouble of asking old questions over again.

fat line: Raster representation of a line that is more than one pixel wide.

feature: A single entity that composes part of a landscape.

field: The contents of one attribute for one record, as written in a file.

field variable: A geographic value that is continuous over space.

figure: The part of a map that is both referenced in the map coordinate system rather than the page layout coordinates and that is the center of the map reader's attention. The figure is contrasted against the ground, or background. For example, on a map of New York State, the state is the figure, and surrounding states, though shown and labeled, are part of the ground and may be toned down.

file: (1) A collection of bytes stored on a computer's storage device. (2) Data logically stored together at one location on the storage mechanism of a computer.

file header: The first part of a file, which contains metadata rather than data.

file server: A computer whose primary function is to store data and make them available on a network as part of a distributed system.

find: A database management operation intended to locate a single record or a set of records or features based on the values of their attributes.

FIPS 173: The federal information processing standard maintained by the USGS and the National Institute of Standards and Technology that specifies a standard organization and mechanism for the transfer of GIS data between dissimilar computer systems. FIPS 173 specifies terminology, features types, and accuracy specifications, as well as a formal file transfer method.

fix: A solution to a software problem or bug. Usually, a section of a computer program or a file to be overwritten to correct the problem, called a "patch."

flat file: A simple model for the organization of numbers. The numbers are organized into a table, with values for variables as entries, records as rows, and attributes as columns.

flatbed scanner: A map input device in which the map is placed on a glass surface and the scanner moves over the map, converting the map into numbers as it moves.

flattening (of an ellipsoid): The ratio of the length of half the short axis of the ellipse to half the long axis of the ellipse, subtracted from 1. The earth's flattening is about 1/300.

flow map: A linear network map that shows, usually by proportionally varying the width of the lines in the network, the amount of traffic or flow within the network.

fonts: A consistent design for the display of the full set of English or other language characters, including special characters such as punctuation and numbers.

format: The specific organization of a digital record.

FORTRAN: An early computer programming language, initially for converting mathematical formulas into computer instructions.

forward/reverse left: Moving along an arc, the identifier for the arc connected in the direction/opposite direction of the arc to the immediate left.

fourth dimension: A common way of referring to time; the first three dimensions determine location in space, the fourth dimension determines creation, duration, and destruction in time.

freeware: Software and data made available on the Internet to any user at zero cost. Immense amounts of freeware, including operating systems and GIS packages, are available.

FTP (File Transfer Protocol): A standardized way to move files between computers. It is a packet switching technique, so that errors in transmission are detected and corrected. FTP allows files, even large ones, to be moved between computers on the Internet, or another compatible network.

fully connected: A set of arcs in which forward and reverse linkages have identically matching begin and end nodes.

functional capability: One of the distinctive processes that a GIS is able to perform as a separate operation or as part of another operation.

functional definition: Definition of a system by what it does rather than what it is.

fuzzy tolerance: Linear distance within which points should be snapped together.

gateway: The entry point to all servers and other computers associated with one project or organization. For example, the U.S. Geological Survey, although spread across the country and throughout many computers, has a single entry point or gateway for its information sources.

GBF (Geographic Base File): A database of DIME records.

general-purpose map: A map designed primarily for reference and navigation use.

generalization: The process of moving from one map scale to a smaller (less detailed) scale, changing the form of features by simplification, and so on.

geocode: A location in geographic space converted into computer-readable form. This usually means making a digital record of the point's coordinates.

geocoding: The conversion of analog maps into computer-readable form. The two usual methods of geocoding are scanning and digitizing.

geodesy: The science of measuring the size and shape of the earth and its gravitational and magnetic fields.

geographic coordinates: The latitude and longitude coordinate system.

geographic information science: (1) Research on the generic issues that surround the use of GIS technology, that impede its implementation, or that emerge from an understanding of its capabilities. (2) The scientific use and study of methods and tools for the capture, storage, distribution, analysis, display, and exploitation of geocoded information.

geographic pattern: A spatial distribution explainable as a repetitive distribution.

geographic property: A characteristic of a feature on earth, usually describable from a map of the feature, such as location, area, shape, distribution, orientation, adjacency, and so on.

geographic search: A find operation in a GIS that uses spatial characteristics as its basis.

geographic(al) information system (GIS): (1) A set of computer tools for analyzing spatial data. (2) A special case of an information system designed for spatial data. (3) An approach to the scientific analysis and use of spatial data. (4) A multi-billion-dollar industry and business.

geographical surface: The spatial distribution traced out by a continuously measurable geographical phenomenon, as depicted on a map.

geography: (1) A field of study based on understanding the phenomena capable of being described and analyzed with a GIS. (2) The underlying geometry and properties of the earth's features as represented in a GIS. (3) The science concerned with all aspects of the earth's surface, including natural and human divisions, the distribution and differentiation of regions, and the role of humankind in changing the face of the earth.

geoid: A complex earth model used more in geodesy than cartography or GIS that accounts for discrepancies over the earth from the reference ellipsoid and other variations due to gravity and so on.

geometric test: A test to establish the spatial relationship between features. For example, a point feature can be given a point-in-polygon test to find whether it is "contained" by an area.

GIF: An industry-standard raster graphic or image format.

GIS/LIS: A U.S. national conference on geographic information and land information systems, sponsored by most GIS professional organizations and held annually.

globe: A three-dimensional model of the earth made by reducing the representative fraction to less than 1 : 1.

GNU: A free software foundation organization that distributes software over the Internet.

goodness of fit: The statistical resemblance of real data to a model, expressed as strength or degree of fit of the model.

GPS (global positioning system): An operational, U.S. Air Force–funded system of satellites in orbits that allow their use by a receiver to decode time signals and convert the signals from several satellites to a position on the earth's surface.

gradient: The constant of multiplication in a linear relationship; that is, the rate of increase of a straight line up or down. See also **slope**.

graduated symbol map: A map type that varies the size of a common geometric symbol to show the amount of an attribute at points or at centroids of areas. For example, cities could be shown with circles of area proportional to population, or census tracts could have a proportional circle divided as a pie chart at a representative point inside the tract.

GRASS: A GIS software package (see Section 8.5).

graticule: The latitude and longitude grid drawn on a map or globe. The angle at which the graticule meets is the best first indicator of what projection has been used for the map.

grid cell: A single cell in a rectangular grid.

grid extent: The ground or map extent of the area corresponding to a grid.

GRID: (1) A United Nations program to assemble, use, and disseminate data sets of global extent of use to United Nations and other agencies. (2) The raster module of the ESRI Arc/Info GIS software.

gridded fishnet map: A map of a three-dimensional surface showing a set of profiles, often parallel to the x, the y, or the viewer's axis so that the surface appears three dimensional, as a raised fishnet viewed in perspective.

ground: The part of the body of the map that is not featured in the figure. This area can include neighboring areas, oceans, and so on. The ground should fall lower than the figure in the visual hierarchy.

GRS80 (Geodetic Reference System of 1980): Adopted by the International Union of Geodesy and Geophysics in 1979 as a standard set of measurements for the earth's size and shape. The length of the semimajor axis is 6,378,137 meters. Flattening is 1/298.257.

GUI (graphical user interface): The set of visual and mechanical tools through which a user interacts with a computer, usually consisting of windows, menus, icons, and pointers.

gypsy moth: A tussock moth introduced into the United States in Boston about 1869. The early stage is a gray-brown, mottled hairy caterpillar that defoliates trees.

harmony: The property by which the elements of a map work together to create a balanced aesthetic whole.

helpline: A telephone service available to software users for help from an expert.

hexadecimal: A counting system based on 16.

hierarchical: System based on sets of fully enclosed subsets and many layers.

hierarchical data model: An attribute data model based on sets of fully enclosed subsets and many layers.

highlight: A way of indicating to the GIS user a feature or element that is the successful result of a query.

histogram: A graphic depiction of a sample of values for an attribute, shown as bars raised to the height of the frequency of records for each class or group of values within the attribute.

HPGL (Hewlett Packard Graphics Language): A device-specific but industry-standard language for defining vector graphics in page coordinates.

HSI: A system of specifying color as three values: hue, saturation, and intensity, respectively.

hue: A color as defined by the wavelength of the light reflected or emitted from the map surface.

hypertext: Textual information in which direct links can be made between related text through "hot links," where pointing to a highlighted term moves the user to the text context for that term in the same or a different document.

hypothesis: A supposition about data expressed in a manner to make it subject to statistical test.

hypsometric map: A map of topography involving a color sequence filling the spaces between successive contours, usually varying from green through yellow to brown.

identify: To find a spatial feature by pointing to it interactively on the map with a pointing device such as a mouse.

image depth: The numbers of bits stored for each pixel in a digital image.

image map: A map that in two dimensions shares many of the characteristics of a map; that is, cartographic geometry, some symbols, a scale and projection, and so on, but is a continuous image taken from an air photo, a satellite image, or a scanner. A scanned paper map used as a backdrop in a GIS becomes an image map.

import: The capability of a GIS to bring data in an external file and in a nonnative format for use within the GIS.

incinerator: A plant designed to burn waste, often producing power as a by-product but primarily for disposal. Often thought to substitute air pollution for ground pollution.

independent variable: A variable on the right-hand side of the equation in a model whose value can range independently of the other constants and variables.

industry-standard format: A commonly accepted way of organizing data, usually advanced by a private organization.

information: The part of a message placed there by a sender and not known by the receiver.

information system: A system designed to allow the user to be delivered the answer to a query from a database.

inset: A map within a map, either at a smaller scale to show relative location, or a larger scale to show detail. An inset may have its own set of cartographic elements such as a scale and graticule.

installation: The step necessary between the delivery of GIS software and its first use, consisting of copying and decompressing files, data, registering licenses, and so on.

installed base: The number of existing implemented systems.

integrated software: Software that works together as part of a common user interface, rather than software that consists of separate programs to be used in sequence.

intensity: The amount of light emitted or reflected per unit area. A map that has high intensity appears bright.

intercept: The value of the dependent variable when the independent variable is zero.

internal format: A GIS data format used by the software to store the data within the program and in a manner unsuitable for use by other means.

Internet: A network of many computer networks. Any computer connected to the Internet can share any of the computers accessible through the network. The Internet shares a common mechanism for communication, called a protocol. Searches for data, tools for browsing, and so on are available to ease the tasks of "surfing" the Internet.

interoperability: The extent to which users, software, and data can move between computer environments without change or retraining. In a fully interoperable GIS, the user interface will look and feel the same in two different environments (say, a microcomputer and a Unix workstation), and the same set of functions will have the same effect on the same data.

interval: Data measured on a relative scale but with numerical values based on an arbitrary origin. Examples are elevations based on mean sea level or coordinates.

in-vehicle navigation system: A navigation aid allowing the driver of a car, pilot of a plane, or navigator of a boat direct assistance during operation. Combinations of GPS, on-board digital maps, GIS functions such as routing, and voice information are common in these systems. Most use outside aids. Those using the sensed motion of the vehicle are called "inertial."

isoline map: A map containing continuous lines joining all points of identical value.

join: To merge both records and attributes for unrelated but overlapping databases.

key attribute: A unique identifier for related records that can serve as a common thread throughout the files in a relational database.

killer app: A computer program or "application" that by providing a superior method for accomplishing a task in a new way becomes indispensable to computer users. Examples are word processors and spreadsheets.

label: Any text cartographic element that adds information to the symbol for a feature, such as the height number label on a contour line.

label placement rules: The set of rules that cartographers use when adding map text, place-names, and labels to features. Some rules are generic to the map as a whole, while others relate to point, line, and area features specifically. Well-designed maps follow the label placement rules and use them to resolve conflicts between the labels, as labels should never be plotted over each other.

label point: A point digitized within a polygon and assigned its label or identifier for use in topological reconstruction of the polygon.

land-cover map: A map showing the type of actual surface covering at a given time. Categories could be grassland, forestlands, cropland, bare rock, and so on.

land-use map: A map showing the human use to which land is put at a given time. Categories could be pasture, forestland, agricultural land, wasteland, and so on.

landmark: A TIGER term for a geographic feature not a part of the census features.

Landsat: A U.S. government satellite program collecting data about the earth's surface in the visible and infrared parts of the spectrum. Two instruments, the multi-spectral scanner (79-meter resolution) and the thematic mapper (30-meter resolution), have been used. Landsat 7 was launched in 1999, with an enhanced thematic mapper that images at 15-meters.

landscape: That part of geographic space showable on a map, including all its features.

latitude: The angle made between the equator, the earth's geometric center, and a point on or above the surface. The south pole has latitude −90 degrees; the north +90 degrees.

layer: A set of digital map features collectively (points, lines, and areas) with a common theme in coregistration with other layers. A feature of GIS and most CAD packages.

learning curve: The relationship between learning and time. A steep learning curve means that much is learned quickly (usually thought to be the opposite). A difficult learning curve is one where learning takes place slowly, over a long period.

least squares: A statistical method of fitting a model, based on minimizing the sum of the squared deviations between the data and the model estimates.

legend: The map element that allows the map user to translate graphic map symbols into ideas, usually by the use of text.

level of measurement: The degree of subjectivity associated with a measurement. Measurements can be nominal, ordinal, interval, or ratio.

line: A one-dimensional (length) map feature represented by a string of connected coordinates.

line feature: A geographic feature recorded on a map as a sequence of locations tracing out a line. A stream is an example.

line thickness: The thickness, in millimeters, inches, or other units, of a line as it appears on a map.

linear relationship: A straight-line relationship between two variables such that the value of the dependent variable is a gradient multiplied by the independent variable plus a constant.

link: The part or structure of a database that physically connects geographic information with attribute information for the same features. Such a link is a defining component of a GIS.

LIS (land information system): The surveying profession's term for GIS where the data are for land ownership.

local area network: An arrangement of computers into a cluster, with network linkages between computers but no external link. Usually, this allows sharing data and software licenses, or the use of a file server.

locate: See **identify**.

location: A position on the earth's surface or in geographic space definable by coordinates or some other referencing system, such as a street address or space indexing system.

logical structure: The conceptual design used to encrypt data into a physical structure.

longitude: The angle formed between a position on or above the earth, the earth's geometric center, and the meridian passing through the center of the observing instrument in Greenwich, England, as projected down onto the plane of the earth's equator or viewed from above the pole. Longitudes range from −180 (180 degrees west) to +180 (180 degrees east).

LUNR (Land Use and Natural Resources Inventory System): An early GIS in New York State.

macro: A command language interface allowing a "program" to be written, edited, and then submitted to the GIS user interface.

magic number: Any number that has a specific value for a specialized need.

map: A depiction of all or part of the earth or other geographic phenomenon as a set of symbols and at a scale whose representative fraction is less than 1 : 1. A digital map has had the symbols geocoded and stored as a data structure within the map database.

map algebra: Tomlin's terminology for the arithmetic of map combination for coregistered layers with rasters of identical size and resolution.

map design: The set of choices relating to how a map's elements are laid out, how symbols such as colors are selected, and how the map is produced as a finished tangible product. The process of applying cartographic knowledge and experience to improve the effectiveness of a map.

map millimeters: A coordinate system based on the dimensions of the map rather than of the features represented on the earth itself, in metric units.

map overlay: Placing multiple thematic maps in precise registration, with the same scale, projections, and extent, so that a compound view is possible.

map projection: A depiction of the earth's three-dimensional structure on a flat map.

map title: The text that identifies the coverage and content of a map. This is usually a major map element, and can be worded to show the map theme or the map's content.

map type: One of the set of cartographic methods or representation techniques used by cartographers to make maps of particular types of data. Data, by their attributes and dimensions, usually determine which map types are suitable in any given map context.

Maptitude: A GIS software package (see Section 8.5).

mask: A map layer intended to eliminate or exclude areas not needed for mapping and analysis.

matrix: A table of numbers with a given number of rows and columns.

mean: A representative value for an attribute, computed as the sum of the attribute values for all records divided by the number of records.

mean center: For a set of points, that point whose coordinates are the means of those for the set.

mean sea level: A local datum based on repeated measurements of sea level throughout all of its normal cycles, such as tides and seasonal change. The basis for elevations on a map.

measurement: A quantitative assessment of a phenomenon.

median: The attribute value for the middle record in a data set sorted by that attribute.

medium: A map medium, for example, is the material chosen on which to produce a map: paper, film, Mylar, CD-ROM, a computer screen, a TV image, and so on.

menu: A component of a user interface that allows the user to make selections and choices from a preset list.

meridian: A line of constant longitude. All meridians are of equal length on the globe.

metadata: Data about data. Index-type information pertaining to the entire data set rather than the objects within the data set. Metadata usually includes the date, source, map projection, scale, resolution, accuracy, and reliability of the information, as well as data about the format and structure of the data set.

metaphor: For a GUI, the physical analogy for the elements with which the user will interact. Many computer GUIs use the desktop as a metaphor, with the elements of a calendar, clock, files and file cabinets, and so on.

metric system: A system of weights and measures accepted as an international standard as the Systeme International d'Unites (SI) in 1960. The metre (meter in the United States) is the unit of length.

microcomputer: A stand-alone computer containing at least a microchip, a keyboard, a display, and some sort of memory, running under an operating system. Typically, microcomputers have separate elements, and extensions such as network links, graphical pointers, and peripheral storage are added to a core system.

military grid: A coordinate system based on the transverse Mercator projection, adopted by the U.S. Army in 1947 and used extensively for world mapping.

MIMO system (map in–map out): A term used to describe a first-generation computer mapping system designed to capture the map by computer and reproduce it.

missing data: Elements where no data are available for a feature or a record.

missing value: A value that is excluded from arithmetic calculations for an attribute because it is missing, not applicable, or is corrupted, and has been signified as such.

mixed pixel: A pixel containing multiple attributes for a single ground extent of a grid cell. Common along the edges of features or where features are ill defined.

MLMIS (Minnesota Land Management System): A very early statewide GIS.

model: A theoretical distribution for a relationship between attributes. A spatial model is a theoretically expected geographic distribution determined by a specified form such as an equation.

modeling: The stage in science when a phenomenon under test is sufficiently well understood that an abstract system can be built to simulate the real system.

modular computer program: A computer program composed of integrated sections of reusable functions rather than a single program.

mosaic: The GIS or digital map equivalent of matching paper maps along their edges. Features that continue over the edge must be "zipped" together and the edge dissolved. To edge-match, maps must be on the same projection, datum, ellipsoid, and scale, and show features captured at the same equivalent scale. See also **edge matching**.

mosaicing: The GIS or digital map equivalent of matching multiple paper maps along their edges. Features that continue over the edge must be "zipped" together, and the edge dissolved. A new geographic extent for the map usually has to be cut or clipped out of the mosaic. For mosaicing, maps must be on the same projection, datum, ellipsoid, and scale, and show features captured at the same scale.

Motif: A graphical user interface standard common on Unix workstations.

multimedia: The use of multiple simultaneous means of communication in a single document, normally including sound, graphics, animation, and hypertext.

multispectral scanner: An instrument carried on the Landsat series of satellites capable of capturing four parts of the spectrum simultaneously for pixels 79 meters on a side.

multitask: The ability of a computer's operating system or GIS to handle more than one process at once; for example, editing and running a command sequence while extracting data from the database and displaying a map.

municipality: An administrative division of geographic space, usually for the purposes of election or service delivery.

Murphy's law: "Anything that can go wrong will go wrong." Long linked to the public use of computers.

NAD27 (North American Datum of 1927): The datum used in the early national mapping of the United States. The Clarke 1866 ellipsoid was used and locations and elevations were referenced to a single point at Meade's Ranch in Kansas.

National GIS Curriculum: An NCGIA-sponsored national college curriculum for GIS, used in many colleges and universities worldwide and with available teaching materials.

National Spatial Data Clearinghouse: A World Wide Web resource that serves as a cross-reference point for the distributed database of all U.S. government public-domain and other geographic information.

NCGIA (National Science Foundation's National Center for Geographic Information and Analysis): A three-university consortium funded to assist in GIS education, research, outreach, and information generation.

neat line: A solid bounding line forming the frame for the visually active part of a map.

network: Two or more computers connected together so that they can exchange messages, files, or other means of communication. A network is part hardware, usually cables and communication devices such as modems, and part software.

network conference group (also network news group, usenet group, List): An Internet resource on which users with similar interests can share broadcast exchanges on information. Several major GIS groups exist, including comp.infosystems.gis; also called GIS-L.

network map: A map that shows as its theme primarily connections within a network, such as roads, subway lines, pipelines, or airport connections.

newsgroup: A discussion area on the Internet for asynchronous many-to-many discussions.

NOAA (National Oceanic and Atmospheric Administration): An agency of the U.S. Department of Commerce that provides digital and other maps for navigation, weather prediction, and for physical features of the United States.

node: (1) At first, any significant point in a map data structure. Later, only those points with topological significance, such as the ends of lines. (2) The end of an arc.

node snap: Instructing the GIS software to make multiple nodes or points in a single node so that the features connected to the nodes match precisely, say at a boundary.

nominal: A level of measurement at which only subjective information is available about a feature. For a point, for example, the name of the place.

normal distribution: A distribution of values symmetrically about a mean with a given variance.

normalize: To remove an effect biasing a statistic; for example, the influence of the size of the sample.

northing: The distance of a point in the units of the coordinate system north of the origin for that system.

null hypothesis: The state opposite to that suggested in a hypothesis, postulated in the hope of rejecting its form and therefore proving the hypothesis.

object-oriented: Computer programming languages and databases that support "objects." Objects are standard "classes" that contain all the properties of an object. As a simple example, an object class could be a point and contain the latitude and longitude of the point, a feature code for the point, such as "radar beacon" and any text necessary to describe the object.

oblate ellipsoid: A three-dimensional shape traced out by rotating an ellipse about its shorter axis.

oblique: A map projection in which the center line of the map is not at right angles to the earth's geographic coordinates, following neither a single parallel nor a meridian.

observation: The process of recording an objective measurement.

Odyssey: A first-generation GIS developed at Harvard to implement the original arc/node vector data structure.

online manual: A digital version of a computer application manual available for searching and examination as required.

OpenLook: A set of GUI specifications agreed upon by several software vendors for an interoperable graphical user interface. Sun's Open Windows is an example of an implementation of OpenLook.

Open/GIS: An active effort to assure interoperability among GIS software packages by specifying a standard set of functions and a common user interface.

operating system: The suite of software programs and utilities necessary for the control and use of a computer, including as a minimum the management of files and use of the computer's processor.

ordinal: A level of measurement at which only relative information is available about a feature, such as a ranking. For a highway, for example, the line may be coded to show a jeep trail, a dirt road, a paved road, a state highway, or an interstate highway, in ascending rank.

origin: A location within a coordinate system where the eastings and northings are exactly equal to zero.

orthophoto map: An image map that is an air photo, corrected for topographic and other effects. A specific type of mapping program, at 1 : 12,000, by the USGS.

OS/2: A microcomputer operating system produced by the IBM Corporation.

overlay: A GIS operation in which layers with a common, registered map base are joined on the basis of their occupation of space.

overlay weighting: Any system for map overlay in which the separate thematic map layers are assigned unequal importance.

page coordinates: The set of coordinate reference values used to place the map elements on the map and within the map's own geometry rather than the geometry of the ground that the map represents. Often, page coordinates are in inches or millimeters from the lower left corner of a standard-size sheet of paper, such as A4 or $8\frac{1}{2}$ by 11 inches.

parallel: A line of constant latitude. Parallels get shorter toward the poles, becoming a point at the pole itself.

parameter: A number, value, text string, or other value required as the consequence of submitting a command to the GIS.

parcel: A land surface partition recognized by law for the purpose of ownership.

patch: A fix to a program or data set involving a sequence of data that are to be overwritten onto an older version.

Pathfinder: A U.S. government prototype effort to use data mining of the older Landsat and other data to provide complete GIS coverages of the United States at different periods.

PC (personal computer): A self-contained microcomputer, providing the necessary components for computing, including hardware, software, and a user interface.

PCMCIA: A credit card–like device interface for microcomputers and other devices, such as GPS receivers, that meets the standards of the Personal Computer Memory Card International Association. PCMCIA cards can act as memory, as connectors to disk drives, as links to other types of devices, and perform many other functions, and they are interoperable across computers.

perfect sphere: A three-dimensional figure traced out by all possible positions of an arc of a fixed radius about a point. A good approximation of the shape of the earth.

permanent map: A map designed for use as a permanent end product in the GIS process.

pheromone: A hormonal substance excreted by an individual that elicits a response in the same species.

physical structure: The mechanical mapping of a section of computer memory onto a set of files or storage devices.

picture symbol map: A map type that uses a simplified picture or geometric diagram at a point to show a feature type. For example, on a reference map, airports could be shown with a small airplane stick diagram, or picnic areas by a picnic table diagram.

pixel: The smallest unit of resolution on a device, often used to show one grid cell at the highest display resolution.

place-name: A text cartographic element that links the language given name to a feature by placing it close to the symbol to which it corresponds, such as a city name as text next to a filled circle.

point: A zero-dimensional map feature, such as a single elevation mark as specified by at least two coordinates.

point feature: A geographic feature recorded on a map as a location. Example: a single house.

point mode: A method of geocoding in semiautomated digitizing, in which one press of the cursor button sends back to the computer only one (the current) tablet location.

polar radius: The distance between the earth's geometric center and either pole.

polygon: A many-sided area feature consisting of a ring and an interior. An example is a lake on a map.

polygon interior: The space contained by a ring, considered part of a polygon.

polygon left: Moving along an arc, the identifier for the polygon adjacent to the left.

polygon right: Moving along an arc, the identifier for the polygon adjacent to the right.

population: The total body of objects from which a sample is taken for measurement.

PostScript: Adobe Corporation's page definition language. An interpreted language for page layout designed for printers but used as an industry standard for vector graphics.

precision: The number of digits used to record a measurement or which a measuring device is capable of providing.

prediction: (1) The scientific ability to forecast the outcome of a process in advance. (2) The ability of a model to provide information beyond that for which measurements are available.

prime meridian: The line traced out by longitude zero and passing through Greenwich, England. The prime meridian forms the origin for the longitude part of the geographic coordinates and divides the eastern and western hemispheres.

proceedings: The formal record of the papers and other prepared presentations at a professional or scientific conference. Usually available to conference attendees, then distributed as a soft-cover book.

professional publication: Books, journals, and other information designed primarily for those using GIS technology as part of their job.

proprietary format: A data format whose specification is a copyrighted property rather than public knowledge.

public domain: Information that has been made available to the general public and is distributed and redistributed without copyright or patent.

quad tree: A way of compressing raster data based on eliminating redundancy for attributes within quadrants of a grid.

query: A question, especially if asked of a database by a user via a database management system or GIS.

query language: That part of a DBMS allowing submission of queries to a database.

***r*-squared:** A common term for the coefficient of determination.

radar mapping system: An active form of remote sensing in which a radar beam is transmitted to earth and the signals reflected are detected and stored. These systems have the distinct advantage of being operable at night, through clouds, and through vegetation, and therefore are used extensively for mapping in the tropics and for mapping terrain.

RAM: That part of a computer's memory designed for rapid access and computation.

random: Having no discernible structure or repetition.

range: The highest value of an attribute less the lowest, in the units of the attribute.

raster: A data structure for maps based on grid cells.

ratio: A level of measurement at which numerical information is available about a feature, based on an absolute origin. For land parcels, for example, the assessed value in dollars would be an example, the value zero having real meaning.

real map: A map that has been designed and plotted onto a permanent medium such as paper or film with tangible form, and is a result of all the design and compilation decisions made in constructing the map, such as choosing the scale, setting the legend, selecting the colors, and so on.

realistic perspective map: A map of a three-dimensional surface showing a colored or shaded image draped over a topographic surface and viewed in perspective.

record: A set of values for all attributes in a database. Equivalent to the row of a data table.

reference map: A highly generalized map type designed to show general spatial properties of features. Examples are world maps, road maps, atlas maps, and sketch maps. Sometimes used in navigation, often with a limited set of symbols and selected data. A cartographic base reference map is often the base layer in a GIS.

relate: A DBMS operation that merges databases through their key attributes to restructure them according to the needs of a user's query, rather than as they are stored physically.

relational DBMS: A database management system based on the relational data model.

relational model: A data model based on multiple flat files for records, with dissimilar attribute structures, connected by a common key attribute.

relative location: A position described solely with reference to another location.

remote log-in: The ability of a computer user to log directly into another computer through a network connection or a modem.

remotely sensed data: Data collected by a sensor that is not in direct contact with the area being mapped. Active remote sensing involves transmitting a beam that is detected after reflection; passive remote sensing simply measures light from the sun being reflected by objects being sensed. Similar instruments for remote sensing can operate from aircraft or satellites.

renumbering: Use of the DBMS to change the ordering or ranges of attributes. Also, especially in raster GISs, to change the numbers within grid cells into categories.

report: A listing of all the values of attributes for all records in a database. A report is often printed as a table, for verification against source material, and for validation by examination.

report generator: That part of a database management system that can produce a listing of all the values of attributes for all records in a database.

representative fraction: The ratio of a distance as represented on a map to the equivalent distance measured on the ground. Typical representative fractions are 1 : 1 million, 1 : 100,000, and 1 : 50,000.

residual: The amount left when the observed value of the dependent variable has subtracted from it that predicted by a model, in units of the dependent variable.

resource management: The intentional control or influence of environmental elements to accomplish particular goals.

restrict: Part of the query language of a DBMS that allows a subset of attributes to be selected out of the flat file.

retrieval: The ability of a database management system or GIS to get back from computer memory records that were stored there previously.

RGB: The system of specifying colors by their red, green, and blue saturations.

ring: A line that closes upon itself to define an area.

rubber sheeting: A statistical distortion of two map layers so that spatial coregistration is accomplished, usually at a set of common points.

run-length encoding: A way of compressing raster data based on eliminating redundancy for attributes along rows of a grid.

sample: A subset of a population selected for measurement.

saturation: The amount of color applied per unit area. Perceptually, saturated colors appear rich or solid, whereas low saturation colors look washed out or pastel-like.

scale: (1) The geographic property of being reduced by a representative fraction. Scale is usually depicted on a map or can be calculated from features of known size. (2) The part of the map display that shows the scale of the map figure as either an expression of values (the representative fraction as a number) or as a graphic, usually a line on the map labeled with an equivalent and whole-number length on the ground, such as 1 kilometer or 1 mile.

scaleless: The characteristic of digital map data in abstract form of being usable and displayable at any scale, regardless of the scale of the map used to geocode the data.

scanning: A form of geocoding in which maps are placed on a surface and scanned by a light beam. Reflected light from every small dot or pixel on the surface is recorded and saved as a grid of digits. Scanners can work in black and white, in gray tones, or in color.

scientific approach: A method for rationally explaining observations about the natural and human world.

scientific visualization: Use of the human visual processing system assisted by computer graphics, as a means of direct analysis and interpretation of information.

search: Any database query that results in successful retrieval of records.

search engine: A software tool designed to search the Internet and WWW for documents meeting the user's query. Examples: Yahoo, Alta Vista, and Google.

secant: A map projection in which the surface used for the map "cuts" the globe at the map's representative fraction. Along this line there is distortion-free mapping of the geographic space. Multiple cuts are possible, for example, on a conic projection.

select: A DBMS command designed to extract a subset of the records in a database.

server: A computer connected to a network whose primary function is to act as a library of information that other users can share.

shareware: Data or software placed in the public domain for distribution, but whose use or support involves the payment of a (usually token) fee to the author.

sift: To eliminate features that are smaller than a minimum feature size.

simulated hill-shaded map: A map in which an apparent shading effect of raised topography is produced by computer (or manually) so that the land surface appears differentially illuminated, as it would in low sun angles naturally.

simultaneous contrast: The tendency for colors at the opposite ends of the primary scale to "jump" when placed together; for example, red and green.

SLF: An early Defense Mapping Agency data format.

sliver: A very small, narrow polygon caused by data capture or overlay error that does not exist on a map.

slope: The constant of multiplication in a linear relationship; that is, the rate of increase of a straight line up or down. See also **gradient**.

snap: Forcing two or more points within a given radius of each other to be the same point, often by averaging their coordinate.

software package: A computer program application.

sort: To place the records within an attribute in sequence according to their value.

spatial data: Data that can be linked to locations in geographic space, usually via features on a map.

Spatial Data Transfer Standard (SDTS): The formal standard specifying the organization and mechanism for the transfer of GIS data between dissimilar computer systems. Adopted as FIPS 173 in 1992, SDTS specifies terminology, feature types, and accuracy specifications as well as a formal file transfer method for any generic geographic data. Subsets for the standard for specific types of data, vector and raster, for example, are called "profiles."

spatial distribution: The locations of features or measurements observed in geographic space.

SPOT (System Proprietaire pour l'Observation de la Terre): A French remote sensing satellite system with 10- and 20-meter resolution and stereo capability.

spreadsheet: A computer program that allows the user to enter numbers and text into a table with rows and columns and then maintain and manipulate those numbers using the table structure.

SQL (Structured Query Language): A standard language interface to relational database management systems.

standard deviation: A normalized measure of the amount of deviation from the mean within a set of values. The mean deviation from the mean.

standard distance: A two-dimensional equivalent of the standard deviation, a normalized distance built from the standard deviations of the eastings and northings for a set of points.

standard parallel: A parallel on a map projection that is secant and therefore distortion free.

state plane: A coordinate system common in utility and surveying applications in the lower 48 United States and based on zones drawn state by state on transverse Mercator and Lambert conformal conic projections.

stepped statistical surface: A map type in which the outlines of areas are "raised" to a height proportional to a numerical value and viewed in apparent perspective. The areas then appear as columns, with a column height proportional to value.

stream mode: A method of geocoding in semiautomated digitizing, in which a continuous stream of points follows a press of the cursor button. This mode is often used for digitizing long features such as streams and coastlines. It can generate many data very quickly, so is often weeded by generalization.

subsetting: Extracting a part of a data set.

SURFACE II: An early computer mapping package from the Kansas Geological Survey.

SYMAP: An early multipurpose computer mapping package.

symbol: An abstract graphic representation of a geographic feature for representation on a map. For example, the feature could be a canal, the symbol a blue line of a given thickness.

symbolization: The full set of methods used to convert cartographic information into a visual representation.

table: (1) An arrangement of attributes and records into rows and columns to assist display and analysis. (2) Any kind of organization by placement of records into rows and columns.

TCP/IP: A network communications protocol that forms the basis of most of the communications on the Internet.

temporary map: A map designed for use as an intermediate product in the GIS process, and not usually subjected to the normal map design sequence.

test of means: Hypothesis test to establish whether two samples with their own means and standard deviations are drawn from the same overall population.

thematic map: A map designed primarily to show a "theme," a single spatial distribution or pattern, using a specific map type.

TIF: An industry-standard raster graphic or image format.

TIGER: A map data format based on zero, one, and two cells, used by the U.S. Census Bureau in street-level mapping of the United States.

TIN: A vector topological data structure designed to store the attributes of volumes, usually geographic surfaces.

tolerance: The distance within which features are assumed to be erroneously located different versions of the same thing.

topographic map: A map type showing a limited set of features but, including at the minimum, information about elevations or landforms. Example: contour maps. Topographic maps are common for navigation and for use as reference maps.

topologically clean: The status of a digital vector map when all arcs that should be connected are connected at nodes with identical coordinates, and the polygons formed by connected arcs have no duplicate, disconnected, or missing arcs.

topology: (1) The property that describes adjacency and connectivity of features. A topological data structure encodes topology with the geocoded features. (2) The numerical description of the relationships among geographic features, as encoded by adjacency, linkage, inclusion, or proximity. Thus, a point can be inside a region, a line can connect to others, and a region can have neighbors. The numbers describing topology can be stored as attributes in the GIS and used for validation and other stages of description and analysis.

toxic release: Release of a toxic substance into the environment, such as the venting of a poisonous gas into the air.

transparent overlay: An analog method for map overlay, where maps are traced or photographed onto transparent paper or film and then overlain mechanically.

transverse: A map projection in which the axis of the map is aligned from pole to pole rather than along the equator.

units: The standardized measurement increments for values within an attribute.

Unix: A computer operating system that has been made workable on virtually every possible computer and has become the operating system of choice for workstations and science and engineering applications.

unsupervised classification: The grouping of pixels by their numerical spectral characteristics without the intervention of direct human guidance.

update: Any replacement of all or part of a data set with new or corrected data.

upward compatibility: The ability of software to move on to a new version with complete support for the data, scripts, functions, and so on, of earlier versions.

U.S. Census Bureau: A division of the Department of Commerce that provides maps in support of the decennial (every 10 years) census of the United States, especially the census of population.

user group: Any formal or informal organization of users of a system that share experiences, information, news, or help among themselves.

user interface: The physical means of communication between a person and a software program or operating system. At its most basic, this is the exchange of typed statements in English or a programlike set of commands.

USGS (United States Geological Survey): An agency of the Department of the Interior and a major provider of digital map data for the United States.

UTM (universal transverse Mercator): A standardized coordinate system based on the metric system and a division of the earth into sixty 6-degree-wide zones. Each zone is projected onto a transverse Mercator projection, and the coordinate origins are located systematically. Both civilian and military versions exist.

validation: The process by which entries placed in records in an attribute data file, and the map data captured during digitizing or scanning, are checked to ensure that their values fall within the bounds expected of them and that their distribution makes sense within the GIS.

value: The content of an attribute for a single record within a database. Values can be text, numerical, or codes.

variance: The total amount of disagreement between numbers. Variance is the sum of all values with their means subtracted and then squared.

vector: A map data structure using the point or node and the connecting segment as the basic building block for representing geographic features.

vector (disease): A disease that requires an intermittent host for transmission, such as the mosquito in malaria.

verification: A procedure for checking the values of attributes for all records in a database against their correct values.

version: An update of software. Complete rewrites are usually assigned entirely new version numbers (version 3), whereas fixes and minor improvements are given decimal increments (version 3.1).

virtual map: A virtual map is one that has yet to be realized as a tangible map; it exists as a set of possible maps. For example, the same digital base map and set of numbers can be an entire series of possible virtual maps, yet only one may be chosen to be rendered as a real map on a permanent medium.

VisiCalc: A spreadsheet package for first-generation microcomputers. Supported data tables in flat files.

visual center: A location on a rectangular map, about 5% of the height above the geometric center, to which the eye is drawn perceptually.

visual hierarchy: The perceptual organization of cartographic elements such that they appear visually to lie in a set of layers of increasing importance as they approach the viewer.

volume: A three-dimensional feature represented by a set of areas enclosing part of a surface. In a GIS, usually the top surface only.

VPF (Vector Product Format): A data transfer standard within DIGEST for vector data.

warping: See **rubber sheeting**.

WGS84 (World Geodetic Reference System of 1984): A higher-precision version of the GRS80 used by the U.S. Defense Mapping Agency in world mapping. A common datum and reference ellipsoid for hand-held GPS receivers.

WIMP: A GUI term reflecting the primary user interface tools available: windows, icons, menus, and pointers.

workstation: A computing device that includes as a minimum a microprocessor, input and output devices, a display, and hardware and software for connecting to a network. Workstations are designed to be used together on local area networks and to share data, software, and so on.

World Data Bank: One of the first digital maps of the world, published in two versions by the Central Intelligence Agency in the 1960s.

World Wide Web (WWW or W3): A distributed database of information stored on servers connected by the Internet and special-purpose software for browsing, searching, and downloading.

X-Windows: A public-domain GUI built on the Unix operating system with computer graphics capabilities, written and supported by the Massachusetts Institute of Technology and the basis of most workstation shareware on the Internet.

zero/one/two cell: TIGER terminology for point, line, and area, respectively.

zip: See **mosaic**.

zone (for a coordinate system): The region over which the coordinates relate with respect to a single origin. Usually, some part of the earth or an administrative unit.

INDEX

Absolute location, 45, 61
Academic research, 26
Accuracy, 61
 of attributes, 120
 of data, 14
 of GPS receivers, 114, 150–156
 of position, 120
 See also Errors
ACSM (American Congress of Surveying and Mapping), 15, 26
ACSM/ASPRS technical meetings, 17, 18
Active data, 202, 226
Activities, 4, 5
Address matching, 84, 106, 124, 210, 216, 226
Address range, 96
Adjacency, 27
Adobe Corporation, 82
Adobe Illustrator, 213
Affine transformations, 208, 227
AGI Source Book, 14
Ahearn, Sean C., 241
Air photography, 6, 101, 115, 116, 139, 270. *See also* Orthophotographs; Satellites
Air temperature, 74
Albers equal area projection, 44
Alexandria digital library, 275
Amazon.com, 14
American Cartographer, 15
American Cartographic Association (Cartography and Geographic Information Society), 36
American Congress of Surveying and Mapping (ACSM), 15, 26
American Society for Photogrammetry and Remote Sensing (ASPRS), 15, 26
American Standard Code for Information Interchange, 67
America Online, 15, 102–103
AML (ARC Macro Language), 141, 215
Analog, 124
Analog-to-digital maps, 100–101
Analysis
 of data, 96, 213, 240
 GIS, 5–6, 27
 slope, 214
 statistical, 149–157
 of terrain, 254, 256, 258–260
 time-series, 231
 See also Spatial analysis
Analytical cartography, 14
Anderson Land Use Classification system, 75
Angle, 244
Animal and Plant Health Inspection Service, 234
Animation, 10, 169–170, 189, 290
Annals of the Association of American Geographers, 13
Anonymous FTP, 285, 297
Anthropology, 6
Anyamba, Assaf, 230–231
Apple II microcomputer, 11, 12
Apple MacIntosh, 204, 215, 281, 276
Arc-Cad, 33
Arc file, 70
ArcGIS, 215
Archeology, 12, 14, 168, 279
Archie, 103, 285
Archiving, 292
Arc/Info, 11, 103, 127, 174, 215, 233, 236, 253, 255, 265, 277
Arc/node, 11, 27, 73–74, 96
ArcPad software, 278
Arcs, 27, 70, 96
 topological structure, 74
ArcView 1, 33, 137, 141, 160, 171, 181, 215–216, 238, 253, 258, 259, 293
ArcVIS, 215
Area, 33, 46, 54–55, 84, 96, 161
 digitizing, 109
 projections and, 43–44, 45
 in raster data, 68
 searching by, 140
 text placement rules for, 183, 184
 in TIGER, 104
 types of maps for, 190, 191
 in vector data, 69
Area event, 5
Area feature, 5, 27, 72, 207
Area qualitative map, 186, 188, 191, 197
Area-ratio, 55
ARGON, 272
Arpanet, 12
Array, 75, 96
ASCII, 67, 71, 72, 82, 85, 96, 118, 205
ASPRS. *See* American Society for Photogrammetry and Remote Sensing
Association of American Geographers, 15, 26
Atlas*GIS, 233
Attribute data, 33, 34–36, 129
 management of, 210
Attributes, 3, 27, 35, 36, 61, 96, 124, 144, 178
 accessing, 128–129
 accuracy of, 120
 creating new, 134–135, 213
 describing, 146–149
 entry of, 115–119
 of geographic feature, 55
 key, 133
 linking with map, 35–36, 147
 searches by, 133–135
 structuring, 71–72
Attribute table, 116, 117
Augmented reality, 278
AutoCad 12, 33, 96, 216
AutoCAD DXF, 81, 82, 205
AutoCarto (International Symposium on Automated Cartography), 17, 27
Autodesk, 80, 82, 96
Autodesk Map, 216–217, 227
Automated digitizing, 110. *See also* Scanning
Automated hill shading, 10
Automatic data processing, 130
Automatic line follower, 110
Automating repetitive tasks, 141
Avenue and Visual Basic, 141
Avenue scripts, 238
Averaging, 152
AVHRR (advanced very high resolution radiometer), 103–104, 115, 254, 255, 265, 274, 297
Azimuth, 244
Azimuthal projection, 41, 42, 44, 61

Bacillus thuringiensis (Bt), 234, 235
Background image, 270, 297
Base layer, 265, 297
Base map, 242, 297
Base of terrain, 140
Basic, 209
Batch, batching, 140, 144, 209, 227
Baylor University, 217
Bearing, 178
Bell curve, 151, 155, 178
Benjamin, Susan, 126–127
Bimodal distribution, 151
BIN. See Building Identification Number
Binary digits, 67
Binary files, 72, 74, 85
Binary images, 212, 213
Biological sciences, GIS and, 233
Bits, 67, 96
Black body calibration, 247
Bliss, Jeffery, 242
Block face, 96

Block level maps, 84
Books, 13–14, 22–24
Boolean operations, 212
Border, 183, 184, 197
Borland, 119
Bortman, Amy L., 252
Bosworth, Mark, 69
Bounding rectangle, 80, 96, 161, 178
Bounding rectangle of the points, 157
Brassel, Kurt, 11
Bridges, 285
Browse, 134, 135, 136, 144
Bruns, Tom, 283
Buffer, buffering, 137–138, 174, 211, 212, 227, 253
Building Identification Number (BIN), 248
Burrough, Peter, 3, 14
Business, 6, 14
    GIS as, 6–7
Business Geographics, 13
Byte, 67, 96

CAD (computer-aided drafting), 75, 96
CADD (computer-aided drafting and design), 110
CALFORM, 10, 202, 227
Caliper Corporation, 220
CAM (computer-assisted mapping), 10, 11, 202, 227
Campbell, John, 102
Canada Geographic Information System (CGIS), 11, 27, 227, 302
CARSI (Center for the Analysis and Research of Spatial Information) lab, 241, 242, 243, 247
CartaLinx, 218
Cartographic convention, 193, 197
Cartographic elements, 183, 192, 197, 213–214
Cartographic Perspectives, 13
Cartographic spaghetti, 73, 96
Cartographica, 13
Cartography, 27, 61
    coordinate systems, 45–54
    GIS and, 2, 36
    map and attribute information, 34–36
    map projections, 40–45
    map scale, 38–40
    parts of maps, 182–184
    shape of the earth, 36–38
Cartography and Geographic Information Society, 36
Cartography and Geographic Information Systems (American Cartographer), 15
Cartography and GIS, 13
Cartography and Graphics, 202
Case, 116
Case studies
    population dynamics of gypsy moth in Michigan, 233–237
    resource management for Channel Islands, 252–255
    traffic accident patterns in Connecticut, 237–241
    "wandering stones" in Death Valley Wilderness Area, 255–260
    World Trade Center rescue and clean-up operations, 241–252
Categories, 117
Cell size, 68
Cellular telephone, 219
    location-based services and, 275–276
Census, census tract, 214. *See also* U.S. Bureau of the Census
Centroid, 158, 178
CGIS (Canadian Geographic Information System), 11, 27, 203, 227, 302
Change, measuring rate of, 169
Channel Islands, resource management in, 252–255
Channel Islands GIS (CIGIS), 252–255
Channel Islands National Marine Sanctuary, 253
Channel Islands National Park, 253
Check plots, 120
Choropleth mapping, 162
Choropleth maps, 10, 27, 55, 144, 186, 190, 191, 192, 197, 213
CHREF, 238
Chrisman, Nick, 7, 8, 73
CIA (Central Intelligence Agency), 10, 202
Clarke, Alexander Ross (Sir), 37
Clarke, Keith, 256
Clarke 1866 ellipsoid, 37, 45
Classes, 55
Clearinghouses, 208
Clump, clumping, 207, 212, 227
Code coordinate, 72
CODES (Crash Outcome Data Evaluation System) GIS, 238–240
Coefficient of determination, 166
COGO, 112, 113, 207
Collaborative GIS, 252–255
Collection modes, 109
Color, map, 192–194
Color table, 85–86, 96
Columns, 3, 71
Commands, as queries, 140. *See also individual commands*
CommunityViz, 180, 181
Compaq iPAQ, 248–249, 278
Complete Geographical Information Systems, 11
Complete Geographic Information Systems, 202
Compression, 227
Compromise projections, 44, 61
CompuServe, 86, 103
Compute operation, 134, 135, 137, 144, 161, 166–167, 178
Computer, 13
Computer-aided drafting (CAD), 75, 96, 110
Computer, Environment and Urban Systems, 13
Computer languages, FORTRAN, 28
Computer mapping, 10, 27
Computer memory, 96
Computer monitor, GIS maps and, 183
Computer networks, finding map data on, 102–103
Computers and Geosciences, 13
Cone, 40, 42
Conferences, 16–18
    proceedings, 25
Conformal projections, 41, 43, 44, 45, 61
Conic projection, 41, 42, 44, 61
Connecticut, traffic accident patterns in, 237–241
Connecticut Department of Transportation (ConnDOT), 238
Connecticut Health Research and Education Foundation, 238
Connectivity, 27
Context-sensitive help, 27
Contiguity, 55, 56, 162, 164
Continuity, 55, 61
Contour interval maps, 189, 197
Contour maps, 55, 127, 197. *See also* Isoline maps
Contours, colors for, 193
Control points, 108, 109, 112, 208
Converge, 178
Cookie cut, 210, 227
Coordinate pair, 45, 55, 61
Coordinates, 45–46
    page, 81, 98, 183, 199
Coordinate systems, 45–54, 61
    determining for map, 108–109
    geographic coordinates, 4, 41, 45, 46–47, 62, 235. *See also* Latitude; Longitude
    map projections and, 207–208
    military grid coordinate system, 46, 48, 50–51, 52, 63
    page, 183
    state plane coordinate system, 46, 48, 51–53, 54, 64
    universal polar stereographic coordinate system, 49, 50, 51
    universal transverse Mercator coordinate system, 46, 48–50, 54, 64, 82, 108, 266
Copyright, data and, 289
Core, 203
CorelDraw!, 86, 213
CORONA, 272, 273, 297
Cost
    of GIS data, 285
    of GIS software, 222
Cost surface, 140
Counterhypothesis, 167

Credits, 183, 184, 197
Criminal justice, 12
Cromley, Ellen, 237
Crosshatching, 193
Cross-variance, 165
Cylinder, 40, 42
Cylindrical projection, 41, 42, 61

Damage assessment, GIS and, 243–247
Data, 35, 61
active, 202, 226
georeferencing, 34–35
missing, 77
ownership of, 289
plotting, 119
redundant, 77
retrieval of, 96, 129, 211–212
sources of future, 268–276
storage of, 208–209
updating, 130–131
See also Map data
Data accuracy, 14
Data analysis, 96, 213
purposes of, 240
Database development, 218
Database dictionary, 72
Database management, 4, 128–133, 203, 210, 216
Database management system (DBMS), 72, 97, 129–132, 144, 227
query interface, 140
relational, 132–133, 229
Database manager, 27, 169, 207, 210
Databases, 27, 35, 36, 61
checking characteristics of, 207
distributed, 284–285
GIS and, 2, 3
linking, 117–118
maintenance of, 130–131
map, 10–11
updating, 130–131, 145
Database systems, 119
Data capture, 205–208
Data collection
in the field, 112–113
in global positioning systems, 114–115
in image and remote sensing, 115
Data definition language, 130, 144
Data definition module, 118
Data dictionary, 96, 118, 124, 130, 144
Data display, 213–214
Data entry, 115–119, 124, 144, 210
errors in, 130
Data-entry module, 118, 124
Data extremes, 149, 150(table), 178
Data formats, 80–87, 96
exchange data formats, 80, 88–90, 96, 202, 227
GIF, 86, 98, 109
industry-standard, 80, 81, 82, 274–275
JPEG, 86, 209
support for, 209
TIF, 86, 99
VPF, 90, 99
See also DEM; DXF; Raster data format; Vector data format
Data layers, 98, 238–239, 240, 265. *See also* Map overlays
Data listing, 120
Data Manipulation Programs, 202
Data model, 96, 129, 130, 144
defined, 129
Data report, 120
Data structures, 27, 96, 129, 227
GIS software and, 214
raster, 75–78, 211, 214, 216
vector, 11, 72–75, 211, 214
Data transfer, 97
standards for, 89–90, 274–275
Datum, 37, 38, 61
DBase IV, 119
Dbase tables, 239
DBF, 205
DBMS. *See* Database management system
DCW (Digital Chart of the World), 90, 104, 275
Death Valley Wilderness Area, nonimpact GIS and GPS in, 256–260
DEC, 204, 276
Decimal degrees (DD), 46–47
Decision making, GIS and, 7
Default, 144
Defense, 14
Defense Mapping Agency, 73, 87, 104, 297
Degrees, minutes and seconds (DMS), 46
Degrees of freedom of the value, 154–155, 156
Delaunay triangulation, 75, 97
Delaware National Map prototype, 271, 272
DEM (digital elevation model), 68, 86, 87, 97, 103, 181, 209, 216, 259, 260, 289
DEM drainage, 255
Demography, 168
Dependent variable, 165, 178
Desert Protection Act (1994), 258
Design loop, 168–169, 192, 197
Design with Nature, 9, 10
Desktop GIS, 12
Desktop mapping, 213, 218, 227
Desktop metaphor, 227, 282
Desktop scanner, 110, 111
Device independence, 204, 227
Difference of means, 178
Differential correction, 154
Differential GPS, 256
Differential mode, 114, 115
DIGEST, 90, 97, 99, 209, 297
standards, 274
Digital cartographic text , 103
Digital Chart of the World (DCW), 90, 104, 275
Digital databases, 252
Digital elevation model (DEM), 68, 86, 87, 97, 103, 181, 209, 216, 259, 260, 289
Digital exchange format (DXF), 82, 97, 205, 209, 216, 227
Digital line graphs (DLGs), 80, 82, 83, 90, 97, 103, 209, 274, 289
Digital map library, 275
Digital orthophotoquads (DOQ), 103, 115, 116, 126–127, 270, 271, 297
Digital raster graphics (DRG), 103, 270–271
Digital terrain analysis, 254
Digitizing, 101, 106–110, 124, 202, 205
Digitizing tablet, 27, 72, 107–108, 124, 280
DIME (dual independent map encoding) coding system, 11, 28, 97
Dimension, 55
Dimensionality, 54–55, 61, 72
Ding, Ji, 242
Direction, 33
maps and, 46
projections and, 43, 45
Display medium, maps and, 183
Dissolve, 206, 227
Distances, projections and, 44
Distance transforms, 174
Distortion, 61
Distributed databases, 284–285
Distributed network, 297
Distribution, 55, 56, 162
DLGs (digital line graphs), 80, 82, 83, 90, 97, 103, 209, 274, 289
DLG-SDTS, 90
DMS (degrees, minutes and seconds), 46
Domain expertise, 286
DOQ (digital orthophotoquads), 103, 115, 116, 126–127, 270, 271, 297
DOS, 220, 276
Dot maps, 185, 186, 191, 197
Dot patterns,193
Double digitized, 97
Douglas-Peucker point elimination, 206, 207
Drainage analysis, 214
DRG (digital raster graphics), 103, 270–271
Dropout, 112, 124
Drum scanner, 110, 112, 124
Dueker, Ken, 4
Dueker's definition of GIS, 4–5, 27, 227
Duplicates, 117
Dvorsky, John, 252
DWG, 216
DX-90 standard, 90, 274, 298
DXF (digital exchange format), 82, 97, 205, 209, 216, 227
Dynamic segmentation, 210, 228

E911, 275
Earth, shape of, 36–38

Earth Data Holdings, 243
Earth Observation System (EOS), 271–272, 273, 298
Earth Science Enterprise data, 272
Earth Science Information Centers, 103
Eastings, 45, 61, 112, 157, 159
 in state plane coordinate system, 52
 in universal transverse Mercator coordinate system, 49
Edge matching, 45, 61, 108, 206, 228
Edge pixels, 77
Editing, 109, 119–120, 124, 205, 210
Education
 in GIS, 14, 18–19, 292
 via Internet, 279
Egenhofer, Max, 282, 283
Elements, cartographic, 183, 192, 197, 213–214
Elevation, 39, 75, 86, 97, 112, 114
 as attribute, 150–156
 surface, 55
Ellipsoid, 37, 38, 39
Emergency management, 280
Emergency Mapping and Data Center (EMDC), 247–249
Encapsulated PostScript, 97, 209
End nodes, 74, 78, 97
Engineering, state plane coordinate system and, 51
English language, as metaphor, 282
Entity by entity, 208, 228
Entomology Spatial Analysis Laboratory, 237
Enumeration map, 84, 97
Environmental management, 252–255
Environmental modeling, 14
EOS (Earth Observation System), 271–272, 273, 298
EOSDIS, 273, 278, 298
Ephemeris data, 114
Epidemiology, 6, 12, 168, 265, 279–280
Equal area projection, 44, 45, 62
Equator, 39, 40(table), 45
Equatorial Mercator projection, 48
Equatorial projection, 43
Equatorial radius, 37, 62
Equirectangular projection, 62
Equivalent projection, 44, 45, 62
ERDAS, 207, 233
ERDAS Imagine, 127, 287
ER-Mapper, 233, 287
EROS Data Center, 103, 104, 273, 275, 297
Error band, 155, 178
Error distributions, 151
Error histogram, 151
Errors
 data entry and, 119–120, 130
 data exchange and, 88
 digitizing, 109, 110
 eliminating data capture, 207
 expected, 97
 GIS and isolating readings causing, 160–161
 GPS measurement, 151
 in macros, 140
 map symbolization, 193
 margin of, 155
 scanning, 119
 standard normal distribution and, 156
 topology and detection of, 78–79
ERS, 272
ESRI, 14, 160, 171, 215, 216, 275, 277, 278
ESRI User Conference, 17
Estes, Jack, 4
ETAK, 102
Ethics, GIS professional, 293
eTrex, 147, 149, 156, 157
European Union, 292
Events, 4–5
Expected error, 155, 178
Extent, of grid, 68

Faber, Brenda G., 180–181
Facilities management, 6
Facilities mapping, 212
FAQ (frequently asked questions), 15, 16, 27
Fat line, 88, 97
Features, 4–5, 97, 116, 144, 161–162
 area, 5, 27, 72, 207
 counting, 211
 line, 4–5, 28, 72, 211
 in map database, 135
 point, 4, 5, 30, 72
Federal Communications Commission, 275
Federal Emergency Management Agency, 247
Federal Geographic Data Committee, 208
Federal government
 data ownership and, 7, 289
 digital map data from, 102
 role in providing GIS technology, 247, 250
Federal Information Processing Standard 173, 90
Field, 55, 71, 97
Field data collection, 101, 112–113, 252, 253
Field note-book, 113
Field sketch, scanning, 111
Field variables, 55, 62
Figure, 183, 184, 198
File, 28, 35, 62, 97, 144
File header, 72, 97
File transfer protocol (FTP), 103, 124, 285
 anonymous, 297
Film negatives, 109, 110
Find, 124, 135, 144
FIPS 173 (Federal Information Processing Standard), 97, 274, 298, 300
First International Advanced Study Symposium on Topological Data Structures for Geographic Information Systems, 74
Fishnet view, 189
Fix, 298
Flag, 72
Flag value, missing, 117
Flat Earth society, 36
Flat file, 68, 70, 71, 116, 117, 118, 124, 144
 relational database management systems and, 132
Flatbed scanner, 124
Flattening, 37, 62
Floating dot, 127
Flow diagram, 282
Flow map, 185–186, 188, 191, 198
Foreign languages, GIS and, 292
Fore Site Consulting, 180
Forest color, 193
Forest cover, 236
Forestry, 6, 214, 236, 279
Format, 28. *See also* Data formats
FORTRAN, 10, 11, 28, 202, 228, 303
Fourth dimension, 5, 28
Freedom of Information Act, 102
Freeware, 222, 269, 298
Frequency, 150
FTP (file transfer protocol), 102, 124, 285, 297
Functional capability, 204–214, 228
Functional definition of GIS, 4, 28, 204–205, 228
Fuzzy tolerances, 79, 206, 228

Gage, Stuart, 233
Garmin Etrex personal navigator, 147, 149, 156, 157
Garmin GPS40, 147, 149
Gateway, 103, 125
Gauss, Johann Carl Friedrich, 48
Gauss conformal projection, 48
Gauss-Kruger projection, 48
GBF (Geographic Base File), 11, 28, 97
GBF/DIME files, 84
Generalization, 206, 228
General-purpose maps, 8, 28, 185, 193
Generate a report, 135
Geocarto International, 13
Geocoding, 28, 35, 62, 101, 125, 203, 205–208, 216, 235
 digitizing, 106–110
 purpose of, 54
 scanning, 110–112
Geodesy, 2, 38, 62
Geodetic reference system (GRS80), 37–38
GeoEurope, 13
Geographer's Desktop, 282, 283, 287
Geographical Review, 10
Geographical surface, 97
Geographical Systems, 13
Geographic base files (GBFs), 11, 28, 97
Geographic continuity, 55
Geographic coordinates, 4, 41, 45, 46–47, 62, 235. *See also* Latitude; Longitude

Geographic database, 147, 148(table)
Geographic Data Committee, 291
Geographic Data Technology, 102
Geographic features. *See* Features
Geographic grid, 54
Geographic information, 35, 36, 54–56
Geographic information science, 2, 6, 28, 298
Geographic Information Systems, 15, 17
Geographic Names Information System (GNIS), 103
Geographic pattern, 28
Geographic properties, 55–56, 62, 162
 searching by, 138, 140
Geographic search, 129, 144
Geographic space, 4–5
Geographic statistics, 161–162
Geography, 2, 6, 28, 35, 62, 280, 303
 searches by, 135–140
Geography departments, GIS education and, 18
Geography Network, 275
Geoid, 38, 39, 62
Geoinformatics, 13
Geology, 6, 12, 14
GeoMedia, 220–221, 222
GeoMedia WebMap, 277
Geometric tests, 213, 228
Geoplace, 16
Georeferencing, 203
Geospatial Information and Technology Association, 15
Geospatial Solutions, 13, 24
GeoVRML, 291
GeoWorld, 13, 25
GIF (Graphics Interchange Format), 86, 98, 209
GIRAS, 73
GIS (geographic information systems)
 as approach to science, 6, 291
 books, 13–14, 22–24
 as business, 6–7
 collaborative, 252–255
 conference proceedings, 25
 conferences, 16–18
 databases and, 2, 3
 data exchange and, 274–275
 definitions of, 1, 2–8
 Dueker's definition of, 4–5, 27, 227
 education in, 14, 18–19, 293
 feature model of, 4–5
 functional definition of, 4, 28, 204–205, 228
 future of, 267–268
 global positioning systems and, 269
 history of, 8–12
 human factors in, 7–8
 image maps and, 270–271
 information sources for, 12–19
 as information system, 4–6
 infrastructure of, 12
 internationalization of, 292
 Internet and, 12, 15–16
 journals and magazines, 13, 24–25
 law and, 288, 292
 location-based services and, 275–276
 map projections and, 45
 measurement and, 8
 object-oriented, 284
 operating systems and, 204
 process definition of, 4
 professional societies for, 15, 26
 remote sensing and, 271–274
 role in society, 7–8
 scalelessness of, 39
 social sciences and, 5, 8
 spatial analysis tools and, 174
 time as data element in, 14
 toolbox definition of, 3–4
 user needs, 285–286
 World Wide Web and, 12, 15–16, 26
GIS analysis, 5–6, 27
GIS Asia/Pacific, 13
GIS Café, 16
GIS functional capabilities, 204–214
 data analysis, 96, 213, 240
 data capture, 205–208
 data display, 213–214
 data management, 210, 216
 data retrieval, 96, 129, 211–212
 data storage, 208–209
GIS groups, 7, 30, 286, 299
GIS interoperability, 287–288, 298
GIS-L, 15, 16, 299
GIS Law, 13, 25
GIS/LIS, 28
GIS maps, displays of, 183
GIS Master Bibliography, 14
GIS Monitor, 16
GISMO option, 174
GIS-Plus GIS, 220
GIS software
 ArcGIS, 215
 ArcView, 1, 33, 137, 141, 160, 171, 181, 215–216, 253, 258, 259, 293
 Autodesk Map, 216–217, 227
 data structures and, 214
 evolution of, 201–203
 future, 281–288
 GeoMedia, 220–221, 222
 GRASS, 12, 13, 15, 127, 207, 217–218, 228, 277
 IDRISI, 11, 207, 218, 219, 233, 235, 277, 287
 issues, 221–223
 MapInfo, 141, 218–220, 276, 277, 278
 Maptitude, 220, 221
 selecting, 214–223
 See also Software
GKS, 203
Global circulation, 290
Global Inventory Monitoring and Modeling Systems (GIMMS) lab, 231
Global Land Information System, 208
Global Map, 275
Global positioning system. *See* GPS
Global scale, 290
Global warming, 290
GML, 291
GNIS (Geographic Names Information System), 103
GNU, 204, 217, 228
GOES, 115, 290
Goodchild, Michael F., 302–303
Goode's homolosine projection, 44
Google, 14
Gopher, 103, 285
Government, map information and, 102
GPS (global positioning system), 6, 12, 37, 62, 101, 209, 243
 data collection and, 112, 113, 114–115
 GIS and, 269
 location-based services and, 275–276
 nonimpact research and, 256, 258
 on-board, 244, 269, 270, 276
GPS card, 275
GPS48, 158
GPS receivers, 113, 114, 115, 147, 149, 156
 handheld, 249–250, 251
 latitude/longitude and, 157–158
GPS-type attribute, 156
Gradient, 178. *See also* Slope
Graduated symbol map, 185, 187, 191, 198
Graphical user interfaces (GUIs), 7, 12, 28, 141, 203, 204, 209, 217, 228, 277, 280, 298
Graphics Interchange Format (GIF), 86, 98, 209
GRASS, 12, 13, 15, 127, 207, 217–218, 228, 277
GrassClippings, 13
Graticule, 41, 62, 183
GRID, 11, 265, 273, 275, 298
Grid cell, 75, 98
Gridded fishnet map, 189,191, 198
Grid extent, 68, 98
Grids
 geographic, 54, 183
 raster 68–69
 vector, 70
Ground, 183, 184, 198
 colors for, 193
Ground locations, 214
Ground Truth, 8
GRS80 (geodetic reference system), 37–38, 45, 62
g-trade (g-commerce), 12, 28
GUIs (graphical user interfaces), 7, 12, 28, 141, 203, 204, 209, 217, 228, 277, 280, 298
Gypsy moth population dynamics, 233–237, 265

Hall, Steven, 291

Handheld GPS receiver, 249–250, 251
Hardware, 3
future developments in, 276–281
Hardware coordinates, 109
Harvard Conference on Topological Data Structures, 17
Harvard Graphics, 203
Harvard Laboratory for Computer Graphics and Spatial Analysis, 11, 202
Header, 72, 97
Health, 14
Help
GIS user, 209
via Internet, 285
Help line, 223, 228
Hewlett-Packard Graphics Language (HPGL), 81, 82
Hexadecimal, 67, 98
Hickman, Melodee, 252
Hierarchical, 98
Hierarchical data model, 144
Hierarchical structure, DBMSs and, 131–132
Hill-shaded relief map, 189, 190, 191
HIS, 198
Histogram, 147, 149, 150–151, 178
History, 14
HIV/AIDS prevention, 238
How to Lie with Maps, 288
HPGL (Hewlett-Packard Graphics Language), 81, 82, 98, 209
Hue, 193, 194, 198
Hue, saturation, and intensity (HSI), 193–194
Human factors, in GIS, 7–8
Hydrological systems, 212
Hydrologic consulting, 32–33
Hypertext, 298
Hypothesis, 167, 178
Hypsometric map, 189, 191, 198

IBM, 11, 180, 204, 248, 299
Ice hypothesis, 257–258
Icons, 282, 287
Identify, 136, 145
IDRISI, 11, 207, 218, 219, 233, 235, 277, 287
IDRISI Project, 230–231
Idrisi32, 282, 283
IDW (inverse distance squared) method, 235–236
IEEE Geosciences, 13
IEEE Transactions on Computer Graphics and Applications, 13
IKONOS, 272
Illuminated contours, 189
Image Alchemy, 86
Image data, 115, 116
Image depth, 85, 98
Image formats, 85–87
Image maps, 189, 191, 198, 270–271
GIS and, 270–271
Image processing functions, 214
Image processing systems, 207
Imagery, 112
IMGRID, 10
Independent variable, 165, 178
Index value, 68
Industry standard format, 98, 281
Information management, 287
Information sources, 12–19
Information system, 28
GIS as, 4–6
Information systems technology, 4, 252
Infoworld, 13
Infrastructure management, 249–250
Infrastructure status, GIS and, 247–249, 250
Inheritance, 284
Inlander, Ethan, 252
Inset, 183–184, 198
Installed base, 7, 28
Instance of a geographic object, 116
Integrated software, 203, 228
Intensity, 193–194, 198
Intercept, 165, 178
Intergraph Corporation, 80, 220, 277
International Geographical Congress, 202
International Geographical Union, 11
International Hydrographic Organization (DX-90), 90, 274, 298
International Journal of Geographical Information Systems, 13
International Journal of Remote Sensing, 13
International Meridian conference, 46
International Steering Committee for Global Mapping, 275
International Union of Geodesy and Geophysics, 62
Internet, 28, 125, 276–277
data access and, 106, 208
data exchange and, 209
development of GIS and, 12
digital map libraries on, 275
distributed databases and, 284–285
EOSDIS and, 273
finding map data on, 103
future data sources and, 269
future role of, 278–279
GIS and, 12, 15–16
location-based services and, 275
standards and, 291–292
U.S. Geological Survey data and, 103–104
See also Web sites; World Wide Web
Internet Explorer, 15
Internet Map Server, 277
Interoperability, 287–288, 298
Interval, 55, 62
In-vehicle navigation system, 244, 269, 270, 296, 298
Inventory of World Topographic Mapping, 102
Inverse distance squared (IDW) method, 235–236
Inversions, 119
Inverted hierarchy, 132
iPAQ, 248–249, 278
Isoline maps, 10, 28, 186, 189, 191, 198. *See also* Contour maps
IWT, Inc, 32

Java, 277
JERS, 272
Jet navigation chart (JNC), 104
Join operation, 133, 134, 137–138, 140, 145
Journal of Cartography, 13
Journals, 13, 25
JPEG format, 86, 209

Kansas Geological Survey, 202
Kapp, Mary, 238
Key, 117
Key attributes, 133, 145
Keyword, 85, 140
Khoros, 287
Killer app, 2, 28, 202, 278

Label, 198
Label placement conventions, 183, 184, 198
Label point, 78, 98
Laboratory for Computer Graphics and Spatial Analysis (Harvard University), 11
Lambert, Johann Heinrich, 48
Lambert conformal conic projection, 43, 44, 51–52, 53, 64
North America, 60
Land-cover digital data, 103
Land-cover index mapping, 274
Land-cover maps, 10, 28
Land information, 28, 220
Landmark, 84, 98
Land planning, 180
Land Remote Sensing Act (1992), 274
Landsat, 68, 115, 116, 231, 265, 273, 298
Landsat 4, 254
Landsat 5, 254
Landsat 7, 77, 116, 265, 272, 273–274, 298
See also Satellites
Landsat Commercialization Act (1984), 274
Landscape, 28
Landscape Ecology, 13
Landscape ecology, 14
Land suitability, 140
Land Use and Natural Resources Inventory System (LUNR), 11, 28
Land-use digital data, 103
Land-use map, 28
LANYARD, 272
Larsen, Nils, 32–33

LAS, 127
LA–the Movie, 290
Latitude, 35–36, 39, 41, 45, 63, 108, 114
  calculating, 157–158, 159
  geocoding, 46
  range of, 149
Law, GIS and, 288, 292
Layers
  data, 98, 238–239, 240, 265
environmental, 254–255
  See also Map overlays
Layer sensitivity, 172
LBSs (location-based services), 275–276
Learning curve, 28
Least squares, 165, 178
LeBlanc, Robert, 241
Legend, 183, 184, 198
Leidner, Alan, 242, 247
Levels, 112
Levels of measurement, 55, 63
Light Detection and Ranging (LIDAR), 243–247
Line activity, 5
Linear relationship, 165, 178
Line features, 4–5, 28, 72
  retrieval of, 211
Line generalization, 206, 207
Line-in-polygon, 140, 213
Line-of-sight calculations, 140
Lines, 54–55, 84, 98, 161
  digitizing, 109
  placement rules for, 183, 184
  in raster data, 68, 71, 76
  searching by, 140
  selecting, 211. *See also* Buffer/buffering
  skeleton, 213
  in TIGER, 104
  types of maps for, 190, 191
  in vector data, 69, 71
Lines that are not ended, 119
Line thickness, 186, 199
Link, 63
Link, between databases, 117–118
Linkspoint Inc., 248
Linux, 204, 217, 276
LIS (land information system), 28
Local area network, 284
Local governments, maps, 102
Locate, 136, 145
Location-based services (LBSs), GIS and, 275–276
Locations, 28, 35–36, 63
  absolute, 45, 61
  GIS and, 54
  number description for, 35–36
  relative, 45, 64
Logical consistency, 120
Long Island, 52, 53
Longitude, 35–36, 39, 41, 45, 63, 108, 114
  calculating, 157–158, 159, 160
  gender ratio and, 165
  geocoding, 46
  range of, 149
Look, 162
LUNR (Land Use and Natural Resources Inventory), 11, 28
Lyme disease, 238

Macro language, 215
Macros, 140, 145, 174, 209, 228
Magazines, 13, 24–25
MAGIC (Map and Geographic Information Center), 241
"Magic number," 85, 98
Maintenance fees, 222
Makower, John, 102
Management, 14
Management information systems, 251
Manuals, 33
Map algebra, 170, 211, 212, 213, 228, 282
MapBasic, 141, 220
Map Catalog 102
Map collar annotation, 54
Map data, 3, 34–36
  accessing, 101–103, 128–129
  checking, 207
  creating new digital, 106
  model, 129
Map databases, 10–11
Map design, 191–194, 199
Map display, 183, 216, 290
  CODES GIS, 241
Map display module, 169
Map elements
  interactive modification of, 213–214
  placement of, 192
Map generalization, 206
Map inches, 109
MapInfo, 141, 218–220, 276, 277, 278
Map library, 101–102, 292
  digital, 275
Map millimeters, 45, 53, 63, 109
Map overlays, 9–10, 29, 138, 139(figs.), 145, 162, 170–174, 211–212, 228
  topologically clean map and, 79
  weighting layers, 30, 171–172
  See also Layers
Mapping error, projection and, 45
Mapping methods, 190
Mapping the Next Millennium, 291
Mapplications, 220
Map projections, 36–45, 40–45, 63
  coordinate systems and, 207–208
  GIS and, 45
MapQuest, 275
Maps, 28, 63, 199
  analog-to-digital, 100–101
  area qualitative, 186, 188, 191, 197
  choropleth, 10, 27, 55, 86, 144, 190, 191, 192, 197, 213
  contour, 55, 127, 197
  contour interval, 189, 197
  defined, 182
  designing, 191–194
  dot, 185, 186, 191, 197
  enumeration, 84, 97
  flow, 185–186, 188, 191, 198
  general-purpose, 8, 28, 185, 193
  generating, 136
  graduated symbol, 185, 187, 191, 198
  gridded fishnet, 189, 191, 198
  hill-shaded relief, 189, 190, 191
  hypsometric, 189, 191, 198
  image, 189, 191, 198, 270–271
  isoline, 10, 28, 186, 189, 191, 198
  land cover, 10, 28
  land-use, 28
  linking with attributes, 35–36, 147
  as metaphor, 282
  network, 185, 191, 199
  orthophoto, 189, 199
  outline, 69
  parts of, 182–184
  pattern and color in, 192–194
  permanent, 182, 199
  picture symbol, 185, 187, 191, 199
  preparation of, 205
  proportional circle, 190
  proportional symbol, 213
  real, 100, 125
  realistic perspective, 189, 190, 191, 199
  reference, 185, 190, 191,199
  representing as numbers, 67–71
  satellite image, 189
  simulated hill-shading, 189, 190, 199
  soil, 10
  solution, 173
  as source of geographic information, 5
  street-level address, 104
  structuring, 72–78
  temporary, 182, 200
  thematic, 8, 10, 30, 185
  topographic, 10, 30, 185, 186, 191, 200, 270–271
  topographic/bathymetric, 253
  unclassed choropleth, 186
  virtual, 100–101, 126
Map scale, 38–40, 82
Maps for America, 102
Map text, 192
Map title, 183, 184, 199
Maptitude, 220, 221, 228
Map type, choosing, 185–191, 199
Map units, 109
Map Use and Analysis, 102
MapXtreme, 277
Margin of error, 155
Marketing, 168, 280
Mask, 210, 228
Master directory, 208
Mathematics, 2

McClean, Gordon, 243
McHarg, Ian, 9, 10
McKenzie, Ceretha, 252
Mean, 147, 152–154, 178
Mean center, 157–159, 161, 179
Mean sea level, 63
Measurement, 29
  field data collection and, 112
  GIS and, 8
  levels of, 55, 63
Measures of central tendency, 152, 154
Median, 152, 154, 179
Medical geography, 237
Medium, 125
Menu, 29, 203, 282
Mercator projection, 43, 44
  North America, 60
Merge, 137
Meridian, 63, 98
Mertes, Leal, A. K., 252
Messina, Paula, 256, 257
Metadata, 208, 228–229, 275, 279, 299
  accessibility of, 285
  standards for, 291
Metaphor, 282, 299
Metric system, 39, 63
Michigan, gypsy moth population dynamics, 233–237
Michigan Department of Agriculture, 234, 236
Michigan Department of Natural Resources, 234, 236
Microcomputers, 11, 12, 277, 299
Microsoft Access databases, 238–239
Microsoft Excel, 119
Microsoft Explorer, 103, 276
Microsoft Terraserver, 275
Microsoft Windows, 12, 141, 203, 204, 218, 220, 276, 281
  Windows for Workgroups, 220
  Windows 95, 204, 220
  Windows NT, 204, 215, 220
  Windows 3.1, 220
Military, standards for data transfer, 90
Military grid coordinate system, 46, 48, 50–51, 52, 63
Miller, Westerly, 147, 150
MIMO (map in–map out) system, 10, 29
Minimum easting, 157
Minimum northing, 157
Minnesota Land Management System (MLMIS), 11, 29
Missing cell, 117
Missing data, 98
Missing flag value, 117
Missing values, 152, 179, 213
Mixed pixel problem, 76–77, 98
Mobility revolution, 6, 277–278
Model, modeling, 6, 29, 168–169, 179, 287
Moderate resolution imagery spectrometer (MODIS), 273
Modular computer program, 29
Modular computer programming languages, 10
Monmonier, Mark, 288
Monochrome, 192–193, 270
Mosaicing, 63, 108, 205, 206
Mosaic program, 103, 229, 285, 303
MOTIF, 204, 229, 276, 287
MSDOS, 204
Multipath signal deflection, 249, 251
Multi-peaked histogram, 151
Multiple membership, 131–132
Multiple translators problem, 89–90
Multiply, 212, 213
Multispectral scanner, 115, 299
Multitasking, 204, 229, 299
Mylar, 109, 110, 125

NAD27 (North American Datum of 1927), 37, 45, 63
NAD83 (North American Datum of 1983), 37–38, 45
NASA, 271, 272, 278
National Airphoto Program, 115
National Center for Geographic Information and Analysis (NCGIA), 12, 18, 19, 29, 293, 302
National Geophysical Data Center, 104
National GIS Curriculum, 29
National Highway Traffic Safety Administration (NHTSA), 238
National Map, 271, 272
National Mapping Division, 82, 103, 126, 127
National Oceanic and Atmospheric Administration (NOAA), 103–105, 115, 125, 253
National Science Foundation, 12, 29
National Spatial Data Clearinghouse, 29, 278, 299
National spatial data infrastructure, 299
National Supercomputing Center, 103
NATO, 90, 274
Natural language interfaces, 287
NCGIA (National Center for Geographic Information and Analysis), 12, 18, 19, 29, 293, 302
Neat line, 183, 184, 199
Neighborhood, 55, 56, 162, 164
Netscape, 15, 103, 276–277, 285, 303
Network analysis, 216
Network conference group, 7, 299
Network maps, 185, 191, 199
Network news group, 15
Networks, 125
  computer, 102–103, 276–277
  construction of, 212
  distributed, 297
  future role of, 278–279
  local area, 284
Newsgroup, 29
News services, 16
New York City GIS, 241–252
New York State, state plane zones, 52–53
New York State Office of Technology, 243
NHTSA (National Highway Traffic Safety Administration), 238
NOAA (National Oceanic and Atmospheric Administration), 103–105, 115, 125, 253
Nodes, 11, 29, 98
Node snapping, 205–205, 229
Nominal, 55, 63
Normal distribution, 151, 179
Normalizing, 159, 179
North America
  geographic center, 158–159
  Mercator projection, 60
North American Datum of 1927 (NAD27), 37, 45, 63
North American Datum of 1983 (NAD83), 37–38, 45
North arrow, 183, 184
Northings, 45, 63, 112, 157, 159
  in state plane coordinate system, 52
  in universal transverse Mercator coordinate system, 48–49
NSFNet, 12
Nuernberger, Andrea, 279
Null hypothesis, 167, 179
Numbers
  to describe locations, 35–36
  maps represented as, 67–71
NYCMap, 241, 242, 247, 248

Object-based data structure, 214
Object-oriented programming systems (OOPSs), 130, 284, 286–287, 299
Oblate ellipsoid, 37, 63. *See also* Ellipsoid
Oblique projection, 41, 43, 63
Observation, 29, 116
Odyssey, 11, 29, 73
On-board global positioning system, 244, 269, 270, 276, 298
ONC (operational navigation chart), 104
One cell, 84, 99, 106
Online manuals, 29, 222–223, 229
On-Star LBS, 276
OOPSs (object-oriented programming systems), 130, 284, 286–287, 299
OPEN/GIS, 287, 299
OpenLook, 204, 276, 287, 299
Open Software Foundation, 204
Open/Systems, 287
Open Windows, 299
Operating systems, 204
Operational navigation chart (ONC), 104
Operator errors, 119
Oracle, 215, 216, 220
Ordinal features, 55, 64
Orientation, 55, 56, 162, 164

Origin, 64
Orthophotographs, 115, 181, 242, 243, 270
Orthophoto map, 189, 199
Orthorectified photography, 243, 244
Orton Family Foundation, 181
OS/2, 204, 299
Oswald, Bruce, 243
Outline map, 69
Overlapping records, 131

Page coordinates, 81, 98, 183, 199
Palm Pilot, 278
Parallel, 41, 64
Parallel processing, 278, 299
Parallel technologies, 7
Parameters, 140, 145
Parcel, 265
Parcel management, 220
Patch, 223, 229, 285, 300
Pathfinder program, 274, 300
Pattern, 55, 56, 161, 162, 164
map, 192–194
PC (personal computer), GIS and, 11, 30, 33, 231
PCMCIA (Personal Computer Memory Card International Association) card, 276, 277, 278, 280, 300
Pearson's correlation coefficient, 166, 179
Perfect sphere, 37, 64
Permanent map, 182, 199
Personal digital assistant, 275
PHIGS, 203
Photogrammetric Engineering and Remote Sensing, 13, 15
Photogrammetrist, 127
Picture symbol map, 185, 187, 191, 199
Pijanowski, Bryan, 233
Pixel, 68, 75, 98
Place-names, 183, 184, 199
Plane, 40, 42
Planimetric map data, 242
Planning, 8, 174, 180–181
Planning Support Systems, 180, 181
Plate Carree, 62
Plot, 162
Plotters, 72, 110
Plotting, 119, 202
Point data sets, 206–207
Point distance to line, 140
Pointers, 282
Point event, 5
Point features, 4, 5, 30, 72
Point-in-polygon, 140, 213
Point mode, 109, 125
Point-nominal, 55
Points, 54–55, 84, 98, 161
bounding rectangle of the, 157
placement rules for, 183, 184
raster system and, 68, 76
searching by, 140
selecting, 211
in TIGER, 104
types of maps for, 190, 191
in vector data, 69
Point select mode, 109
Point-to-line distance, 213
Polar radius, 37, 64
Polar regions, 49
Polyconic projection, 45
Polygon attribute table, 70
Polygon features, retrieval of, 211
Polygon information, 11
Polygon interior, 98
Polygons, 78, 98
in arc/node model, 73–74
construction of, 74
detecting errors and, 119
listing attributes for, 137
selecting, 211. *See also* Buffers/buffering
sliver, 170
POLYVRT, 73
Pope, Brian, 238
Population, 155, 179
Population density, 134–135
Population dynamics, of gypsy moths, 233–237
Positional accuracy, 119
Positional error, 120
PostScript, 81, 82, 86, 97, 98, 209, 281
PPGIS (Public Participation GIS), 7
Precision, 64, 120
Prediction, 6, 30, 168–169, 179
Prime meridian, 45, 46, 47, 64
Principles of Geographical Information Systems for Land Resources Assessment, 14
Privacy issues, 276, 288
Private sources, for map data, 102
Problem solving, GIS and, 5–6
Procedural tree, 9
Proceedings, 25, 30
Process definition of GIS, 4
Professional publications, 14, 30
Professional societies, GIS, 15, 26
Profiles, 90
Programming languages, 209, 217
FORTRAN, 10, 11, 28, 202, 228, 303
Projection systems, 33
Proportional circle map, 190
Proportional symbol map, 213
Proportional symbol mapping, 191
Ptolemy, 36
Public access, to CODES GIS, 240–241
Public domain, 300
data ownership and, 289
satellite data in, 272–274
Public health, GIS and, 14, 237–241
Public works, 220

Quadrangle maps, 154
Quad tree, 77–78, 99, 209, 214
Query(ies), 5, 30, 134, 137, 145
in CODES GIS, 239
map as vehicle for, 211
of network, 212
Query interface, 140–141
Query language, 131, 145
Quicksurf, 33

Racetrack Playa, 256–260
Radar mapping system, 300
RADARSAT, 115, 272
Rain events, rock movement and, 257–259
RAM (random access memory), 278
Random distribution, 151
Range, 179
of latitude, 149
of values, 117
Raster data format, 55, 85–87, 99, 170, 209, 229, 265
conversion to vector, 88, 209
digital orthophotquads and, 270
population dynamic study and, 235–236
remote sensing data and, 274
vs. vector, 282–284
Raster data model, 68–69, 70
Raster data set, editing, 205
Raster data structure, 75–78, 214, 216
map overlay and, 211
Ratio values, 55, 64
Real estate, 280
Realistic perspective, 189, 190, 191, 199
Real maps, 100, 125
Recode, recoding, 135, 137, 212
Records, 3, 30, 35, 36, 64, 71, 116–117, 147
overlapping, 131
residuals and, 166–167
Red, green, blue triplets (RGB), 193, 199
Redistricting, 212, 214
Reference map, 185, 190, 191, 199
Relate command, 133, 145
Relational database management systems, 132–133, 229
Relational database managers, 203
Relational model, 145
Relative locations, 45, 64
Remote sensing, 6, 76, 101, 115, 127, 207, 231, 252, 300
GIS and, 271–274
Remote Sensing Review, 13
Renumbering, 134, 145, 229
Reordering, 131
Report, 125
Report generator, 133, 145
Representative fraction, 38–39, 40(table), 64
Residual mapping, 166–167
Residuals, 166, 179
Resolution, scanning and, 111–112
Resource exploration, 32–33
Resource management, 180, 233, 236–237, 266

Channel Islands, 252–255
Restrict operation, 134, 135, 145
Retrieval, 96, 129, 145, 210, 211–213
RGB (red, green, blue triplets), 193, 199
Richards, Jim, 283
River plumes, 254, 255
Road color, 193
Robinson projection, 44, 61
Rock trails, 256–260
Rows, 71
r-squared, 166, 179
Rubber sheeting, 208, 229
Run-length encoding, 77, 99, 209

Sample, 155, 179
Santa Clara River, 254, 255
Santa Cruz Island Reserve, 253
SAS (Statistical Analysis System), 174
SASGRAPH, 203
Satellites, 115, 209, 214, 272–273
    image maps, 189
    spy data, 272, 273
    See also Landsat
Saturation, 193, 194, 199
Scale, 38–40, 46, 55, 56, 64, 82, 120, 162, 164, 183, 184, 199
    scanning and, 111–112
Scaleless, 39, 64
Scaling errors, 119
Scanners, 88, 280
    desktop, 110, 111
    drum, 110, 112, 124
Scanning, 76, 101, 110–112, 125
Scatter plot, 165, 166
Scenario development, 251
Science, GIS and, 6, 167, 291
Scientific approach, 30
Scientific visualization, 289–290, 300
SCITEX workstations, 127
SDTS (Spatial Data Transfer Standard), 80, 90, 91, 99, 209, 229, 274, 287, 291, 300
Search engines, 15, 30
Search(es), 131, 134, 145
    by attribute, 133–135
    by geography, 135–140
Secant conic projection, 41, 42, 64
Select-by-attribute, 211
Selective availability, 114
Select operation, 134, 137–138, 145
Self-support, 7
Semiautomated digitizing, 107. *See also* Digitizing
Server, 125
Shading, on maps, 192–193
Shape, 33, 46, 55, 56, 161, 162, 164
    numbers to describe, 212
    projections and, 43, 44
Shareware, 204, 222, 269, 279, 300
Sharp, Robert P., 258
Show attributes, 135
Show records, 135
Shuttle Radar Topographic Mapping mission, 272
Sift, 212, 229
Silicon Graphics, 276
Simple hierarchy, 132
Simulated hill shading, 189, 190, 199
Simultaneous contrast, 194, 200
Sinusoidal projection, 44
SIR (shuttle imaging radar), 272
Size, 55, 56, 162
Skeleton lines, 213
SLF, 99
Sliver polygons, 170
Slivers, 74, 79, 99, 119, 207
Slope, 140, 178, 179, 213
Slope analysis, 214
SmallTalk, 209, 286
Smallworld GIS, 284
Smart Places, 180
Snap, snapping, 78, 99, 109, 207
Social science, 5, 8
Social theory, 14
Software, 3
    fixes, 285
    maintenance, 223
    packages, 10, 30
    research, 286–287
    self-configuring, 284
    upgrading, 222, 223
    See also GIS software
Soil Conservation Service (Natural Resource Conservation Service), 10
Soil maps, 10
Solution map, 173
Sort, sorting, 131, 135, 136–137, 145, 179
SPANS, 78, 127
Spatial, defined, 3
Spatial analysis, 6, 162–174
    environmental layers, 254–255
    map overlays and, 170–174
    prediction, 168–169
    residual mapping, 166–167
    testing spatial model, 165–166
    tools, 174, 252
    U.S. gender ratios, 162–165
Spatial data, 3, 30
    hierarchy for, 73
Spatial databases, 3
Spatial Data Clearinghouse, 16, 208
Spatial data handling conferences, 17
Spatial data infrastructure, 277
Spatial Data Transfer Standard (SDTS), 80, 90, 91, 99, 209, 229, 274, 287, 291, 300
Spatial description, 157–162
    geographic features and statistics, 161–162
    mean center, 157–159
    standard distance, 159–161
Spatial distribution, 30
Spatial information, 3
Spatial News, 16
Spatial Odyssey, 14
Spatial searching, 136
Sphere, 37, 38, 39
Spheroid, 37
Spikes, 79, 119
S-Plus, 174
SPOT satellite series, 115, 272, 300
Spreadsheet, 30, 119, 213, 229
SPSS (Statistical Package for the Social Sciences), 174
Spy satellite data, 272, 273
SQL (Structured Query Language), 141, 145, 216, 220
Square the difference, 154
Stadia, 112
Standard deviation, 154–155, 179
Standard deviation of the mean, 156
Standard distance, 159–161, 179
Standard exchange formats, 209
Standard normal distribution, 155–156
Standard parallel, 41, 42, 53, 60, 64
Standards, 287, 291–292
Standard transfer formats, 89–90, 274–275
Star, Jeffrey, 4
State governments, maps, 102
State plane coordinate system, 46, 48, 51–53, 54, 64
Statistical analysis, 149–157
Statistical graphics, 203
Statistical software, GIS and, 174
Statistical testing, 155–157
Statistical tools, 169
Steinberg, Richard, 248
Stella, 287
Stepped statistical surface,186, 188, 191, 200
Stereo plotters, 127
Straightness index, 161
Stream mode, 109, 125
Street address, 45. *See also* Address matching
Street-level address maps, 104
Structured Query Language (SQL), 141, 145, 216, 220
Student's T, 156
Subsetting, 131, 145
Sun, 276
Sunblade, 276
Support, 209, 223, 285
Surface, partitioning, 213
Surface elevation, 55
SURFACE II, 10, 202, 229
Surveying, 6, 14
Surveying and Land Information Systems, 15
"Surveys for Planning," 9
Sweden, national grid, 53
SYMAP, 10, 11, 202, 229
Symbol, 183, 192, 200